Advances in Highly Correlated Systems

RIVER PUBLISHERS SERIES ELECTRONIC MATERIALS, CIRCUITS AND DEVICES

The "River Publishers Series in Electronic Materials, Circuits and Devices" is a series of comprehensive academic and professional books which focus on theory and applications of advanced electronic materials, circuits and devices. This includes analog and digital integrated circuits, memory technologies, system-on-chip and processor design. Also theory and modeling of devices, performance and reliability of electron and ion integrated circuit devices and interconnects, insulators, metals, organic materials, micro-plasmas, semiconductors, quantum-effect structures, vacuum devices, and emerging materials. The series also includes books on electronic design automation and design methodology, as well as computer aided design tools.

Books published in the series include research monographs, edited volumes, handbooks and textbooks. The books provide professionals, researchers, educators, and advanced students in the field with an invaluable insight into the latest research and developments.

Topics covered in this series include:

- Analog Integrated Circuits
- Data Converters
- Digital Integrated Circuits
- Electronic Design Automation
- Insulators
- Integrated circuit devices
- Interconnects
- Memory Design
- MEMS
- Nanoelectronics
- Organic materials
- Power ICs
- Processor Architectures
- Quantum-effect structures
- Semiconductors
- Sensors and actuators
- System-on-Chip
- Vacuum devices

For a list of other books in this series, visit www.riverpublishers.com

Advances in Highly Correlated Systems

Editors

Parvathy Nancy
Mahatma Gandhi University, India

Saju Joseph
Mahatma Gandhi University, India

Wilfrid Prellier
French National Centre for Scientific Research, France

Sabu Thomas
Mahatma Gandhi University, India

Nandakumar Kalarikkal
Mahatma Gandhi University, India

River Publishers

Routledge
Taylor & Francis Group
NEW YORK AND LONDON

Published 2025 by River Publishers
River Publishers
Broagervej 10, 9260 Gistrup, Denmark
www.riverpublishers.com

Distributed exclusively by Routledge
605 Third Avenue, New York, NY 10017, USA
4 Park Square, Milton Park, Abingdon, Oxon OX14 4RN

Advances in Highly Correlated Systems / by Parvathy Nancy, Saju Joseph, Wilfrid Prellier, Sabu Thomas, Nandakumar Kalarikkal.

Routledge is an imprint of the Taylor & Francis Group, an informa business

ISBN 978-87-7022-411-6 (hardback)
ISBN 978-87-7004-207-9 (paperback)
ISBN 978-87-7004-221-5 (online)
ISBN 978-87-7004-214-7 (master ebook)
DOI: 10.1201/9781003337102

Contents

Preface

The realm of highly correlated systems occupies a central and progressively essential place in modern physics and materials science. In these complex systems, the behaviour of individual particles is inseparably correlated with the collective dynamics of the whole, generating emergent phenomena that challenge classical and mean-field descriptions. The intricate interplay of charge, spin, orbital, and lattice degrees of freedom in such materials gives rise to a variety of quantum phases, including unconventional superconductivity, Mott insulating states, multiferroicity, colossal magnetoresistance, and topologically protected excitations.

The conception of this book, *Advances in Highly Correlated Systems*, was born from the realisation that we are witnessing a transformative era in the scientific understanding and design of correlation-driven events. The chapters presented herein reflect a diversity of scholarly approaches from quantum field theoretic constructs and many-body numerical simulations to state-of-the-art experimental investigations and materials-specific analyses, each contributing uniquely to the expansion of our comprehension of collective behaviour in highly correlated systems.

The book mostly focuses on the design, development, and application of theoretical formalisms that transcend conventional single-particle frameworks. Scientific approaches grounded in quantum dynamical considerations, extended cluster models, and real-material simulations converge to describe the subtleties of electronic localization, correlation-induced phase transitions, and symmetry-breaking instabilities. These contributions provide not only small steps but also fresh conceptual angles for investigating the entangled quantum states arising in connected materials.

The experimental viewpoints on two-dimensional quantum materials, low-dimensional heterostructures, and multiferroic oxides, where correlation effects appear in tunable and technologically promising ways, are equally compelling. The precise control of lattice strain, chemical substitution, layer stacking, and external fields has led to unprecedented control over magnetic, ferroelectric, and topological properties. In addition to being fundamentally interesting, these studies point the way towards the development of next-generation quantum materials for energy conversion, memory devices, and spintronics.

This book stands out for its focus on interdisciplinary connections. Some chapters provide insights into the quantum dynamics of complex oxide nanostructures and plasma environments in astrophysical contexts, extending the discussion of correlated systems beyond conventional condensed matter. This scope serves as an example of how correlation phenomena are universal, enriching the intellectual framework within which we approach quantum matter.

In my capacity as editor, I firmly believe that this volume not only documents important scientific advancements but also serves as a catalyst for further exploration in this dynamically evolving field. I hope the concepts, models, and findings discussed in this book will stimulate new theoretical understandings, experimental approaches, and material innovations that will thereby propel the field towards new horizons of understanding.

Dr Parvathy Nancy
Editor
Advances in Highly Correlated Systems

List of Figures

List of Tables

List of Contributors

Bhadrapriya, B. C., *School of Pure and Applied Physics, Mahatma Gandhi University, India; International and Inter University Centre for Nanoscience and Nanotechnology, Mahatma Gandhi University, India*

Bose, Bosely Anne, *School of Pure and Applied Physics, Mahatma Gandhi University, India; International and Inter University Centre for Nanoscience and Nanotechnology, Mahatma Gandhi University, India*

Chaplot, Samrath L., *Solid State Physics Division, Bhabha Atomic Research Centre, India; Homi Bhabha National Institute, India*

Chakroborty, Brahmananda, *High Pressure & Synchroton Radiation Physics Division, Bhabha Atomic Research Centre, Trombay, Mumbai-400085, India; Homi Bhabha National Institute, Mumbai*

Gupta, Gaurav Kumar, *Technion-Israel Institute of Technology, Israel*

Gupta, Mayanak K., *Solid State Physics Division, Bhabha Atomic Research Centre, India*

Haldar, Prosenjit, *Laboratoire de Physique Théorique, IRSAMC, Universitéde Toulouse, CNRS, UPS, France*

Jani, Mona, *Department of Physical Sciences and Nanotechnology, Yachay Tech University, San Miguel de Urcuqui, Ecuador*

Jayalakshmy, M. S., *International and Inter University Centre for Nanoscience and Nanotechnology, Mahatma Gandhi University, India*

Joseph, Saju, *International and Inter University Centre for Nanoscience and Nanotechnology, Mahatma Gandhi University, India*

Kala, M. S., *St. Teresa's College, Ernakulam, India*

Kalarikkal, Nandakumar, *International and Inter University Centre for Nanoscience and Nanotechnology, Mahatma Gandhi University, India; School of Pure and Applied Physics, Mahatma Gandhi University, India*

Lakshmy, Seetha, *International and Inter University Centre for Nanoscience and Nanotechnology, Mahatma Gandhi University, India*

Mathew, George, *School of Physics, IISER Thiruvananthapuram, India*

Mayeen, Anshida, *School of Pure and Applied Physics, Mahatma Gandhi University, India*

Michael, Manesh, *Department of Physics, Bharata Mata College, India*

Mishra, Sanjay K., *Solid State Physics Division, Bhabha Atomic Research Centre, India*

Mittal, Ranjan, *Solid State Physics Division, Bhabha Atomic Research Centre, India*; *Homi Bhabha National Institute, India*

Mohan, Saj, *Department of Materials Science and Metallurgical Engineering, Indian Institute of Technology Hyderabad, India*

Nair, Anju K., *Mar Thoma College, Tiruvalla, India*

Nancy, Parvathy, *School of Energy Materials, Mahatma Gandhi University, India; School of Pure and Applied Physics, Mahatma Gandhi University, India*

Pillai, V. P. Mahadevan, *Department of Optoelectronics, University of Kerala, India*

Praveena, M. G., *Department of Physics and Centre for Research, St. Teresa's College, Mahatma Gandhi University, India*

Ramadurai, Ranjith, *Department of Materials Science and Metallurgical Engineering, Indian Institute of Technology Hyderabad, India*

Sajna, M. S., *Department of Optoelectronics, University of Kerala, India*

Sebastian, Sijo, *Department of Physics, St. Aloysius College, India*

Shemim, S. S., *Thangal Kunju Musaliar College of Engineering, India*

Shilpa, S., *International School of Photonics, Cochin University of Science and Technology, India*

Sreedevi, Kiran, *School of Energy Materials, Mahatma Gandhi University, India*

Swalihu, Muhammed, *International and Inter University Centre for Nanoscience and Nanotechnology, Mahatma Gandhi University, India; Department of Physics, Catholicate College, India*

Trivedi, Ravi, *Department of Physics, Karpagam Academy of Higher Education, Coimbatore 641021, Tamil Nadu, India; Centre for Computational Physics, Karpagam Academy of Higher Education, Coimbatore 641021, Tamil Nadu, India*

Thomas, Paulose, *Mar Thoma College for women Perumbavoor, India*

Thomas, Sabu, *School of Energy Materials, Mahatma Gandhi University, India; International and Inter University Centre for Nanoscience and Nanotechnology, Mahatma Gandhi University, India; School of Nano Science and Nano Technology, Mahatma Gandhi University, India*

Unnikrishnan, N. V., *School of Pure and Applied Physics, Mahatma Gandhi University, India*

Venugopal, Chandu, *School of Pure & Applied Physics, Mahatma Gandhi University, India*

Vimal, G., *St. Dominic's College* Kanjirappally, *India*

List of Abbreviations

2D	Two-dimensional
AF	Antiferromagnetic
AFM	Atomic force microscopy
AFP	Antiferroparamagnetic
AHE	Anomalous Hall effect
AIAO	All-in-all-out
BCS	Bardeen, Cooper, and Schrieffer
BD	Bond disproportionated
BFO	Bismuth ferrite
BG	Boron doped Graphene
BNG	Boron nitrogen doped graphene
BP	Black phosphorus
BSCCO	Bismuth strontium calcium copper oxide
BST	$Ba_{1-x}Sr_xTiO_3$
BZ	Brillouin zone
CCDW	Commensurate charge-density-wave
CDMFT	Cluster dynamical mean field theory
CMOS	Complementary metal oxide semiconductor
CMR	Colossal magnetoresistance
CNT	Carbon nano tubes
CPA	Coherent potential approximation
CTQMC	Continuous-time quantum Monte Carlo
CV	Crystal violet
CVD	Chemical vapor deposition
DCA	Dynamical cluster approximation
DFT	Density functional theory
DM	Dzyaloshinskii–Moriya
DMFT	Dynamical mean field theory
DMRG	Density matrix renormalization group
DMS	Dilute magnetic semiconductor
DOS	Density of states
ED	Exact diagonalization

EDAX	Energy dispersive X-ray analysis
EF	Energy in fermilevel
EG	Ethylene glycol
EIC	Electrostatic ion cyclotron
FE	Ferroelectric
FeRAM	Ferroelectric memories
FeSC	Ferro superconductor
FET	Field-effect transistors
FKM	Falicov–Kimball model
FM	Ferromagnetic
GO	Graphene oxide
HF	Hartree–Fock
HF	Heavy-fermion
HFQMC	Hirsch-Fye Quantum Monte Carlo
HOMO	Highest occupied molecular orbital
HRTEM	High resolution transmission electron microscope
HS	High twist state
IA	Ion-acoustic
IBSCS	Iron-based superconductors
IMT	Insulator-to-metal transition
IPT	Iterated perturbation theory
IQHE	Integer quantum Hall effect
IR	Infrared
IREE	Inverse Rashba–Edelstein effect
ITRS	International technology roadmap for semiconductors
JT	Jahn–Teller
KE	Kinetic energy
LB	Long bond
LBCO	$La_{2-x}Ba_xCuO_4$
LCMO	Lanthanum calcium manganite
LDA	Local density approximation
LDOS	Local density of states
LGD	Landau–Ginzburg–Devonshire
LIB	Lithium–ion battery
LS	Low spin
LSCO	$La_{2-x}Sr_xCuO_4$
LSDA	Local spin density approximation

LSMO	$La_{0.7}Sr_{0.3}MnO_{3\pm\delta}$
LUMO	Lowest unoccupied molecular orbital
MBE	Molecular beam epitaxy
MC	Monte Carlo
ME	Magnetoelectric
MERAM	Magnetoelectric random access memory
MESO	Magnetoelectric, spin−orbit
MIT	Metal–insulator transition
ML	Multilayer
MM	Magnetoelectric multiferroics
MOR	Methanol oxidation process
MR	Magnetoresistance
MRAM	Magnetic memory devices
NCCDW	Nearly commensurate charge-density-wave
NCF	$Ni_{1-x}Co_xFe_2O_4$
NG	Nitrogen dopped Graphene
NIR	Near Infrared
NP	Nanoparticle
NRG	Numerical renormalization group
OER	Oxygen evolution reaction
OLED	Organic light emitting diode
ORR	Oxygen reduction reaction
OS	Optical spectroscopy
PES	Photoemission spectroscopy
PFM	Piezoresponse force microscopy
PFN	$Pb(Fe_{1/2}Nb_{1/2})O_3$
PFW	$Pb(Fe_{1/2}W_{1/3})O_3$
PPMS	Physical property measurement system
Pt	Platinum
PVDF	Polyvinylidene fluoride
PZT	$Pb(ZrTi)O_3$
QHE	Quantum hall effect
QSL	Quantum spin liquid
RDM	Reduced density matrix
RDMFT	Reduced density matrix functional theory
RE	Rare-earth
RF	Radiofrequency
rGO	Reduced GO
RSM	Reciprocal space mapping

RVB	Resonating valence bond
SAXS	Small-angle X-ray spectrum
SB	Short bond
SC	Superconductivity
SCE	Strongly correlated
SEM	Scanning electron microscopy
SERS	Surface enhanced raman spectroscopy
SES	Strongly correlated systems
SG	Sulfur-doped graphene
SIC	Self-interaction corrected
SNG	Syngas
SOC	Spin–orbit coupling
SOFC	Solid oxide fuel cell
SPA	Specific power absorption
STM	Scanning tunneling microscope
TBG	Twisted bilayer graphene
TDGL	Time-dependent Ginzburg–Landau
TEM	Transmission electron microscope
TF	Thomas–Fermi
TGA	Thermogravimetric analysis
TI	Topological insulator
TMAH	Tetramethyl ammonium hydroxide
TMD	Transition metal dichalcogenide
TMDC	Transition metal dichalcogenide
TMI	MI transition temperature
TMO	Transition metal oxide
UC	Unit cell
UPS	Ultraviolet photoelectron spectroscopy
VBC	Valence-bond crystal
vdW	Van der Waals
VSM	Vibrating sample magnetometer
WAXS	Wide-angle X-ray spectrum
WRC	Wurtz-type reductive coupling
WSM	Weyl semi-metal
XMR	Extreme magnetoresistance
YBCO	Yttrium barium copper oxide

1

Dynamical Mean Field Theory: Application to the Strongly Correlated Materials

Prosenjit Haldar

Laboratoire de Physique Théorique, IRSAMC,
Université de Toulouse, CNRS, UPS, France

Abstract

In this chapter, we present a bird's-eye view of dynamical mean field theory (DMFT) and its extensions to highly correlated systems. We first demonstrate the elementary introduction to DMFT on a single site with detailed derivations and then a solution of the simple interacting models: the Hubbard model and the Falicov–Kimball model on infinite dimensional Bethe lattice using DMFT. Thereafter, we focus on two kinds of non-local extensions of single-site DMFT - (a) cluster dynamical mean field theory (CDMFT) on real space and (b) dynamical cluster approximation (DCA) on momentum space. Finally, we conclude the chapter with the application of DMFT combining with density functional theory (DFT) in the real-material systems (LDA+DMFT).

Keywords: DMFT, CDMFT, DCA, LDA+DMFT, Hubbard model, Falicov–Kimball model

1.1 Introduction

The numerical implementation of highly correlated systems is a fascinating branch of condensed matter theory nowadays. But, no numerical technique is sufficient enough to solve many-body systems in the highly correlated region. The methods like exact diagonalization are restricted to problems

DOI: 10.1201/9781003337102-1

with a limited number of sites as the number of basis of the Hilbert spaces becomes exponentially large with the number of sites, whereas perturbative approaches are limited only in the weakly coupled region. On the other hand, the quantum Monte Carlo method can be applied to a large system, but it is not applicable at low temperatures owing to the sign problem. The density matrix renormalization group (DMRG) method is bounded in the low-dimensional systems.

A convenient numerical technique to treat the electronic systems in the strongly interacting region is DMFT. The development of the theory was pioneered by Muller-Hartmann and Metzner and Vollhardt [12], while the comprehensive insights were flourished by Georges, Kotliar, and co-workers [5]. The DMFT provides a good approximation in the high-dimensional lattice, whereas it is insufficient in the low-dimensional lattice. Though DMFT has some limitations, it emerges to be the prominent technique to understand systems in the highly interacting regime with a myriad of applications [9]. Cluster extensions of the DMFT are crowd-pleasing approximation schemes for low-dimensional systems with local interaction to access their equilibrium properties in the thermodynamic limit.

1.2 DMFT on Single Lattice Site

The elementary scheme of DMFT was lent from the Weiss mean-field theory of magnetism [6] in classical physics. It considers a single lattice site, and the consequence of all the other lattice sites is included in a self-consistently determined "effective medium" [5]. Unlike the Weiss mean-field theory, in DMFT, we cannot determine the environment by a static field but must include a dynamic "bath function" ($\Delta_i(\omega)$), which takes into consideration the dynamics of the electrons on a given site due to the environment. So, DMFT formalism maps the tedious full lattice problem to an effective single-site problem. The problem then effectively brings down to solving a quantum impurity problem add-on with a self-consistently determined bath function $\Delta(\omega)$. The exact form of DMFT equation will vary for interaction, disorder, or both.

Some of the important physical aspects of DMFT are:

- DMFT is an approximate formalism where the problems of infinite length scale are mapped to a single lattice site with infinite time scale. In the limit of infinite coordination number, the spatial correlation vanishes, and the DMFT becomes exact.

- DMFT shows the Fermi liquid characteristics of metal both in the low-temperature and the incoherent regimes.
- One can evaluate local self-energy $\Sigma_{ii}(\omega)$ from a local Green function, and one can also calculate the effective mass of the quasi-particle as $m^* = 1 - (\delta\Sigma_{ii}/\delta\omega)_{\omega=0}$ and the inelastic scattering rate $\tau_{\text{in}}^{-1}(\omega) = -Im\Sigma_{ii}(\omega = 0)$. These quantities display spatial fluctuations [13] with disorder, and so there is a need of appropriate distribution function to characterize them.
- As DMFT is unable to take into account spatial fluctuations, it cannot demonstrate the phenomena that are born out of long wavelength spatial fluctuations.

1.3 Introduction to the Hubbard Model

The simplest microscopic model to describe the strong local correlation in the electronic systems is the single-band Hubbard model. The Hamiltonian of this model is

$$\hat{H} = -\sum_{\langle i,j \rangle \, \sigma} t_{ij}\hat{c}^{\dagger}_{i\sigma}\hat{c}_{j\sigma} + \mu \sum_{i,\sigma} \hat{n}_{i\sigma} + \sum_{i} U\, \hat{n}_{i\uparrow}\, \hat{n}_{i\downarrow}. \tag{1.1}$$

Here, $\hat{c}^{\dagger}_{i}, \hat{c}_{j}$ are the creation and annihilation operators of electrons with spin indices σ (where, $\sigma \in [\uparrow, \downarrow]$) at sites i and j, respectively, and satisfying the fermionic anti-commutation relation:

$$\{\hat{c}^{\dagger}_{i\sigma}, \hat{c}_{j\sigma}\} = \delta_{ij}, \qquad \{\hat{c}_{i\sigma}, \hat{c}_{j\sigma}\} = 0, \qquad \{\hat{c}^{\dagger}_{i\sigma}, \hat{c}^{\dagger}_{j\sigma}\} = 0 \tag{1.2}$$

where $\hat{n}_{i\sigma} = \hat{c}^{\dagger}_{i\sigma}\hat{c}_{i\sigma}$ is the fermionic number operator at site i with spin σ. The first term in the Hamiltonian, $\hat{H}$, describes the hopping between nearest-neighbor sites i and j with amplitude t, the second term represents the onsite chemical potential μ, and the final terms represent onsite interaction between $\uparrow$-electron and $\downarrow$-electron with strength U at site i. The Hubbard model has an exact solution only in $d = 1$ dimensions using Bethe ansatz. So we need a controlled approximation scheme for this model in $d = 2, 3$.

1.3.1 Solution of the Hubbard model using DMFT by cavity method

The infinite-dimensional ($d \rightarrow \infty$) limit provides the basis for the construction of DMFT where the full lattice problem is mapped to an effective

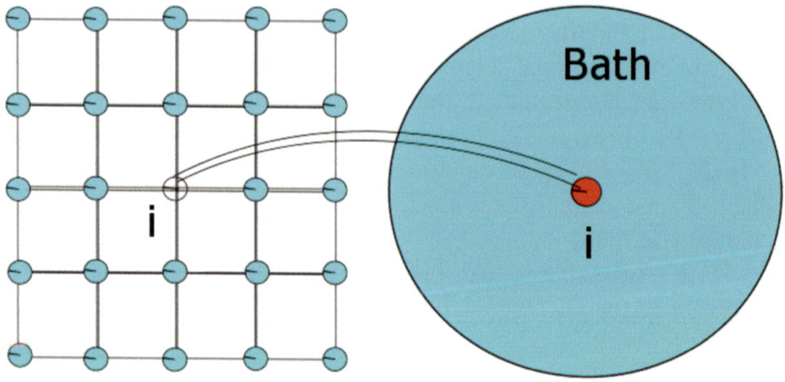

Figure 1.1 Within DMFT formalism, we map the full lattice problem into a single-site (say $'i'$) problem that is embedded by an effective medium. Creating a cavity with bonds removed and mapping to an effective bath. Figure taken from [8].

single-site impurity Anderson model with local propagator and self-energy, which are purely dynamical (function ω); as in this approach, all the spatial fluctuations are frozen, but local quantum fluctuations remain intact – this helps in describing the correlation effects.

The first derivation of the dynamical mean field theory is known under the name of "cavity method." In the cavity method, one removes a lattice site (Figure 1.1) together with its bonds from the rest of the lattice. The remaining lattice containing the cavity is now replaced by an effective dynamical bath. The bath is coupled via a hybridization to the cavity. The resulting problem then reduces to an effective single-impurity Anderson model with degrees of freedom of the bath represented by an appropriate hybridization function, to be determined self-consistently. Consider the partition function with Hubbard action $\mathcal{S}$ in Grassmann variables $\hat{c}^{\dagger}_{i\sigma}$, $\hat{c}_{i\sigma}$,

$$\mathcal{Z} = \int \prod_{i,\sigma} \mathcal{D}\hat{c}^{\dagger}_{i\sigma} \mathcal{D}\hat{c}_{i\sigma}\, exp[-\mathcal{S}]. \tag{1.3}$$

The Hubbard action is given by

$$\mathcal{S} = \int_0^{\beta} d\tau \Big[\sum_{i\sigma} \hat{c}^{\dagger}_{i\sigma}(\tau) \Big(\partial_\tau - \mu\Big) \hat{c}_{i\sigma}(\tau) - \sum_{ij,\sigma} t_{ij} \hat{c}^{\dagger}_{i\sigma}(\tau)\, \hat{c}_{j\sigma}(\tau) + \sum_i U \hat{c}^{\dagger}_{i\uparrow}(\tau)\, \hat{c}_{i\uparrow}(\tau)\, \hat{c}^{\dagger}_{i\downarrow}(\tau)\, \hat{c}_{i\downarrow}(\tau) \Big] \tag{1.4}$$

where β is the inverse temperature and τ is the imaginary time in Matsubara formalism. The action is split into three parts, $\mathcal{S} = \mathcal{S}_0 + \Delta\mathcal{S} + \mathcal{S}^{(0)}$, where $\mathcal{S}_0$ contains only variables on a site, say $i = 0$. $\Delta\mathcal{S}$ contains the hopping between site 0 and other sites of the lattice $i \neq 0$ and $\mathcal{S}^{(0)}$ denotes the lattice with cavity, i.e., site 0 and bonds removed

$$\mathcal{S}_0 = \int_0^\beta d\tau \Big[\sum_\sigma \hat{c}^\dagger_{0\sigma}(\tau) \Big(\partial_\tau - \mu\Big) \hat{c}_{0\sigma}(\tau) + U \hat{c}^\dagger_{0\uparrow}(\tau)\, \hat{c}_{0\uparrow}(\tau)\, \hat{c}^\dagger_{0\downarrow}(\tau)\, \hat{c}_{0\downarrow}(\tau) \Big] \tag{1.5}$$

$$\Delta\mathcal{S} = -\int_0^\beta d\tau \sum_{i,\sigma} t_{i0} \hat{c}^\dagger_{i\sigma}(\tau)\, \hat{c}_{0\sigma}(\tau) + t_{0i} \hat{c}^\dagger_{0\sigma}(\tau)\, \hat{c}_{i\sigma}(\tau) = -\int_0^\beta d\tau\, \Delta\mathcal{S}(\tau) \tag{1.6}$$

$$\mathcal{S}^{(0)} = \int_0^\beta d\tau \Big[\sum_{i\neq 0,\sigma} \hat{c}^\dagger_{i\sigma}(\tau) \Big(\partial_\tau - \mu\Big) \hat{c}_{i\sigma}(\tau) - \sum_{ij\neq 0,\sigma} t_{ij} \hat{c}^\dagger_{i\sigma}(\tau)\, \hat{c}_{j\sigma}(\tau) + \sum_{i\neq 0} U \hat{c}^\dagger_{i\uparrow}(\tau)\, \hat{c}_{i\uparrow}(\tau)\, \hat{c}^\dagger_{i\downarrow}(\tau)\, \hat{c}_{i\downarrow}(\tau) \Big]. \tag{1.7}$$

Separating site 0 and integrating out rest of the lattice degrees of freedom, the partition function describes an effective action $\mathcal{S}_{\text{eff}}$,

$$\begin{aligned} \mathcal{Z} &= \int \prod_\sigma \mathcal{D}\hat{c}^\dagger_{0\sigma} \mathcal{D}\hat{c}_{0\sigma} \exp[-\mathcal{S}_{\text{eff}}] \\ &= \int \prod_\sigma \mathcal{D}\hat{c}^\dagger_{0\sigma} \mathcal{D}\hat{c}_{0\sigma} \exp[-\mathcal{S}_0] \times \int \prod_{i\neq 0,\sigma} \mathcal{D}\hat{c}^\dagger_{i\sigma} \mathcal{D}\hat{c}_{i\sigma} \exp[-\mathcal{S}^{(0)}] \exp[-\Delta\mathcal{S}]. \end{aligned} \tag{1.8}$$

The second integral represents tracing out all degrees of freedom of lattice, except at site 0. Expanding $\exp[\int_0^\beta d\tau\, \Delta\mathcal{S}(\tau)]$,

$$\mathcal{Z} = \int \prod_\sigma \mathcal{D}\hat{c}^\dagger_{0\sigma} \mathcal{D}\hat{c}_{0\sigma} \exp[-\mathcal{S}_0] \times \int \prod_{i\neq 0,\sigma} \mathcal{D}\hat{c}^\dagger_{i\sigma} \mathcal{D}\hat{c}_{i\sigma} \exp[-\mathcal{S}^{(0)}] \left(1 + \int_0^\beta d\tau\, \Delta\mathcal{S}(\tau) + \frac{1}{2!} \int_0^\beta d\tau_1 \int_0^\beta d\tau_2 \Big[\mathcal{T}_\tau \Delta\mathcal{S}(\tau_1) \Delta\mathcal{S}(\tau_2)\Big] + \cdots \right). \tag{1.9}$$

The imaginary time ordering is coming as $\hat{c}_{i\sigma}$ are τ dependent. Define cavity partition function $\mathcal{Z}^{(0)} = \int \prod_{i\neq 0} \mathcal{D}\hat{c}^\dagger_{i\sigma} \mathcal{D}\hat{c}_{i\sigma} \exp[-\mathcal{S}^{(0)}]$ and ensemble average taken with respect to $\mathcal{S}^{(0)}$ as

$$\langle \mathcal{A} \rangle_{(0)} = \frac{1}{\mathcal{Z}^{(0)}} \int \prod_{i \neq 0} \mathcal{D}\hat{c}^{\dagger}_{i\sigma} \mathcal{D}\hat{c}_{i\sigma}\, \mathcal{A} \exp[-\mathcal{S}^{(0)}]. \tag{1.10}$$

Eqn (1.9) becomes,

$$\begin{aligned} \mathcal{Z} = \int \prod_{\sigma} \mathcal{D}\hat{c}^{\dagger}_{0\sigma} \mathcal{D}\hat{c}_{0\sigma} \exp[-\mathcal{S}_0]\, \mathcal{Z}^{(0)} \Big(1 + \int_0^{\beta} d\tau \, \langle \Delta\mathcal{S}(\tau) \rangle_{(0)} \\ + \frac{1}{2!} \int_0^{\beta} d\tau_1 \int_0^{\beta} d\tau_2 \langle \mathcal{T}_{\tau} \Delta\mathcal{S}(\tau_1) \Delta\mathcal{S}(\tau_2) \rangle_{(0)} + \cdots \Big). \end{aligned} \tag{1.11}$$

All the odd terms vanish as they are average over single Fermionic operators. The second term now,

$$\begin{aligned} &\frac{1}{2!} \int_0^{\beta} d\tau_1 \int_0^{\beta} d\tau_2 \Big\langle \mathcal{T}_{\tau} \Delta\mathcal{S}(\tau_1) \Delta\mathcal{S}(\tau_2) \Big\rangle_{(0)} \\ &= \frac{1}{2!} \int_0^{\beta} d\tau_1 \int_0^{\beta} d\tau_2 \Big\langle \mathcal{T}_{\tau} \sum_{i\sigma} \left[t_{i0} \hat{c}^{\dagger}_{i\sigma}(\tau_1) \hat{c}_{0\sigma}(\tau_1) + t_{0i} \hat{c}^{\dagger}_{0\sigma}(\tau_1) \hat{c}_{i\sigma}(\tau_1) \right] \\ &\quad \times \sum_{j\sigma'} \left[t_{j0} \hat{c}^{\dagger}_{j\sigma'}(\tau_2) \hat{c}_{0\sigma'}(\tau_2) + t_{0j} \hat{c}^{\dagger}_{0\sigma'}(\tau_2) \hat{c}_{j\sigma'}(\tau_2) \right] \Big\rangle_{(0)} \\ &= \frac{1}{2!} \int_0^{\beta} d\tau_1 \int_0^{\beta} d\tau_2 \Big\langle \sum_{ij,\sigma\sigma'} \mathcal{T}_{\tau} \big[t_{i0} t_{0j} \hat{c}^{\dagger}_{i\sigma}(\tau_1) \hat{c}_{0\sigma}(\tau_1) \hat{c}^{\dagger}_{0\sigma'}(\tau_2) \hat{c}_{j\sigma'}(\tau_2) \\ &\quad + t_{0i} t_{j0} \hat{c}^{\dagger}_{0\sigma}(\tau_1) \hat{c}_{i\sigma}(\tau_1) \hat{c}^{\dagger}_{j\sigma'}(\tau_2) \hat{c}_{0\sigma'}(\tau_2) \big] \Big\rangle_{(0)}. \end{aligned}$$

The other two terms give $\left[\int_0^{\beta} d\tau \sum_{i,\sigma} t_{i0} \hat{c}^{\dagger}_{i\sigma}(\tau) \hat{c}_{0\sigma}(\tau) \right]^2$, which is zero after τ integration. The two terms in the bracket are equal, as the first term for $\tau_1 > \tau_2$ becomes

$$= \int_0^{\beta} d\tau_1 \int_0^{\beta} d\tau_2 \sum_{\sigma\sigma'} \hat{c}_{0\sigma}(\tau_1) \sum_{ij} t_{i0} t_{0j} \langle \hat{c}^{\dagger}_{i\sigma}(\tau_1) \hat{c}_{j\sigma'}(\tau_2) \rangle_{(0)} \hat{c}^{\dagger}_{0\sigma'}(\tau_2).$$

And for $\tau_2 > \tau_1$, the second term becomes

$$= \int_0^\beta d\tau_2 \int_0^\beta d\tau_1 \sum_{\sigma\sigma'} \hat{c}_{0\sigma'}(\tau_2) \sum_{ij} t_{0i} t_{j0} \langle \hat{c}^\dagger_{j\sigma,}(\tau_2) \hat{c}_{i\sigma}(\tau_1) \rangle_{(0)} \hat{c}^\dagger_{0\sigma}(\tau_1). \tag{1.12}$$

Here, τ_1 and τ_2 being dummy variables, we can switch them so that the two terms above are the same. Similarly, for $\tau_1 < \tau_2$, the first term and for $\tau_1 > \tau_2$, the second term are the same. Taking the hopping amplitude for hopping into and out of the cavity are taken to be same for same i, i.e., $t_{0i} = t_{i0}$, we get

$$\begin{aligned} &\frac{1}{2!}\int_0^\beta d\tau_1 \int_0^\beta d\tau_2 \Big\langle \mathcal{T}_\tau \Delta\mathcal{S}(\tau_1) \Delta\mathcal{S}(\tau_2) \Big\rangle_{(0)} \\ &= \int_0^\beta d\tau_1 \int_0^\beta d\tau_2 \sum_{\sigma} \hat{c}^\dagger_{0\sigma}(\tau_1) \sum_{ij} t_{i0} t_{j0} \langle \mathcal{T}_\tau \hat{c}_{i\sigma}(\tau_1) \hat{c}^\dagger_{j\sigma}(\tau_2) \rangle_{(0)} \hat{c}_{0\sigma}(\tau_2) \\ &= -\int_0^\beta d\tau_1 \int_0^\beta d\tau_2 \sum_{\sigma} \hat{c}^\dagger_{0\sigma}(\tau_1) \sum_{ij \neq 0} t_{i0} t_{j0} G^{(0)}_{ij}(\tau_1 - \tau_2) \hat{c}_{0\sigma}(\tau_2) \end{aligned} \tag{1.13}$$

where $G^{(0)}_{ij,\sigma}(\tau_1 - \tau_2) = -\langle \mathcal{T}_\tau \hat{c}_{i\sigma}(\tau_1) \hat{c}^\dagger_{j\sigma}(\tau_2) \rangle_{(0)}$. Similarly, all the higher terms can be evaluated and $\mathcal{S}_{\text{eff}}$ can be written as

$$\begin{aligned} \mathcal{S}_{\text{eff}} = &\mathcal{S}_0 + \sum_{n=1}^{\infty} \sum_{i_1 \cdots i_n} \int d\tau\, \eta^\dagger_{i_1}(\tau_{i_1}) \cdots \eta^\dagger_{i_n}(\tau_{i_n})\, \eta_{j_1}(\tau_{j_1}) \cdots \eta_{j_n}(\tau_{j_n}) \\ &G^{(0)}_{i_1 \cdots j_n,\sigma}(\tau_{i_1} \cdots \tau_{i_n}, \tau_{j_1} \cdots \tau_{j_n}) \end{aligned} \tag{1.14}$$

where $\eta_i = t_{i0}\hat{c}_{i0}$. Now t_{ij} in $d \to \infty$ limit goes like $\frac{1}{\sqrt{d}}$ and $G^{(0)}_{i_1 j_1}$ goes $\sim \frac{1}{(\sqrt{d})^{|i_1 - j_1|}}$. For $n = 1$, two sums over $i_1 \neq j_1$ give $\sim d^2$. Two factors of t gives $\frac{1}{d}$. The two-point propagator G^0_{ij} goes $\sim \frac{1}{d}$, as i_1 and j_1 are both NN of site 0 and the shortest lattice step between them is at least 2, i.e., $|i_1 - j_1| = 2$. Hence, the total contribution for $n = 1$ is $\mathcal{O}(d^0)$. For $i_1 = j_1$ also the $n = 1$ term is $\mathcal{O}(d^0)$ for $d \to \infty$. For the $n = 2$ term, four sums over distinct i_1, i_2, j_1, j_2 give d^4, and four factors of t_{ij} give $\frac{1}{d^2}$. The four-point Green's function $G^0_{i_1,i_2,j_1,j_2}(\tau_{i_1}\tau_{i_2}, \tau_{j_1}\tau_{j_2})$ goes like $\frac{1}{(\sqrt{d})^{|i_1-i_2|}} \frac{1}{(\sqrt{d})^{|i_1-j_1|}} \frac{1}{(\sqrt{d})^{|i_1-j_2|}} \sim \frac{1}{d^3}$. The net result is $\frac{1}{d}$. Similarly, all the higher terms vanish, and only the two-point propagator contributes. The effective action now becomes

$$\mathcal{S}_{\text{eff}} = \int_0^\beta d\tau \sum_\sigma \hat{c}_{0\sigma}^\dagger(\tau)\Big(\partial_\tau - \mu\Big)\hat{c}_{0\sigma}(\tau) + U\int_0^\beta d\tau \hat{c}_{0\uparrow}^\dagger(\tau)\,\hat{c}_{0\uparrow}(\tau)\,\hat{c}_{0\downarrow}^\dagger(\tau)\,\hat{c}_{0\downarrow}(\tau)$$
$$+ \int_0^\beta d\tau_1 \int_0^\beta d\tau_2 \sum_\sigma \hat{c}_{0\sigma}^\dagger(\tau_1) \sum_{ij\neq 0} t_{i0}t_{j0}G_{ij,\sigma}^{(0)}(\tau_1-\tau_2)\hat{c}_{0\sigma}(\tau_2). \quad (1.15)$$

Now the above single-site effective dynamics can be compared to the Weiss effective action with free Weiss effective field $\mathcal{G}_0(\tau_1 - \tau_2)$, which is introduced as,

$$\mathcal{S}_{\text{eff}}^{\text{Weiss}} = -\int_0^\beta d\tau_1 \int_0^\beta d\tau_2 \sum_\sigma \hat{c}_{0\sigma}^\dagger(\tau_1)\mathcal{G}_{0\sigma}^{-1}(\tau_1-\tau_2)\hat{c}_{0\sigma}(\tau_2) + \int_0^\beta d\tau\, U\hat{c}_{0\uparrow}^\dagger(\tau)\,\hat{c}_{0\uparrow}(\tau)\,\hat{c}_{0\downarrow}^\dagger(\tau)\,\hat{c}_{0\downarrow}(\tau) \quad (1.16)$$

where $\mathcal{G}_{0\sigma}(\tau_1-\tau_2)$ denotes the effective amplitude for a fermion to be created at site 0 at time τ_1 (coming from the external bath) and being destroyed at time τ_2 (going back to the bath).

$$\mathcal{G}_{0\sigma}^{-1}(\tau_1-\tau_2) = -\Big(\frac{\partial}{\partial\tau} - \mu\Big)\delta_{\tau_1\tau_2} - \Delta_{0\sigma}(\tau_1-\tau_2) \quad (1.17)$$

where $\Delta_{0\sigma}(\tau_1 - \tau_2) = \sum_{ij\neq 0} t_{i0}t_{j0}G_{ij\sigma}^{(0)}(\tau_1 - \tau_2)$ is the hybridization function. In frequency space, Eqn (1.17) becomes

$$\mathcal{G}_{0\sigma}^{-1}(i\omega_n) = i\omega_n + \mu - \sum_{ij\neq 0} t_{i0}t_{j0}G_{ij,\sigma}^{(0)}(i\omega_n) = i\omega_n + \mu - \Delta_{0\sigma}(i\omega_n). \quad (1.18)$$

Now for a general lattice, the cavity Green's function $G_{ij\sigma}^{(0)}$ and full lattice, Green's function $G_{ij\sigma}$ are given by (Hubbard, 1964),

$$G_{ij\sigma}^{(0)} = G_{ij\sigma} - G_{i0\sigma}G_{00\sigma}^{-1}G_{0j\sigma} \quad (1.19)$$
$$\text{where,}\quad G_{00\sigma} \equiv G_{\sigma,\text{imp}}(i\omega_n) = -\big\langle \hat{c}_\sigma(i\omega_n)\hat{c}_\sigma^\dagger(i\omega_n)\big\rangle_{S_{\text{eff}}^{\text{Weiss}}(\mathcal{G})}$$

where "imp" is for impurity site. Inserting Eqn (1.19) in (1.18) and Fourier transforming into $\boldsymbol{k}$-space, one finds the Dyson equation as

$$G_{\sigma,\text{imp}}^{-1}(i\omega_n) = \mathcal{G}_{0\sigma}^{-1}(i\omega_n) - \Sigma_\sigma(i\omega_n) = i\omega_n + \mu - \Delta_\sigma(i\omega_n) - \Sigma_\sigma(i\omega_n). \quad (1.20)$$

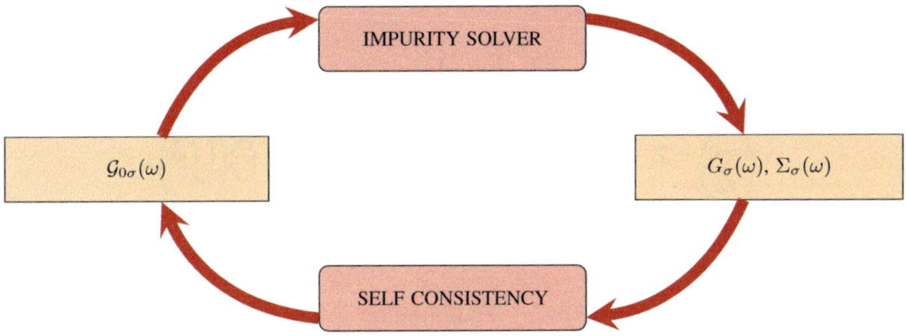

Figure 1.2 Flowchart of DMFT algorithm. For a chosen $\mathcal{G}_{0\sigma}(\omega)$, one can solve the impurity problem and calculate $G_\sigma(\omega)$ (interacting Green's function) and $\Sigma_\sigma(\omega)$ (self-energy). A new $\mathcal{G}_{0\sigma}(\omega)$ (bath Green's function) will produce using the self-consistency condition. After sufficient number of iteration loops, the convergence is achieved. Figure taken from [8].

Now, DMFT requires $G_{\sigma,\text{imp}}(i\omega_n)$ to be the same with the site-diagonal interacting lattice Green's function given as,

$$G_\sigma(i\omega_n) = \int_{-\infty}^{\infty} d\epsilon \, \frac{\rho(\epsilon)}{i\omega_n - \epsilon + \mu - \Sigma_\sigma(i\omega_n)}. \tag{1.21}$$

The analytical continuation $i\omega_n \rightarrow \omega + i\delta$ gives the real frequency (time) full lattice local Green's function $G_\sigma(\omega)$. Eqn (1.19), (1.20) and (1.21) form a closed set of equations for DMFT. An iterative solution is: start with some Weiss field $\mathcal{G}_\sigma$, obtain $G_{\sigma,\text{imp}}$ from Eqn (1.19) and determine Σ_σ from Eqn (1.20), and then calculate G_σ from Eqn (1.21) (Figure 1.2). If $G_{\sigma,\text{imp}} = G_\sigma$, then we have the solution, and if not, then obtain a new $\mathcal{G}_\sigma$ using Eqn (1.20) and repeat until convergence is reached.

1.3.2 Single-impurity anderson model (SIAM)

Due to a purely local interaction term U, the dynamics of the Hubbard model remains difficult even in $d \rightarrow \infty$ as the frequencies are coupled. There are a number of numerical methods known as impurity solvers, for solving single-site Anderson impurity model, such as for finite temperatures, the Hirsch-Fye quantum Monte Carlo (HFQMC) algorithm, and continuous-time quantum Monte Carlo (CTQMC) and for zero temperature, exact diagonalization (ED), numerical renormalization group (NRG) and the density matrix renormalization group (DMRG). One of the first methods used to solve the impurity Hamiltonian is the iterated perturbation theory (IPT).

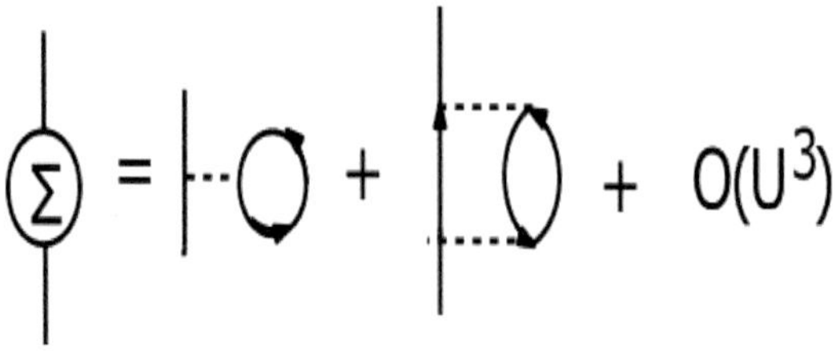

Figure 1.3 Self-energy of the impurity Hamiltonian using perturbation theory up to the second order of U.

Although simplistic in nature, IPT gives a qualitatively right description of the metal–insulator transition in the Hubbard model on infinite-dimensional lattices (on the quantitative level, it tends to underestimate correlations though). In IPT, one iteratively solves the impurity solver using the second-order perturbation theory in Hubbard interaction U to approximate the impurity self-energy (Figure 1.3). For the half-filling (particle-hole symmetric case), it reads as

$$\Sigma(i\omega_n) \approx \frac{U}{2} + U^2 \int_0^\beta d\tau \mathcal{G}_0(-\tau)^2 \mathcal{G}_0(-\tau). \tag{1.22}$$

Here, we have neglected any spin–flip terms in the expression of the self-energy. We omitted the spin index for convenience. The first term in the self-energy expression is the Hartree–Fock contribution $U/2$. This term cancels with a term from $\mathcal{G}_0(\text{in})^1$ when the self-energy is substituted into the Dyson equation. For this reason, we neglect the Hartree contribution from both functions.

The IPT is successful by the fact that it not only becomes exact in the weak coupling limit (by construction) but also reproduces an atomic-limit expression for (in) as U grows large.

1.3.3 Results and discussions

The DMFT equations should produce correct non-interacting and atomic limits, which are as follows: (i) In the non-interacting case, $U = 0$; thus, $\Sigma_\sigma(i\omega_n) = 0$ and so Eqn (1.21) gives $G_\sigma(i\omega_n) = \mathcal{G}_\sigma(i\omega_n)$. (ii) In atomic limit, $t_{ij} = 0$, i.e., $\epsilon_{\mathbf{k}} = 0$. $\rho(\epsilon) = \delta(\epsilon)$. From Eqn (1.21) $G_\sigma(i\omega_n) = \left[i\omega_n + \mu - \Sigma_\sigma(i\omega_n)\right]^{-1}$. So, from Eqn (1.20) $\mathcal{G}_\sigma^{-1}(i\omega_n) = i\omega_n + \mu$, i.e., $\mathcal{G}_\sigma^{-1}(\tau) = -\partial_\tau + \mu$, which agrees with Eqn (1.18) as $\Delta_\sigma = 0$ for $t_{ij} = 0$. We present the result for intermediate U using DMFT with IPT as an impurity

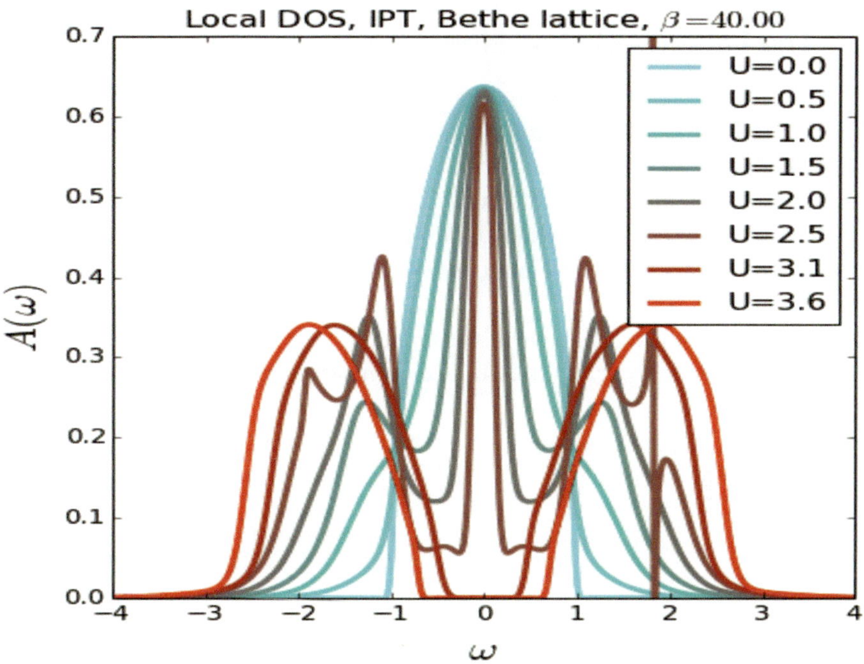

Figure 1.4 The local density of states ($A(\omega)$) for different interaction strengths U in the Hubbard model using IPT+DMFT. At the critical point, the $A(\omega)$ vanishes at $\omega = 0$. The MIT is driven by the electron correlations. The result is produced using the TRIQS software [14].

solver (Figure 1.4). The local density of states is calculated from the local Green's function as, $A(\omega) = -\frac{1}{\pi}G_\sigma(\omega)$. The $G_\sigma(\omega)$ is calculated from the $G_\sigma(i\omega_n)$ using the Padé approximation. At half filling, the $A(\omega)$ vanishes at $\omega = 0$ at a critical value of U that manifests the metal–insulator transition (MIT). As the MIT is driven by strong electron–electron correlations, it is called Mott metal-to-insulator transition. In the metallic phase, the self-energy shows the Fermi-liquid behavior ($Im\Sigma(\omega) \sim -\omega^2$) in the low frequency region with Kondo peak at $\omega = 0$. At the critical U, the Kondo peak suddenly disappears, giving the insulating phase. In this phase, in the gap, the imaginary part of the self-energy vanishes except with a singularity at $\omega = 0$. At the critical U, the real part of the self-energy shows discontinuity at $\omega = 0$. The MIT is first-order as it exhibits the coexistence region.

1.4 The Falicov–Kimball Model (FKM)

In 1963–65, Hubbard and Gutzwiller introduced the Falicov–Kimball model (FKM) for the first time as a simplification of the Hubbard model. In 1969, the FKM model [4] was introduced with a few extra terms so that it can capture the metal–insulator phase transitions in rare-earth materials and transition-metal compounds. Experimental data suggested that this transition is due to the interactions between electrons in two electronic states: non-localized states (itinerant electrons) and states that are localized around the sites corresponding to the metallic ions of the crystal (static electrons). The model is exactly solvable analytically in infinite dimensions, first shown by Brandt and Mielsch [1, 2, 3].

The FKM model describes the interaction between spinless conduction electrons and localized electrons or ions. The Hamiltonian of this model is

$$H = t\sum_{\langle ij\rangle}(c_j^\dagger c_i + c_i^\dagger c_j) + \sum_i \epsilon_f f_i^\dagger f_i - \mu\sum_i (c_i^\dagger c_i + f_i^\dagger f_i) + U\sum_i c_i^\dagger c_i f_i^\dagger f_i \tag{1.23}$$

where c_i is the annihilation operator and $c_i^\dagger$ is the creation operator of the conduction electron at site i, respectively, the kinetic term t is for c-electrons, the site energy ϵ_f for f-electrons, the chemical potential μ, on site interaction between c and f electrons is U. It is worth mentioning that the solution of the $d = \infty$ Hubbard model has been approximated by several authors with the solution of the infinite-dimensional FKM model. Here statistical averaging is used to determine the number density of localized f electrons. Considering spinfull Hubbard model with one spin (up or down spins) electron as frozen or static and, in this background, the other spin electrons are moving, then the Hubbard model and the FKM model are equivalent.

It is to mention that the FKM model has one-to-one correspondence (isomorphic) to the binary disorder model. The localized f-electrons act like a binary disorder. Although in the binary disorder model the disorder average is already fixed, i.e., disorder is quenched-like, whereas in FKM model, the average f-electrons, i.e., $\langle n_f\rangle$ is statistically calculated, i.e., disorder is annealed-like. So one may assume the FKM model as a binary disorder model with annealed averaging over $\langle n_f\rangle$ in place of quenched averaging.

1.5 Exact Solution of the Falicov–Kimball Model using DMFT

The Falicov–Kimball model can be solved exactly in an infinite dimension using DMFT. Separating site 0 and integrating out rest of the lattice degrees of freedom, the partition function describes an effective action $\mathcal{S}_{\text{eff}}$,

$$\mathcal{Z} = \int \prod_{\sigma} \mathcal{D}\hat{c}_{0\sigma}^{\dagger} \mathcal{D}\hat{c}_{0\sigma} \exp[-\mathcal{S}_{\text{eff}}] \tag{1.24}$$

with the effective action given as

$$\begin{aligned} \mathcal{S}_{\text{eff}} = -\int_0^{\beta} d\tau \int_0^{\beta} d\tau' \hat{c}^{\dagger}(\tau) \mathcal{G}_0^{-1}(\tau - \tau') \hat{c}(\tau') \\ + U \int_0^{\beta} d\tau \hat{c}^{\dagger}(\tau) \hat{c}(\tau) \hat{f}^{\dagger}(\tau) \hat{f}(\tau) \\ + \int_0^{\beta} d\tau \hat{f}^{\dagger}(\tau) (\frac{\partial}{\partial \tau} + \epsilon_f - \mu) \hat{f}(\tau). \end{aligned} \tag{1.25}$$

For the FKM model, the two Hilbert subspace of f-electrons $f^{\dagger}f = 0$ and $f^{\dagger}f = 1$ evolve independently under the dynamics of the c-electrons. Hence, we can separate out the partition function for $n_f = 0$ and $n_f = 1$ subspaces with

$$\mathcal{Z} = \mathcal{Z}_0 + \mathcal{Z}_1. \tag{1.26}$$

The $\mathcal{Z}_0$ corresponds to $f^{\dagger}f = 0$ sector and $\mathcal{Z}_1$ corresponds to $f^{\dagger}f = 1$ sector. The impurity Green's function obtain as

$$G(i\omega_n) = -\langle c^{\dagger} c \rangle_{S_{\text{eff}}} = \frac{w_0}{\mathcal{G}_0^{-1}(i\omega_n)} + \frac{w_1}{\mathcal{G}_0^{-1}(i\omega_n) - U} \tag{1.27}$$

with $w_0 = \frac{\mathcal{Z}_0}{\mathcal{Z}} = 1 - w_1$ and

$$w_1 = \left\{ 1 + \exp\left(\beta(\epsilon_f - \mu) + \sum_n \{\ln[\mathcal{G}_0^{-1}(i\omega_n)] - \ln[\mathcal{G}_0^{-1}(i\omega_n) - U]\} e^{i\omega_n 0^+}\right) \right\}^{-1}. \tag{1.28}$$

By differentiating the partition function w.r.t. the ϵ_f, we can show that w_1 is the average of the $f-$electrons per lattice site. The conduction electron self-energy can also be written using Dyson's equation as

$$\Sigma_n(G_n) = \frac{U}{2} - \frac{1}{2G_n} + \sqrt{\left(\frac{U}{2} - \frac{1}{2G_n}\right)^2 + w_1 \frac{U}{G_n}} \tag{1.29}$$

where both w_1, Σ_n are functions of G_n. If w_1 is constant, the result will be similar to the coherent potential approximation (CPA) self-energy of random binary alloy.

One can compute a lattice green function using the Dyson equation. For the lattice green function, the Dyson equation is

$$(i\omega_n + \mu - \Sigma_n^{ii}(G_n^{ii}))G_n^{il} - \sum_m t_{im} G_n^{ml} = \delta_{il}. \tag{1.30}$$

One can assume Σ^{ii} as of site independent for $d = \infty$, i.e., the self-energy is local. Therefore, in the Fourier space (k-space), the lattice green function is

$$G_n(k) = \frac{1}{i\omega_n + \mu - \Sigma_n - \epsilon_k}. \tag{1.31}$$

Thus, the local lattice green function becomes

$$G_n^{ii} = \sum_k \frac{1}{i\omega_n + \mu - \Sigma_n - \epsilon_k} = \int_{-\infty}^{\infty} d\epsilon \frac{\rho_0(\epsilon)}{i\omega_n + \mu - \epsilon - \Sigma(i\omega_n)}. \tag{1.32}$$

1.6 Results and Discussions

Considering a Bethe lattice of infinite coordination number ($d = \infty$) and half-bandwidth unity, we have computed the local Green function $G(\omega)$ using DMFT formalism. The self-consistency loop has converged upto an error of 10^{-5}. The formula of spectral density of any lattice is: $\rho(k,\omega) = -\frac{1}{\pi} ImG(k,\omega)$. For a Bethe lattice of infinite dimension (or infinite coordination number), we know that both Green function and self-energy are local. So, the spectral function or the density of states (DOS) of the infinite-dimensional Bethe lattice is given as, $\rho(\omega) = -\frac{1}{\pi} ImG(\omega)$.

We plot the local DOS using DMFT for different disorder strengths (U) in the particle–hole symmetric case (half filling) as shown in Figure 1.5. The plot exhibits a clear continuous metal–insulator transition as a function of U. The critical value of U is $U_c = 1.01$ (cyan curve). It is worth mentioning the behavior of the self-energy in the metallic state as well as the insulating state. In the metallic state, self-energy displays non-Fermi-liquid characteristics. The imaginary part of the self-energy displays the quadratic behavior with a wrong sign, and at zero temperature, it does not go to zero in the limit $\omega = 0$. The real part of the self-energy shows a linear behavior. In the insulating phase, the imaginary part diverges while the real part shows a sharp discontinuity at $\omega = 0$. As a result, the DOS vanishes at $\omega = 0$.

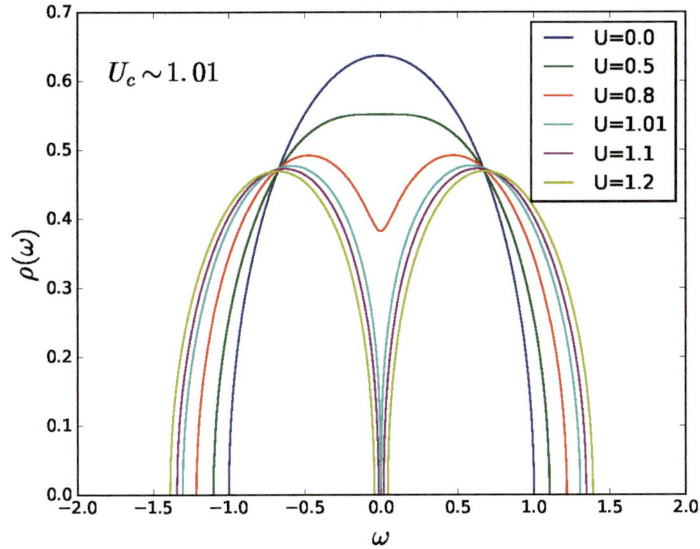

Figure 1.5 For binary disorder model with $p - h$ symmetry case, the local density of states (LDOS) shows a continuous metal–insulator transition at $U_c \approx 1.0$ (cyan curve). Figure taken from [8].

Some important characteristics of the Falicov–Kimball model are as follows.

- f-electrons acts like a disorder scatterer that provides a finite lifetime even at the Fermi surface, causing the break down of the Fermi-liquid theory. Therefore, the system behaves like a non-Fermi-liquid even in the metallic region. Although, for small U, we describe the system as a "dirty Fermi liquid." "Dirty Fermi liquid" refers to the destabilization of the Fermi liquid phase in weakly disorder systems at a very low temperature.
- Contrasting the Hubbard model that exhibits two distinct phases, this model shows three different phases: (i) dirty Fermi liquid metal, (ii) non-Fermi-liquid metal, and (iii) insulator.
- At a low temperature, the $d = \infty$ FKM exhibits charge density wave (CDW) ordering phase transition at half-filling in all bipartite lattices with the increasing interaction (U).
- It is worth mentioning that there is an X-ray edge singularity in the f-electron spectral function. The detailed discussion of f-electron spectrum part is not presented in this chapter.

1.7 Cluster Extensions of DMFT

In DMFT formalism, a lattice with infinite coordination number, the lattice model has an exact mapping with the Anderson impurity model, and there is self-consistency between the lattice Green function and the impurity Green function. Unfortunately, for a lattice with finite coordination number, the mapping is an approximation. Within this approximation, we consider non-local lattice self-energy (momentum-dependent) as local self-energy, i.e., $\Sigma(k, \omega a) \approx \Sigma(\omega)$. Therefore, we neglect the inter-sites correlation. But to describe certain physical phenomena such as the super-exchange in the Hubbard model, it is important to incorporate the inter-sites correlation. DMFT formalism in such situations are not suitable at all.

Many ideas have been proposed recently to extend DMFT formalism so that inter-site correlations can be incorporated. In these extended formalisms, the self-energy is non-local (momentum dependent), i.e., $\Sigma(\vec{K}, \omega)$. With the increase of $\vec{K}$ values, the complexity of the problem increases.

In this chapter we discuss two different versions of the extended DMFT formalism - i) CDMFT (cluster dynamical mean field theory) and ii) DCA (dynamical cluster approximations). These two methods are described schematically in Figures 1.6(a) and Fig 1.6(b).

1.7.1 Cluster dynamical mean field theory (CDMFT)

In this technique, as shown in Figure 1.6(a), we treat a real space cluster of sites as impurity and it is embedded by a cluster of baths. For an infinitely

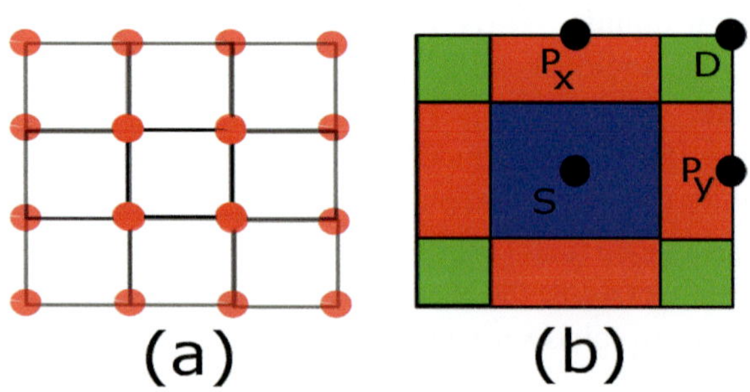

Figure 1.6 Two widely used extensions of DMFT: (a) real space cluster DMFT (CDMFT) on 2d square lattice with four sites square as an impurity and (b) momentum space cluster DMFT (DCA) [15] on 2d square lattice with four cluster momentum plaquette (S, P_x, P_y, and D). Figure taken from [8].

extended lattice, a large subset of the lattice is chosen as a cluster, which respects all the symmetries of the original lattice. For example, 2×2 or 4×4 cluster would be suitable choices for the two-dimensional lattice. Then the full lattice is converted to a super-lattice, treating the cluster as a basis of the super-lattice. The super-lattice possesses all the symmetries that the original lattice also has. Now, consider the Anderson impurity problem on the super-lattice by choosing one site as the impurity, while rest acts as the bath. In CDMFT, the impurity green function ($\hat{G}_{\mathrm{imp}}(\omega)$) and the bath function ($\hat{\Delta}(\omega)$) are in matrix form (the dimension is equal to the number of sites in each cluster) and they are connected through the self-consistent matrix equations.

Let us convert an N-site lattice into an N_c-sites super-lattice where each site is a cluster of sites of the original lattice. Let the lattice vector of the cluster from the origin in the super-lattice be $\tilde{\mathbf{x}}$. Then a lattice point $\mathbf{x}$ in the original lattice is given as, $\mathbf{x} = \tilde{\mathbf{x}} + \mathbf{X}$, where $\mathbf{X}$ is the vector from the cluster to the lattice point. The Brillouin zone (BZ) of the original lattice in the reciprocal space is split into N_c number of super-lattice Brillouin zones with $\mathbf{k} = \tilde{\mathbf{k}} + \mathbf{K}$, where $\mathbf{k}, \tilde{\mathbf{k}}$ are the reciprocal lattice vectors in the original lattice BZ and super-lattice BZ, $\mathbf{K}$ in cluster momentum.

The hopping and the self-energy terms can be divided into two parts: intra-cluster and inter-cluster. Then one has

$$\begin{aligned} \mathbf{t}(\tilde{\mathbf{x_i}} - \tilde{\mathbf{x_j}}) &= \mathbf{t_c}\delta_{\tilde{\mathbf{x_i}},\tilde{\mathbf{x_j}}} + \delta\mathbf{t}(\tilde{\mathbf{x_i}} - \tilde{\mathbf{x_j}}) \quad (1.33) \\ \mathbf{\Sigma}(\tilde{\mathbf{x_i}} - \tilde{\mathbf{x_j}}, i\omega_n) &= \mathbf{\Sigma_c}(\mathbf{i\omega_n})\delta_{\tilde{\mathbf{x_i}},\tilde{\mathbf{x_j}}} + \delta\mathbf{\Sigma}(\tilde{\mathbf{x_i}} - \tilde{\mathbf{x_j}}, i\omega_n) \quad (1.34) \end{aligned}$$

where all bold quantities are of size $N_c \times N_c$. One can expand the Green function as

$$\begin{aligned} \mathbf{G}(\tilde{\mathbf{x_i}} - \tilde{\mathbf{x_j}}, i\omega_n) &= \mathbf{g}(i\omega_n)\delta_{\tilde{x}_i,\tilde{x}_j} \\ &+ \mathbf{g}(i\omega_n)\sum_l [\delta\mathbf{t}(\tilde{\mathbf{x_i}} - \tilde{\mathbf{x_j}}) \\ &+ \delta\Sigma(\tilde{x}_i - \tilde{x}_j, i\omega_n)\mathbf{G}(\tilde{\mathbf{x_i}} - \tilde{\mathbf{x_j}}, \mathbf{i\omega_n})] \quad (1.35) \end{aligned}$$

where the Green function g restricted to the cluster is defined as

$$\mathbf{g} = [(i\omega_n + \mu) - \mathbf{t}_c - \mathbf{\Sigma}_c(i\omega_n)]^{-1}. \quad (1.36)$$

This equation in momentum space simplifies to

$$\mathbf{G}(\tilde{\mathbf{k}}, i\omega_n) = \mathbf{g}(i\omega_n) + \mathbf{g}(i\omega_n)[\delta\mathbf{t}(\tilde{\mathbf{k}}) + \delta\mathbf{\Sigma}(\tilde{\mathbf{k}}, i\omega_n)]\mathbf{G}(\tilde{\mathbf{k}}, i\omega_n). \quad (1.37)$$

The self-energy is local in the DMFT formalism. Similar to DMFT, in CDMFT, we consider the cluster self-energy on the super-lattice as local, i.e., $\delta\mathbf{\Sigma} = 0$; so we can write

$$\begin{aligned} \mathbf{G}(\tilde{\mathbf{k}}, i\omega_n) &= \mathbf{g}(i\omega_n) + \mathbf{g}(i\omega_n)\delta\mathbf{t}(\tilde{\mathbf{k}})\mathbf{G}(\tilde{\mathbf{k}}, i\omega_n) \\ &= [\mathbf{g}^{-1}(i\omega_n) - \delta\mathbf{t}(\tilde{\mathbf{k}})^{-1}]^{-1}. \end{aligned} \tag{1.38}$$

The cluster Green function is in the "coarse gaining" form as

$$\bar{\mathbf{G}}(i\omega_n) = \frac{N_c}{N}\sum_{\tilde{\mathbf{k}}} \mathbf{G}(\tilde{\mathbf{k}}, i\omega_n) \tag{1.39}$$

and the non-interacting cluster Green function in CDMFT is

$$\mathcal{G}^{\mathbf{0}}(i\omega_n) = [\bar{\mathbf{G}}^{-1}(i\omega_n) + \mathbf{\Sigma}_c(i\omega_n)]^{-1}. \tag{1.40}$$

1.7.2 Dynamical cluster approximation (DCA)

In the DCA technique [15], the Brillouin zone (BZ) of the lattice is divided equally into N_c number of sub-zones (shown in Figure 1.6(b)). It is chosen so that the center of each sub-zone is the reciprocal lattice vector of the real space cluster. The function defining the subzones is given as

$$\phi_{\mathbf{K}}(\mathbf{k}) = \begin{cases} 1, & \text{if } \mathbf{k} \text{ is in patch } \mathbf{K} \\ 0, & \text{otherwise.} \end{cases} \tag{1.41}$$

We restrict the sum over momentum inside the $\mathbf{K}_{th}$ sub-zone only. The approximation of this technique is chosen in such a way that self-energy is constant on each sub-zone but different in different sub-zones, i.e.,

$$\Sigma(\mathbf{k}, i\omega_n) \approx \Sigma_c(\mathbf{K}, i\omega_n). \tag{1.42}$$

As we know that any self-energy is a function of the Green function, the self-energy on each cluster is also a function of the coarse-grained Green function, i.e., $\Sigma_c(\mathbf{K}, i\omega_n) = \Sigma_c[\bar{G}(\mathbf{K}, i\omega_n)]$. Now the coarse-grained Green function $\bar{G}(\mathbf{K}, i\omega_n)$ is given as

$$\begin{aligned} \bar{G}(\mathbf{K}, i\omega_n) &= \frac{N_c}{N}\sum_{\mathbf{k}} \phi_{\mathbf{K}}(\mathbf{k}) G(\mathbf{k}, i\omega_n) \\ &= \frac{N_c}{N}\sum_{\mathbf{k}} \phi_{\mathbf{K}}(\mathbf{k}) \frac{1}{i\omega_n - \epsilon_k + \mu - \Sigma^{\mathrm{DCA}}(\mathbf{k}, i\omega_n)}. \end{aligned} \tag{1.43}$$

The DCA self-energy $\Sigma^{\mathrm{DCA}}(\mathbf{k}, i\omega_n)$ is

$$\Sigma^{\mathrm{DCA}}(\mathbf{k}, i\omega_n) = \sum_{\mathbf{K}} \phi_{\mathbf{K}}(\mathbf{k})\Sigma_c(\mathbf{K}, i\omega_n). \tag{1.44}$$

The corresponding Weiss function (non-interacting Green function) is

$$\mathcal{G}^0(\mathbf{K}, i\omega_n) = [\bar{G}^{-1}(\mathbf{K}, i\omega_n) + \Sigma_c(\mathbf{K}, i\omega_n)]^{-1}. \tag{1.45}$$

This Weiss function $\mathcal{G}^0(\mathbf{K}, i\omega_n)$ is used as input in the cluster impurity solver. The impurity solver provides the new interacting cluster Green function $G_c(\mathbf{K}, i\omega_n)$. From the interacting cluster Green function $G_c(\mathbf{K}, i\omega_n)$ using Dyson's equation, we compute the cluster self-energy $\Sigma_c(\mathbf{K}, i\omega_n)$ as

$$\Sigma_c(\mathbf{K}, i\omega_n) = \mathcal{G}_0^{-1}(\mathbf{K}, i\omega_n) - G_c^{-1}(\mathbf{K}, i\omega_n). \tag{1.46}$$

The new cluster self-energy is used as input in eqn (1.42) and the iteration is continued till self-consistently is reached. The self-consistency loop stops as soon as the $\Sigma_c(\mathbf{K}, i\omega_n)$ converges.

1.8 Beyond Model Hamiltonian: LDA + DMFT

Till now, we have considered the single band model Hamiltonians like the Falicov–Kimball model and the Hubbard model. The DMFT technique can also be applied to the multi-orbital Hubbard model. Marrying density functional theory (DFT) band structure calculation, we can apply DMFT at ab-initio calculations for correlated materials. The technique like LDA+DMFT method allows us to treat one part of the system that is expected to exhibit the weakly interacting effects while another part – like d- or f-shells that show strong correlation effects. In the weakly interacting region, we can apply the local density approximation (LDA), and in the strongly correlated region, we can apply the DMFT. Kotliar et al. [9] provide a detailed review on this subject. Owing to the favorable scaling technique of the new impurity solvers with temperature enables us to include lower temperatures, many orbitals, and stronger interactions than before. The LDA+DMFT technique can be applied to solve any lattice problem with density–density interactions within the DMFT approximation. For example, in metals like Cerium for which the physics is well-understood in high-temperature limits, various LDA+DMFT techniques with the Hirsch-Fye QMC impurity solver have been performed previously. The results are very much impressive.

1.9 Conclusion

In brief, we can safely conclude that the DMFT framework provides a powerful quantitative description of the highly correlated fermionic systems. Although the approximation becomes exact in the infinite-dimensional lattice, it can be applicable in 2D or 3D systems along with LDA approximation. The technique is further in the developing stage and we present a very short introduction of it. The interested reader may go through an introductory article [10], a detailed review [5], a PhD thesis, as well as lecture notes [7, 8, 9] on DMFT and the cluster DMFT [11].

References

[1] U. Brandt and C. Mielsch. Thermodynamics and correlation functions of the falicov-kimball model in large dimensions. *Zeitschrift für Physik B Condensed Matter*, 75(3):365–370, Sep 1989.

[2] U. Brandt and C. Mielsch. Thermodynamics of the falicov-kimball model in large dimensions ii. *Zeitschrift für Physik B Condensed Matter*, 79(2):295–299, Jun 1990.

[3] U. Brandt and C. Mielsch. Free energy of the falicov-kimball model in large dimensions. *Zeitschrift für Physik B Condensed Matter*, 82(1):37–41, Feb 1991.

[4] L. M. Falicov and J. C. Kimball. Simple model for semiconductor-metal transitions: Smb_6 and transition-metal oxides. *Phys. Rev. Lett.*, 22:997–999, May 1969.

[5] Antoine Georges, Gabriel Kotliar, Werner Krauth, and Marcelo J. Rozenberg. Dynamical mean-field theory of strongly correlated fermion systems and the limit of infinite dimensions. *Rev. Mod. Phys.*, 68:13–125, Jan 1996.

[6] N. Goldenfeld. *Lectures on Phase Transitions and the Renormalization Group*. Frontiers in physics. Avalon Publishing, 1992.

[7] Emanuel Gull. *Continuous-time quantum Monte Carlo algorithms for fermions*. PhD thesis, 2008.

[8] Prosenjit Haldar, Study of quantum transport at the Metal-insulator transition in Falicov-Kimball Model within Alloy analogy, HBNI, Th133, 2017.

[9] G. Kotliar, S. Y. Savrasov, K. Haule, V. S. Oudovenko, O. Parcollet, and C. A. Marianetti. Electronic structure calculations with dynamical mean-field theory. *Rev. Mod. Phys.*, 78:865–951, Aug 2006.

[10] Gabriel Kotliar and D. Vollhardt. Strongly correlated materials: Insights from dynamical mean-field theory. *Physics Today*, 57:53, 2004.

[11] Thomas Maier, Mark Jarrell, Thomas Pruschke, and Matthias H. Hettler. Quantum cluster theories. *Rev. Mod. Phys.*, 77:1027–1080, Oct 2005.

[12] Walter Metzner and Dieter Vollhardt. Correlated lattice fermions in $d = \infty$ dimensions. *Phys. Rev. Lett.*, 62:324–327, Jan 1989.

[13] E Miranda and V Dobrosavljević. Disorder-driven non-fermi liquid behaviour of correlated electrons. *Reports on Progress in Physics*, 68(10):2337, 2005.

[14] Olivier Parcollet, Michel Ferrero, Thomas Ayral, Hartmut Hafermann, Igor Krivenko, Laura Messio, and Priyanka Seth. Triqs: A toolbox for research on interacting quantum systems. *Computer Physics Communications*, 196:398 – 415, 2015.

[15] E. Pavarini, E. Koch, Dieter Vollhardt, and Alexander Lichtenstein. The lda+dmft approach to strongly correlated materials modeling and simulation vol. 1. 2011.

2

Multiferroic Perspectives on Strain, Structure, and Properties

Ranjith Ramadurai[1] and Saj Mohan[1]

[1]Department of Materials Science and Metallurgical Engineering, Indian Institute of Technology Hyderabad, India

Abstract

This chapter briefly introduces multiferroic materials and the underlying physics behind the coupling between order parameter of these materials. Multiferroic materials being the most promising candidate of the correlated materials are classified into two broad classes based on the origin of ferroelectric polarization. Bismuth ferrite being the only known naturally existing room temperature magnetoelectric multiferroic material so far has been the major focus of the researchers around the world, in an attempt to develop the magnetoelectric multiferroic devices. The later part of the chapter introduces the physics and applications of the bismuth ferrite (BFO) thin films and the recent advancement in strain-induced effects on structure and properties of BFO epitaxial layers. The chapter describes various means of strain engineering and its use as a method to introduce and magnify some physical properties for device applications.

Keywords: Multiferroics, Magnetoelectric, Strain engineering, Bismuth ferrite, Epitaxial thin films

2.1 Introduction

The Maxwell theory of relativity has shown the correlated nature of phenomena such as the electric and magnetic fields in the early century. However, the cross-coupling between these was considered to be negligible as the origin of these parameters is mutually exclusive; the electric charges of the electrons

DOI: 10.1201/9781003337102-2

and ions are responsible for the charge effects and the spin of the electrons, and the exchange interactions between them define the magnetic properties. Later, Pierre Curie, Landau, and Lifshitz have predicted the coupling of the electric and magnetic ordering in materials; however, the effect of interaction between the respective order parameters was expected to be weak. It was Dzyaloshinskii who predicted and Astrov who experimentally observed the linear magnetoelectric coupling in Cr_2O_3. Later, it was in 1994 that H. Schmid coined the term multiferroic, considering the growing popularity of research in this field. The multiferroics are those multifunctional materials that simultaneously possess more than one cross-coupled ferroic ordering like ferroelectric (FE), magnetic, ferroelastic, or ferrotoroidicity. The order parameters in these materials are often coupled to each other and facilitate us to manipulate one using the other. The key research interest shown by the researchers all over the world is due to the possibility of controlling the magnetic spin orientation using the electric field or by changing the polarization direction on application of the magnetic field. In addition, certain materials might undergo a change in the crystal structure on application of the magnetic or electric fields, which then can be a potential candidate for the sensor and actuator applications. The major focus of the worldwide research is on the magnetoelectric materials though, aiming at various applications. The term ferroic describes the materials' vectorial property, which can be changed from one orientational state to another on the application of the external force. This orientational state also exhibits hysteresis behavior under the application of the external applied force. The components can be strain, polarization, magnetization, etc. The external force can be magnetic field, electric field, or mechanical stress.

2.1.1 Ferroics ordering and multiferroics

The four different ferroic ordering in the material could be explained as follows:

a. Ferroelectric (FE) materials possess spontaneous polarization that is switchable and can be changed by an external electric field; the materials with antiferroelectric orderings have permanent spontaneous polarization that cancels out in the crystallographic unit cell.
b. Ferromagnetic (FM) materials are those that have a spontaneous magnetization that can change in response to an external magnetic field. Similar to ferroelectric materials, antiferromagnetic materials also have spontaneously generated magnetic ordering, but they cancel one another within magnetic unit cells.

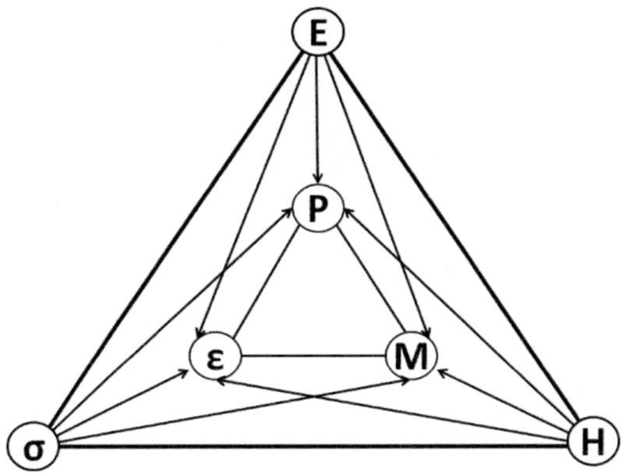

Figure 2.1 Interrelation between order parameters in multiferroics.

c. Ferroelastic materials are those that exhibit spontaneous strain and can be altered by the application of external applied stress.
d. Ferrotoroidic materials are those that display spontaneous order parameter and defined by the curl of magnetization or polarization. The order parameter is also switchable by application of the respective external field.

The magnetoelectric property is the phenomenon by which the polarization in a material is influenced by the applied magnetic field and vice versa [1]. As mentioned before, the multiferroic magnetoelectric material is the primary focus of research to develop materials with strong cross-coupling between parameters. The interrelationship between the order parameters are illustrated in the Heckmann diagram in Figure 2.1.

An FE material is one that can possess at least two different stable orientational polarization states in the absence of an external electric field. These polarization states should be able to switch from one state to another by the application of an external electric field because they are structurally identical but differ in the direction of polarization. The origin of the ferroelectricity in the material can be explained by the non-centrosymmetric crystal structure of the material, which is one of the fundamental criteria for the existence of FE behavior. Out of the 32-point group symmetries, only 21 of them possess a non-centrosymmetric nature, and FE materials belong to the class of materials that shows piezoelectric effect with a unique polar axis for the spontaneous polarization. However, FE materials are the subclass of the piezoelectric and pyroelectric materials. The FE materials undergo a phase

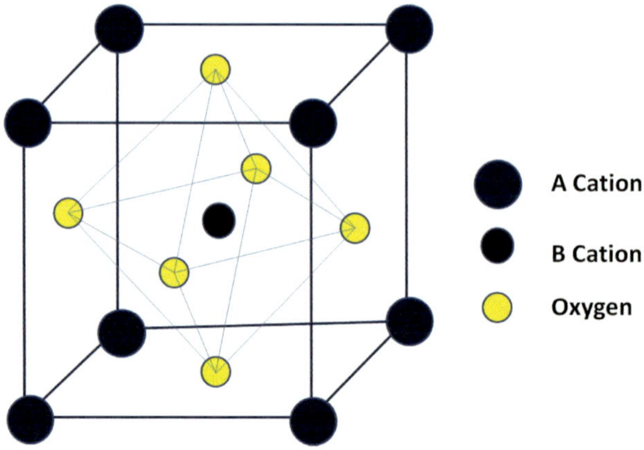

Figure 2.2 Ideal perovskite structure (e.g., $BaTiO_3$).

transition from a high-temperature centrosymmetric non-polar phase to the non-centrosymmetric polar phase below a certain temperature called Curie temperature (T_C)

The most widely studied FE materials are the materials with a perovskite structure, with the general formula ABO_3. In this structure, A and B represent the cations and O is the oxygen forming an octahedral cage with the B-cation. The origin of ferroelectricity is different in different perovskite materials. The ideal perovskite structure is illustrated in Figure 2.2. The perovskite structure deviates from the ideal cubic structure and stabilizes in the low symmetry structures. The degree of deviation and the stability of the perovskite are determined by the Goldschmidt tolerance factor, which is given by

$$\tau = \frac{(r\mathrm{A} + r\mathrm{O})}{\sqrt{(2\,(r\mathrm{B} + r\mathrm{O})\,)}} \tag{2.1}$$

where r_A, r_B, and r_O are the ionic radii of A, B, and O ions respectively. This is derived from the geometry of a cube in which the atoms are of such sizes that they touch each other and hence the face diagonal of the cell would be $\sqrt{2}$ times of the unit cell length. For the cubic structure, in the ideal case, the tolerance factor is 1.

2.1.2 Classification of multiferroics

The origin of magnetism in the materials is caused by the uncompensated spins of the half-filled *d* shell or *f* shell electrons in the materials. The

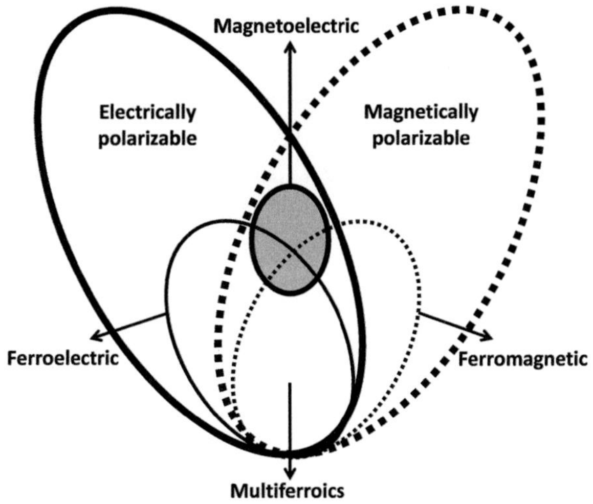

Figure 2.3 Venn diagram representing the cross-coupling in the electrically and magnetically polarizable materials.

exchange interactions between these spins give rise to a specific type of magnetic ordering in the system. There can be different kinds of magnetic ordering like FM ordering, antiferromagnetic ordering, ferrimagnetic ordering, etc., depending upon the type of the interactions present in the system. But the origin of the ferroelectricity is mainly due to the d^0 configuration of the rattling B-cation present in the perovskite (ABO_3) system and/or presence of $6s^2$ lone pair effect in lead and bismuth type ferroelectrics. The plausible formation of magnetoelectric multiferroics materials by cross-coupling of the magnetically and electrically polarizable materials is illustrated using the Venn diagram in Figure 2.3.

Multiferroic materials were classified into two major types based on the origin of FE polarization in a given material.

***Type-I multiferroics*:** The multiferroic materials are classified as the type-I multiferroic where the origin of ferroelectricity and magnetism has different sources and are usually independent of each other, with weak cross-coupling between them. In these materials, the polarization and magnetism appear at rather different temperatures and possess large polarization. They are again classified into four types depending on the origin of ferroelectricity in the material.

a. Multiferroic mixed perovskites:

One of the ways to address the contraindicative presence of magnetic and electric ordering, that is the d^0 vs. d^n problem, is to substitute/dope d^n cations in or different from the d^0 site. In these systems, the ferroelectricity and magnetism originate from different cations; however, as discussed before, the cross-coupling between them could be weak, similar to systems like $PbFe_{1-x}Nb_xO_3$, etc.

b. Ferroelectricity due to lone pair:

In materials like $BiFeO_3$, $BiMnO_3$, and $PbTiO_3$, the stereo chemically active $6s^2$ lone pair electrons that do not participate in the chemical bonding give rise to a lattice distortion and displace the B-cation, which gives rise to high polarization. The ordering of these lone pair electrons and the structural distortion due to it is the origin of ferroelectricity.

c. Ferroelectricity due to charge ordering:

In the charge ordered systems, materials with different transition metal cations and/or with different oxidation state and bond lengths can lead to a polarization and effectively give rise to ferroelectricity in such systems. In materials like $Pr_{0.5}Ca_{0.5}MnO_3$, $TbMn_2O_5$, $CaCoMnO_6$, and $RMnO_3$, the ferroelectricity is caused because of the charge ordering.

d. Geometric ferroelectricity:

In the case of the materials like $YMnO_3$, the tilting of the hard MnO_5 block creates the spontaneous polarization. The oxygen atom moves close to the relatively small Y ions as a result of the tilting, which only serves to promote compact packing, as shown in Figure 2.4.

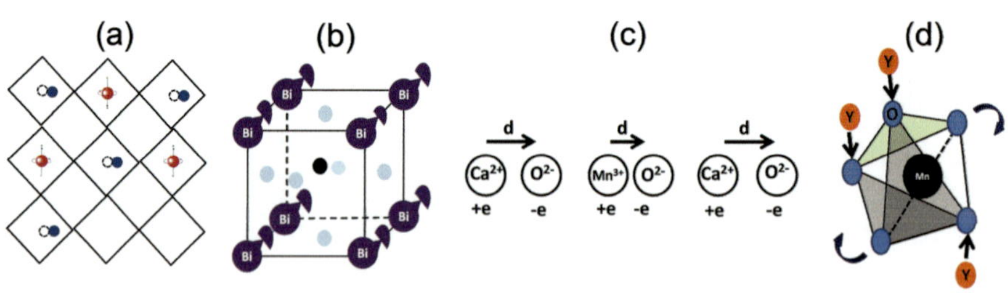

Figure 2.4 Different mechanisms for ferroelectricity in type-I multiferroics. (a) Mixed perovskite. (b) Ferroelectricity due to lone pair. (c) Charge ordering. (d) Geometric ferroelectricity (figure recreated for detailing similar to ref.[2])

***Type-II Multiferroics*:** In this class of the materials, a certain magnetically ordered state and a specific type of magnetic orderings are required for the ferroelectricity to exist [3, 4]. For example, $TbMnO_3$, $TbMn_2O_5$, etc., exhibit the ferroelectricity only in the magnetically ordered state below the lower transition temperature at 28 K. The ferroelectricity occurs only in the magnetically ordered state, where the spin rotation plane is perpendicular to the propagational vector. In other cases, the polarization is zero. The inverse D-M interaction occurs and induces a lattice distortion which results in a polarization that appears only in the magnetically ordered state. There are two subclasses for the type-II multiferroics.

a. *Spiral type-II multiferroics:*

In the presence of external magnetic field, the magnetic spin cycloid will flip and accordingly the polarization will flip, as the 90° phase difference should be maintained between the two.

b. *Collinear type-II multiferroics:*

In this type of multiferroics, the ferroelectricity appears as a result of the exchange striction in the material, when the material is magnetically ordered. Ca_3CoMnO_6 is an example for this type of material. The distortion and thereby breaking of the material's inversion symmetry is caused by differences in the lengths of the ferromagnetic and antiferromagnetic bonds, leading to spontaneous polarization.

2.2 Bismuth Ferrite Structure and Properties

Bismuth ferrite (BFO) is the only single-phase room temperature multiferroic that has been discovered so far. Because of its scope in utilizing its multiferroic property for the device applications, the researchers all over the world have been doing immense research in understanding its structure, property, and applications. BFO is formed with the equal composition of Bi_2O_3 and Fe_2O_3. The formation of the impurity phases like bismuth oxide, iron oxide, and other iron-rich phases like $Bi_2Fe_4O_9$ and the bismuth-rich phase $Bi_{25}FeO_{39}$ is commonly observed in the synthesis of bismuth ferrite. Also, BFO can decompose at higher applied voltages into its substituent components [5]. These impurity phases are the cause for the material's leaky nature and could also influence the magnetic properties.

The crystal structure of BFO is the rhombohedral distorted perovskite R3c structure with a pseudocubic unit cell parameter of $a = 3.965$ Å and the rhombohedral angle of 89.4°. It can alternatively be represented as a hexagonal structure with the lattice parameters $a_{hex} = 5.58$ Å and $c_{hex} = 13.9$ Å. The Gold Schmidt tolerance factor "τ" for BFO is 0.88, with Fe^{3+} in six-fold coordination with high spin state. When the tolerance factor is less than 1, the octahedral has to buckle to fit into the unit cell and will have the oxygen octahedral tilt of $11° - 14°$ around the [111] axis, changing the bond angle of Fe-o-Fe to $154° - 156°$. BFO is a Type-I multiferroic in which the bismuth $6s^2$ lone pair induced ferroelectricity and (Fe-O-Fe) are in commensurate spin cycloidal G-type antiferromagnetic ordering.

At a higher temperature of 825 °C, the β phase of the BFO forms with volume contraction, which is the ferroelectric to paraelectric phase transition that most groups claim to be centrosymmetric in nature [6, 7]. There is another phase transition at 925 °C into the ϒ phase of BFO, which is also a centrosymmetric cubic structure. There are some reports claiming the β phase being orthorhombic, monoclinic, tetragonal, etc. But most recent neutron diffraction studies show the orthorhombic nature of the crystal structure [8]. There are reports of a high-pressure-induced phase transition at 10 Gpa into an orthorhombic symmetry at room temperature, which is similar to the β phase proposed by many groups [4]. The pressure-induced rhombohedral–orthorhombic–cubic phase transition is similar to the temperature-induced phase transition. Another way of the phase transition at room temperature reported is the chemical pressure induced, where the bismuth ions are substituted by the lanthanum ion, and at around 0.75% substitution, the structure changes to a orthorhombic structure [9, 10]. Since the lone pair $6s^2$ electrons of the bismuth ions are the source of ferroelectricity in BFO, non-FE cation substitution in the bismuth site will hamper the polarization in the material.

The displacement of the bismuth ions relative to the FeO_6 octahedra causes the ferroelectric ordering in BFO [11]. The symmetry allows the polarization direction in the <111> pseudocubic direction, forming eight possible polarization variants. In the early studies a small P_r was observed in the bulk ceramics, but the unprecedented large polarization in the epitaxial thin films were reported around 60 μC/cm^2 [12] for the film grown in (001) direction and around 100 μC/cm^2 for the (111) oriented film. However, the major drawbacks in BFO thin films are the high leakage current and low resistivity, caused by the defects and/or non-stoichiometry in the BFO compound. On an attempt to bring down the leakage problems, a report shows many attempts in introducing the substitution in the A/B sites. La, Ca, Tb, Sm, Nd, and Gd ions are used for partial substitution into the A-site, which improves the ferroelectricity and enhance the spin structure [13–14]. Substitution of the ions like Sc, Mn, Ti, Cr, and Nb into the B-site will strongly reduce the oxygen vacancies and the leakage current and stabilize the spin structure [15–16]. The dielectric constant of the BFO in (001) oriented films is ∼100, which is smaller than the conventional piezoelectric materials like PZT, BST BTO, etc., which causes the limited application approach with relatively smaller d_{33} values [17].

The magnetic ordering in the BFO is G-type antiferromagnetic with all the spins aligned ferromagnetically within the (111) plane and antiferromagnetically aligned to the adjacent (111) plane. Each Fe^{3+} spin is surrounded by the anti-parallel spins of the nearest Fe neighbor. Owing to the intrinsic structural distortion, Fe-O-Fe bond angle changes and forms a canting in the system due to the Dzyaloshinskii–Moriya (D-M) interaction. Thus, the magnetoelectric coupling in BFO is coupled through the lattice by the inverse D-M interaction. This forms an incommensurate spin cycloidal structure of 62–64 nm along the direction [10I], as shown in Figure 2.5. The magnetic Neel temperature is ∼643 K and the spin cycloidal arrangement can be altered at a lower temperature [18].

The magnetoelectric coupling is weak in the case of BFO because the origin of magnetism and ferroelectricity are from different sites. The magnetoelectric coupling in the BFO enables the electric-field-induced change in the magnetism and vice versa and manifests for the multiferroic-based device applications. The change in the magnetic easy plane can be manipulated by applying an electric field, which switches the polarization direction by 71 ° in epitaxial thin films [20]. There are also reports of the disappearance of the

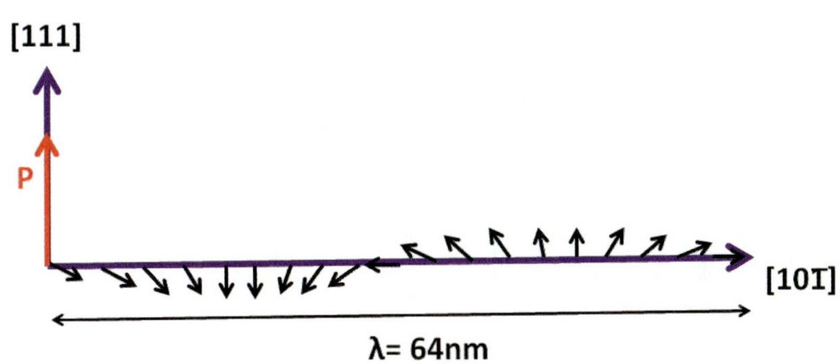

Figure 2.5 Schematic of the spin cycloidal antiferromagnetic arrangement in BFO. The spins are confined within the plane defined by the polarization vector [19].

spin cycloid in the epitaxial thin films and forming a homogeneous G-type antiferromagnetic ordering [21]

2.2.1 Domains and domain walls in BFO

The domains and domain wall studies are crucial from both the fundamental science and application perspectives. The domains scale up with the sample dimension, and, for thin films, the domain wall density will be very high. The rhombohedral BFO has its polarization direction along any of the body diagonal; so there exist only three different kinds of domain walls, 71 °, 109 °, and 180 °. The formation of the type of domain walls is shown in Figure 2.6.

The FE domain shape of the epitaxial BFO varies from the stripped domains [22, 23] for the BFO grown on $SrRuO_3$ buffered $SrTiO_3$, $TbScO_3$, and $DyScO_3$ to irregularly shaped fractal structure/mosaic structure [24] as reported by different groups. The irregularly shaped domain walls are energetically not favorable; however, they are commonly found in some studies. These domain walls possess unique local symmetries and characteristics. The conductivity of 180 ° domain walls is comparatively higher than the 109 ° and 71 ° domain walls [25]. There are two reasons explaining this effect. First, the polarization normal to these domain walls is not uniform throughout the walls and could provide an electrostatic depolarization field that could draw charge carriers in favor of conduction. Second, the band gap of the 180 ° and 109 ° domain walls is comparatively less than the domain regions. This is associated with the Fe-O-Fe bond angle, which regulates orbital overlapping and the material distortion. Within the domain wall, the local polarization suppression exists, which leads to a volume expansion by canceling the

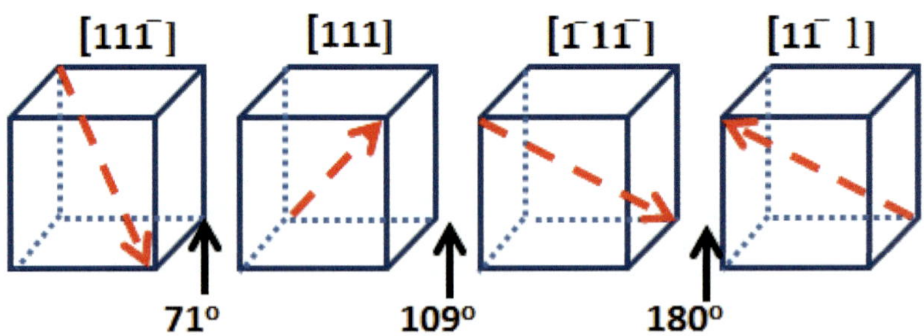

Figure 2.6 Illustration of different FE domain formation and possible domain walls in the BFO.

spontaneous strain. As a result, the bond angle straightens, increasing the orbital overlap and resulting in a reduction in the band gap. The absolute value of polarization is minimum in the middle of the 180 ° domain wall (P = 0), intermediate in the 109 ° domain wall ($P = 0.57P_0$), and maximum for the 71 ° domain wall ($P = 0.82P_0$). So the change in volume is highest in the 180 °, then 109 °, and the smallest in the 71 ° domain walls [26].

2.3 Strain Tuning in Bismuth Ferrite

Due to its potential multiferroic application possibilities at room temperature, the researchers all over the world have explored various promising properties of BFO such as the magnetoelectric coupling, integration to field effect transistor (FET) devices, resistive switching memory devices, and various other areas of spintronics. The room temperature structure of the BFO is rhombohedral with R3c symmetry in bulk ceramics with rhombohedral lattice parameter of 3.965 Å and a rhombohedral angle of 0.6 °. The symmetry allows the BFO to have eight equally possible spontaneous polarization variants under no strain condition, which can be denoted as $R_1 = +^-$ [111], $R_2 = +^-$ $[\overline{11}1]$, $R_3 = +^-$ $[\overline{111}]$, and $R_4 = +^-$ $[1\overline{11}]$. The previous studies on the bulk BFO under the high-temperature, high-pressure, and strain conditions reveal the symmetry transitions from lower rhombohedral to higher orthorhombic structure. The influence of epitaxial strain on the structure and property of the BFO have been studied immensely especially on different substrates like $LaAlO_3$(LAO(001)), $YiAlO_3$(YAO(001)), and $SrTiO_3$(STO(001)). These different studies have shown that the epitaxial BFO, grown on different substrates with different lattice strains, mimics the transformation of the crystal structure from low symmetry rhombohedral phase to the higher symmetry tetragonal, monoclinic and orthorhombic symmetries. The epitaxially grown BFO on different substrates like STO(001) will have a change in the c/a ratio, reveals change in symmetry towards the tetragonal symmetry, and is expected to have strong (001) out-of-plane polarization components. The high-resolution reciprocal space mapping (RSM) and the piezoresponse force microscopy (PFM) studies reveal the presence of other polarization invariants than the out-of-plane components, suggesting the presence of distorted monoclinic phase at higher thickness growth of BFO on different substrates. There is much experimental and theoretical evidence based on the manipulation of the ferroelectric domains through substrate-induced changes in epitaxial strain conditions or different film orientations, for the tuning of the structural and polarization variants. These structural changes were also evident when the BFO thin films were employed along with a chemical substitution. These studies also reveal that the epitaxial strain-driven rhombohedral to tetragonal structure transformation happens through two different types of monoclinic Mc- and MA-type distortions. Such strain-driven structural transformations were plausible by varying the thickness and with temperature-induced structural transformations of the BFO layer grown on different substrates.

The morphotropic phase boundary regions in the materials like $Pb(Zr_xTi_{(1-x)})O_3$ (PZT) enhance the piezoelectric properties via the mechanism of polarization rotation and polarization extension. Similar effects can also be induced in the BFO if such phase coexistence can be induced in the system through epitaxial strain. If such a series of phase transformation and phase coexistence in the BFO is induced through epitaxial strain, it leads to the similar effects of morphotropic phase-boundary-driven materials. This will give rise to enhanced electromechanical response in the BFO thin films. The tetragonal symmetry allows the system to have six equally possible spontaneous polarization variants denoted by C+ [001] and C− [001ŕ] types of polarization variants. The polarization vector in the monoclinic phase can lie in any of the 24 equally possible different directions and lie in the mirror plane with unrestricted rotation. The ferroelectric and piezoelectric properties of the materials are defined by their polarization variants, and the crystal structure can be enhanced in the materials by introducing the phase instability and coexistence of multiple phases using the epitaxial strain. Thus, introducing the advantages of the morphotropic phase boundary like the behavior using the strain-driven structural transitions, without the disadvantages of thermal instability usually associated with the composition and temperature-driven systems, can be used to tune its overall electromechanical property.

In this section, we address the growth and characterization of the epitaxial BFO thin films with different strain conditions on a single crystal substrate. Different phases of the BFO are realized, viz. rhombohedral, tetragonal, monoclinic, and a possible mixed-phase region on a single crystal substrate. We utilize a template-assisted strain-tuning approach to induce the different structural and polarization invariants in the system. As the materials with morphotropic phase boundary regions, BFOs with different phases are expected to have enhanced electromechanical response. The epitaxial strain from the substrate is tuned by varying the template layer ($La_{0.7}Sr_{0.3}MnO_{3\pm\delta}$ – LSMO) thickness, to realize the different structural phases like a fully strained tetragonal-like structure, fully relaxed rhombohedral structure, and a mixed-phase BFO realized on the LAO (001) substrate. With the help of X-ray diffraction, reciprocal space mapping, and piezoresponse force microscopy technique, we try to explain the strain relaxation behavior and functional property correlation. The mixed-phase BFO is realized when the LSMO layer thickness is within 10 nm, and as the thickness increases to 20 nm, the BFO is strain-relaxed to stabilize in the rhombohedral phase. The super tetragonal phase is stabilized at a very low thickness of 2 nm of LSMO on the LAO substrate. A phase field model to investigate the domain morphologies in the epitaxial thin film in these samples profile was created with the experimental parameters, which undergoes the aforementioned structural transformations. The schematic of the layers grown on the strain-tuning LSMO layer is shown in Figure 2.7.

The reciprocal space mapping results show that the ultra-thin template layer of LSMO (2 nm) results in complete transfer of the strain from the LAO substrate to the BFO to stabilize it in a super tetragonal phase. The super tetragonal phase has an out-of-plane lattice parameter of 4.64 Å and an in-plane lattice parameter of 3.79 Å, while accommodating an in-plane compressive strain of 4.3%. The large c/a ratio observed in this study is in correlation with other studies published with tetragonal-like structures of BFO epilayers. The out-of-plane tensile strain experienced by the BFO as a result of the in-plane compressive strain is as huge as 16.99% [27]. The mixed-phase BFO is stabilized when the 20-nm BFO is grown on a 10-nm LSMO layer. The epitaxial strain configurations are adapted by the thickness of the LSMO, causing partial relaxation of epitaxial strain in the BFO layer and resulting in the formation of a mixed phase as shown in the reciprocal space mapping in Figure 2.8. The complete strain relaxation of the BFO layer occurs when the LSMO thickness increases to 20 nm. The BFO layer then stabilizes to the rhombohedral symmetry. The details of the structural analysis could be found elsewhere [28]. BFO, under strained conditions,

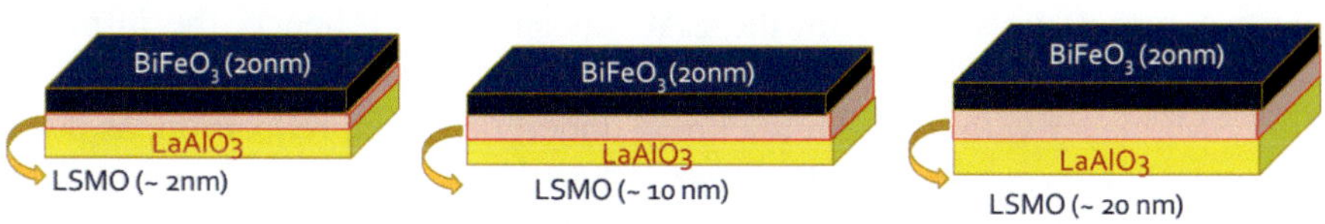

Figure 2.7 The configuration of the different layers with varying thicknesses of the strain-tuning layer, with a constant BFO thickness (tetragonal (left), mixed (middle), and rhombohedral (right)).

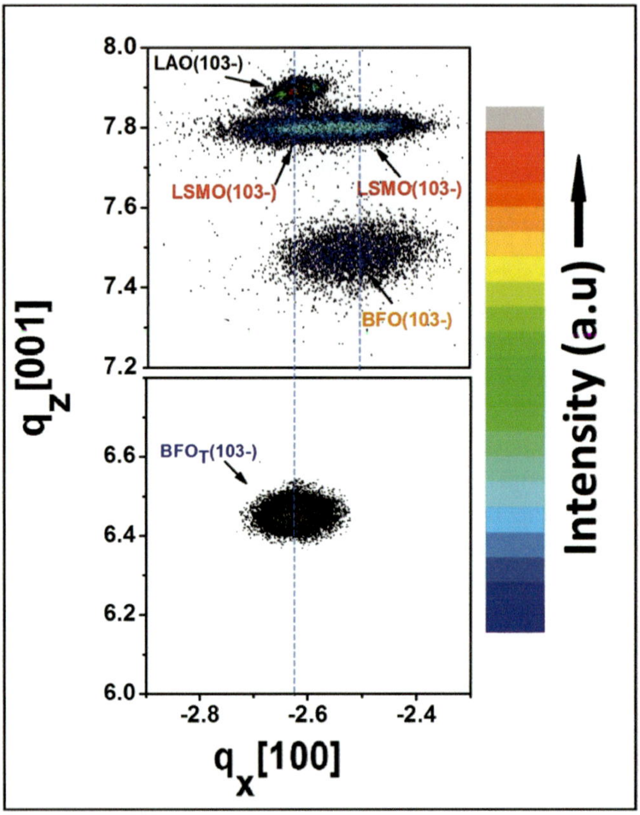

Figure 2.8 RSM of asymmetric (103-) reflection of a mixed-phase BFO thin film.

usually stabilizes to a macroscopic tetragonal symmetry but takes on an internal monoclinic distortion, which permits various plausible polarization components to be observed [16].

The fully strained tetragonal BFO film will accommodate the biaxial in-plane compressive strain from the LAO substrate, and, hence, the in-plane lattice constant is coherent with the substrate. For the mixed-phase BFO, the LSMO layer with a 10-nm thickness undergoes strain relaxation, as does the top BFO layer. The LSMO layer is known to possess different crystal symmetries with respect to their compositional phase diagram.

However, when BFO layers are grown to relatively large thicknesses (~80 nm), they undergo strain relaxation and super tetragonal phase changes to a monoclinic distorted structure as shown in Figure 2.9 [16, 33]. The

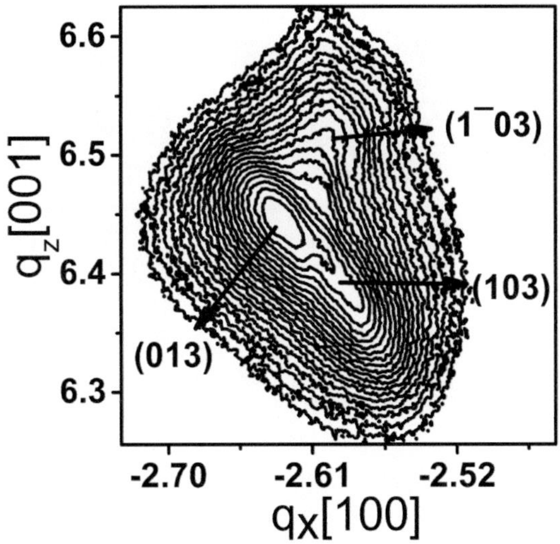

Figure 2.9 Reciprocal space mapping of the asymmetric (103-) plane showing monoclinic distortion in the BFO/LAO (001) substrate.

asymmetric (103-) reflection of the 80-nm thick BFO film shows clear splitting of monoclinic distortion. We come to the conclusion that the BFO film with a higher thickness, in order to reduce the strain energy, possesses a structural distortion even when it is fully strained on the LAO substrate. This indicates the presence of polarization variants in the film even at a lower thickness.

Thus, the different structural symmetries, such as tetragonal, mixed (tetragonal + rhombohedral), and rhombohedral symmetries, are plausible to achieve in the same substrate by utilizing a strain-tuning layer. Such a variation in the crystal structure is expected to affect the FE domain patterns. A piezoresponse force microscope (PFM) is used to study the ferroelectric/piezoelectric properties, domain patterns of the BFO thin films with different crystal symmetries, and strain configurations. The strain-tuning layer of LSMO is serving as a conduction bottom electrode for this measurement. Figures 2.10(a), (b), (d)–(e), (g), and (h) represent the in-plane and out-of-plane domain images (phase) for the super tetragonal, mixed-phase, and rhombohedral phases of a BFO thin film. The PFM provides simultaneous measurement of topography imaging and the ferroelectric domain imaging. By applying a small AC signal through the tip of the conductive AFM tip, the inverse piezoresponse from the sample is simultaneously measured. Both the in-plane and the out-of-plane responses can be simultaneously mapped using

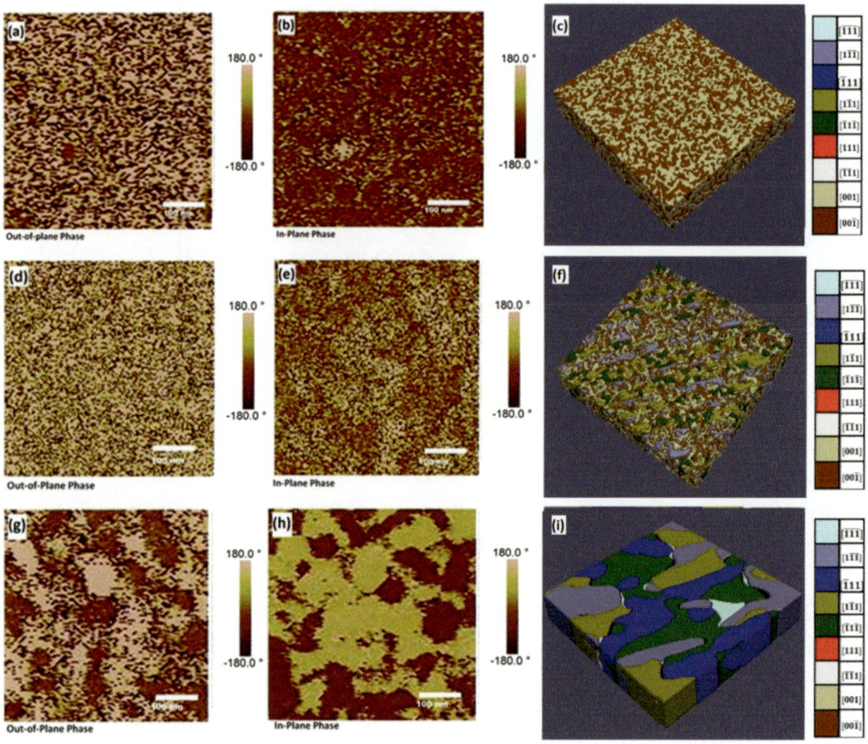

Figure 2.10 (a), (b), (d), (e), (g), and (h) Experimental observation from PFM with in-plane and out-of-plane ferroelectric domain orientations for fully strained, partially strained, and fully relaxed BFO thin films, respectively. (c), (f) and (i) The phase field modeling of domain morphology under the respective strain conditions.

the PFM. More information about the PFM and the analysis techniques could be found elsewhere [29]. We are utilizing the phase contrast images of the PFM studies for further discussion, where the phase contrast images are the finger print for the FE domain orientation.

In the super tetragonal phase of BFO(001), it is evident from the magnified phase images that the major out-of-plane components are dominant with sharp domain boundaries. The phase contrast images of the mixed-phase BFO(001) have comparatively enhanced in-plane components with curved domain boundaries. The rhombohedral BFO thin film, grown fully relaxed on the LAO(001) substrate, has both in-plane and out-of-plane components with curved domain patterns having comparatively large domain size.

The theoretical calculations using the phase field approach were employed to understand the promising strain effects in the structure and property correlation of this study. We incorporated the experimental results and calculated strain values to predict the functional properties of the system under study. The model considers the different strain conditions employed in the experimental strain engineering using the LSMO layer thickness as the substrate effect. The piezoelectric properties are calculated theoretically, considering the LSMO as the bottom electrode. The model considers the BFO as a single crystal with different strain configurations from the substrate for the study. The aim is to understand the evolution of the polarization under different strain conditions and to extract the physics of functional response under external electric fields. Extensive studies have been reported using phase field models to retrieve the ferroelectric domain evolution and domain dynamics in thin films [28, 32]. The phase field simulations are based on the numerical solutions of the time-dependent Ginzburg–Landau (TDGL) equation, which takes into account the long-range electrostatic and elastic interactions.

The phase field model based on the Landau–Ginzburg–Devonshire (LGD) phenomenological theory uses a spontaneous polarization vector "P" as the order parameter to describe the domain configuration and the thermodynamic energies of a ferroelectric $BiFeO_3$ single crystal. The total free energy (F_{total}) of a $BiFeO_3$ thin film consists of the bulk free energy (F_{Bulk}), the gradient or the domain wall energy (F_{grad}), the elastic energy (F_{elast}), and the electric energy (F_{elect}) as follows:

$$F_{\text{total}} = F_{\text{bulk}} + F_{\text{grad}} + F_{\text{elast}} + F_{\text{elec}} \tag{2.2}$$

For the theoretical calculations of the BFO thin films under the strained conditions, a model size of $128\Delta \times 128\Delta \times 20\Delta$, with Δ being the grid size in nm, is considered. A 20-nm-thick BFO film is considered for the calculation. For the modeling, a vacuum layer of around 40 nm was considered as the boundary condition, and the semi-implicit Fourier spectral technique was applied to solve the TDGL equation. Careful consideration is taken for substrate properties such as misfit angle, and the presence of terrace-like features may result in different polarization variants and reduced crystallographic symmetry. For the theoretical calculations, we have considered a misfit angle of 0.5°, and the misfit strain is applied by appropriately setting up a homogeneous strain in the film. The simulated domain pattern and calculations show good correlation with the experimental results, and

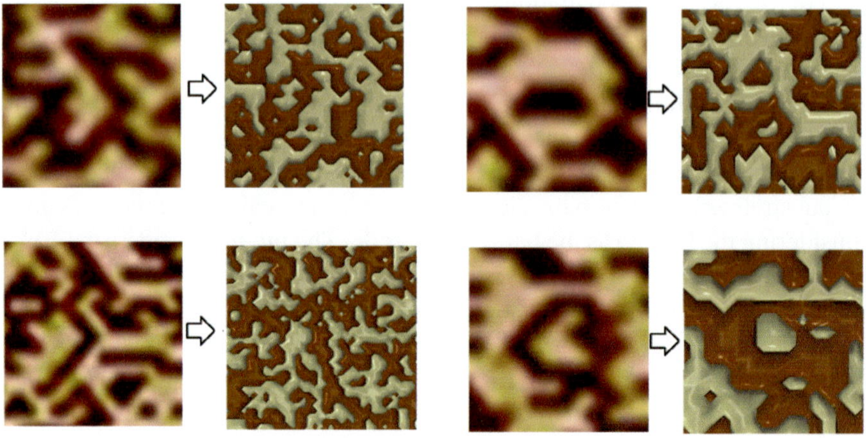

Figure 2.11 The experimental and simulated FE out-of-plane domain patterns in highly strained BFO epitaxial thin films.

both the experimental phase contrast images and the simulated patterns are shown in Figure 2.10 for comparison. In order to emphasize more on the correlation between the theory and experiments, the cropped and magnified view of specific regions of domain pattern is shown in Figure 2.11. Thus, tuning of epitaxial strain in BFO could alter the structure from a super tetragonal phase to a rhombohedral phase via a mixed-phase behavior. Most importantly, such a transformation could be achieved by altering the strain conditions experienced by BFO. From the application perspective, the mixed-phase films are expected to possess a relatively larger piezoelectric coefficient (d_{33}) due to its analogy to the morphotropic phase boundary of FE systems.

Table 2.1 shows the experimentally observed values of the piezoelectric coefficients of BFO epilayers ($\sim$20 nm) with various structures. It could evidently be seen that the mixed-phase film has almost 250% increment in the piezoelectric values for a film of same thickness. Such an enhancement in piezoelectric behavior is similar to a morphotropic phase boundary

Table 2.1 Piezoelectric coefficient (d_{33}) values of BFO epilayers thin films ($\sim$20 nm) with varying crystal structures.

Tetragonal	Rhombohedral	Mixed phase
$\sim$12 pm/V (experimental)	$\sim$16 pm/V (experimental)	$\sim$44 pm/V (experimental)

effect observed in substitutional compounds with mixed structure. Thus, it is important to note that epitaxial strain is one of the key factors that could facilitate engineering various properties of multiferroic systems, more specifically in BFO epilayers. Most importantly, strain tuning using additional layers could facilitate achieving a required crystal structure in a given substrate for a specific thickness and still obtain superior properties.

References

[1] Spaldin NA, Fiebig M. The Renaissance of Magnetoelectric Multiferroics. Science. 2005; 309: 391-392. DOI:10.1126/science.1113357

[2] Khomskii D. Classifying multiferroics: Mechanisms and effects Physics.2009; 2: 20. DOI:10.1103/Physics.2.20.

[3] Kimura T, Goto T, Shintani H, Ishizaka K, Arima T, Tokura Y. Magnetic control of ferroelectric polarization. Nature. 2003 ; 426 : 55–58. DOI~: 10.1038/nature02018

[4] Hur N, Park S, Sharma P, Ahn JS, Guha S, Cheong S-W. Electric polarization reversal and memory in a multiferroic material induced by magnetic fields. Nature. 2004 ; 429 : 392–395. DOI~:10.1038/nature02572

[5] Lou XJ, Yang CX, Tang TA, Lin YY, Zhang M, Scott JF. Formation of magnetite in bismuth ferrrite under voltage stressing. Appl. Phys. Lett. 2007; 90 (26): 262908. DOI:10.1063/1.2752724

[6] Haumont R, Kreisel J, Bouvier P. Raman scattering of the model multiferroic oxide BiFeO3: effect of temperature, pressure and stress. Phase Transit. 2006; 79(12): 1043-1064. DOI:10.1080/01411590601067342

[7] Haumont R, Kreisel J, Bouvier P, Hippert F. Phonon anomalies and the ferroelectric phase transition in multiferroic BiFeO3 . Phys. Rev. B 2006; 73: 132 101. DOI:10.1103/PhysRevB.73.132101

[8] Arnold DC, Knight KS, Morrison FD, Lightfoot P. Ferroelectric-paraelectric transition in BiFeO 3: crystal structure of the orthorhombic β phase.., Phys. Rev. Lett. 2009; 102(2): 027602. DOI:10.1103/PhysRevLett.102.027602

[9] Gabbasova ZV, Kuz'Min MD, Zvezdin AK, Dubenko IS, Murashov VA, Rakov DN, Krynetsky IB. Bi1− xRxFeO3 (R= rare earth): a family of novel magnetoelectrics. Phys. Lett. A 1991; 158(9): 491-498. DOI:10.1016/0375-9601(91)90467-M

[10] Zalesskii AV, Frolov AA, Khimich TA, Bush AA. Composition-induced transition of spin-modulated structure into a uniform antiferromagnetic state in a Bi1 – xLaxFeO3 system studied using 57 Fe NMR. Phys. Solid State. 2003;45: 141-145. DOI:10.1134/1.1537425

[11] Chu YH, Martin LW, Holcomb MB, Ramesh R. Controlling magnetism with multiferroics. Mater. Today. 20007; 10: 16-23. DOI:10.1016/S1369-7021(07)70241-9

[12] Wang JB, Neaton JB, Zheng H, Nagarajan V, Ogale SB, Liu B, Viehland D, Vaithyanathan V, Schlom DG, Waghmare UV, Spaldin NA. Epitaxial BiFeO3 multiferroic thin film heterostructures. science. 2003; 299(5613): 1719-22. DOI:10.1126/science.1080615

[13] Yang SY, Seidel J, Byrnes SJ, Shafer P, Yang CH, Rossell MD, Yu P, Chu YH, Scott JF, Ager Iii JW, Martin LW. Above-bandgap voltages from ferroelectric photovoltaic devices. Nature nanotech. 2010; 5(2): 143-147. DOI:10.1038/nnano.2009.451

[14] Singh VR, Garg A, Agrawal DC. changes in chemical solution deposited lanthanum doped bismuth ferrite thin films. Appl. Phys. Lett. 2008; 92 (15): 152905. DOI:10.1063/1.2901017

[15] Cheng ZX, Wang XL, Kimura H, Ozawa K, Dou SX. La and Nb codoped BiFeO3 multiferroic thin films on LaNiO3/Si and IrO2/Si substrates. Appl. Phys. Lett. 2008; 92 (9): 092902. DOI:10.1063/1.2890068

[16] Wang Y, Nan CW. Enhanced ferroelectricity in Ti-doped multiferroic BiFeO3 thin films. Appl. Phys. Lett. 2006; 89 (5): 052903. DOI: 10.1063/1.2222242 Lett. 2009; 94 (11): 112904. https://doi.org/10.1063/1.3098408

[17] Wang Y, Nan CW. Enhanced ferroelectricity in Ti-doped multiferroic BiFeO3 thin films. Appl. Phys. Lett. 2006; 89 (5): 052903. DOI:10.1063/1.22222424

[18] Fischer P, Polomska M, Sosnowska I, Szymanski M. Temperature dependence of the crystal and magnetic structures of BiFeO3. J. Phys. C: Solid State Phys.1980; 13(10):1931. DOI:10.1088/0022-3719/13/10/012

[19] Lebeugle D, Colson D, Forget A, Viret M, Bataille AM, Gukasov A. Electric-field-induced spin flop in BiFeO3 single crystals at room temperature. Phys. Rev. Lett. 2008; 100(22): 227602. DOI:PhysRevLett.100.227602

[20] Lebeugle D, Colson D, Forget A, Viret M, Bataille AM, Gukasov A. Electric-field-induced spin flop in BiFeO3 single crystals at room

temperature. Phys. Rev. Lett.2008; 100(22): 227602. DOI:10.1103/PhysRevLett.100.227602

[21] Zhao T, Scholl A, Zavaliche F, Lee K, Barry M, Doran A, Cruz MP, Chu YH, Ederer C, Spaldin NA, Das RR. Electrical control of antiferromagnetic domains in multiferroic BiFeO3 films at room temperature. Nature Mater. 2006; 5: 823–829. DOI:10.1038/nmat1731.

[22] Chu YH, Zhao T, Cruz MP, Zhan Q, Yang PL, Martin LW, Huijben M, Yang CH, Zavaliche F, Zheng H, Ramesh R. Ferroelectric size effects in multiferroic BiFeO3 thin films. Appl. Phys. Lett. 2007; 90 (25): 252906. DOI:10.1063/1.2750524

[23] C. M. Folkman, S. H. Baek, H. W. Jang, C. B. Eom, C. T. Nelson, X. Q. Pan, Y. L. Li, L. Q. Chen, A. Kumar, V. Gopalan, S. K. Streiffer; Stripe domain structure in epitaxial (001) BiFeO3 thin films on orthorhombic TbScO3 substrate. Appl. Phys. Lett. 2009; 94 (25): 251911. DOI:10.1063/1.3152009

[24] Catalan G, Béa H, Fusil S, Bibes M, Paruch P, Barthélémy A, Scott JF. Fractal dimension and size scaling of domains in thin films of multiferroic BiFeO3. Phys. Rev. Lett. 2008; 100(2): 027602. DOI: 10.1103/PhysRevLett.100.027602

[25] Seidel J, Martin LW, He Q, Zhan Q, Chu YH, Rother A, Hawkridge ME, Maksymovych P, Yu P, Gajek M, Balke N. Conduction at domain walls in oxide multiferroics. Nature Mater. 2009; 8: 229–234. DOI:10.1038/nmat2373.

[26] Catalan G, Scott JF. Physics and Applications of Bismuth Ferrite. Adv. Mater., 21: 2463-2485. DOI:10.1002/adma.200802849

[27] Zeches RJ, Rossell MD, Zhang JX, Hatt AJ, He Q, Yang CH, Kumar A, Wang CH, Melville A, Adamo C, Sheng G. A strain-driven morphotropic phase boundary in BiFeO3. Science. 2009; 326(5955): 977-80. DOI:10.1126/science.1177046

[28] Saj Mohan MM, Bandyopadhyay S, Jogi T, Bhattacharya S, Ramadurai R; Realization of rhombohedral, mixed, and tetragonal like phases of BiFeO3 and ferroelectric domain engineering using a strain tuning layer on LaAlO3(001) substrate. J. Appl. Phys. 2019; 125 (1): 012501. DOI: 10.1063/1.5054372

[29] Soergel E, Piezoresponse force microscopy (PFM) J. Phys. D Appl. Phys. 44, 464003 (2011). DOI:10.1088/0022-3727/44/46/464003

3

Electronic and Spintronic Devices using Strongly Correlated Two-dimensional Semiconducting Transition Materials

Mona Jani[1], Kiran Sreedevi[2], and Lakshmipriya Ravindran[2]

[1]Department of Physical Sciences and Nanotechnology, Yachay Tech University,
[2]School of Energy Materials, Mahatma Gandhi University, India

Abstract

The strong electron correlations in semiconducting two-dimensional (2D) transition materials offer rich structural and electronic properties due to the energy gap in the spectrum's optical range and, thus, form the heart of modern electronics. The important features of these materials are the bandgap, which involves a passage from a direct gap (in a single layer) to an indirect gap (in multilayer) samples, indicating the vitality of interlayer coupling; electrons in 2D crystals have a staggered honeycomb lattice structure, which is inversion symmetric, and possess extra valley degrees of freedom (namely orbital, lattice, layer, valley, etc.), in addition to charge and spins. These features open up possibilities to manipulate and tune these material properties for future electronic, spintronic, as well as valleytronic devices. Devices like field-effect transistors, optoelectronics: van der walls heterostructures, batteries, supercapacitors, solar cells, photocatalysis, and sensors are described herewith. An overview of the structure along with distinct properties of the emerging 2D transition materials, including numerous experimental methods implemented to synthesize them, is addressed. We conclude in conjunction with perspectives, new aspects, and prospects in this captivating a class of material.

Keywords: 2D Transition materials, Topological insulators, optoelectronics, spintronics, Semiconducting heterostructures

DOI: 10.1201/9781003337102-3

3.1 Introduction

The formation of a new aspect of quantum matter by putting the fundamental elements together has sparked tremendous interest in the research related to condensed matter physics lately [1]. Amongst various materials studied so far, the two-dimensional (2D) structures have become a focal point for researchers [2–4]. The studies on 2D materials initially started after the empirical realizations on exfoliated graphene. Graphene is a simple monolayer of carbon crystal forming 2D hexagonal lattice with a geometric structure [5]. Graphene is an exemplary model due to the dynamics of fermions described by the Dirac equation in 2Ds; the charge carriers represent relativistic particles with rest mass to be zero along with an effective speed comparable to the speed of light [6]. The lower energy excitations are indicated by a four-component Dirac fermion that originates on the topology and geometry of the hexagonal lattice [7]. As a result of graphene's high carrier mobility, specific surface area, quantum hall effect, and extra-ordinary optical, electronic, and thermal properties, this material finds application in high-speed transistors, optical communications, supercapacitors, hybrid materials, sensing chemicals, as well as sequencing of DNA [8–13]. However, the toxicity present in the graphene-related materials is upsurged through the fabrication method, oxygen containing groups, and their sheet size [14]. The Fermi liquid parameters (such as correlations and disorder in their effective mass increment) and the local factors (electronic compressibility of doped graphene) are modulated by effective electron−electron interactions [15, 16]. Theoretical calculations related to the effective mass and the corresponding spin susceptibility of electron configurations for graphene are scaffolded as Landau's Fermi liquid theory [15], whose core is the quasiparticle concept. Furthermore, simultaneous inclusions of charge and spin densities, in addition to the variations exceeding the random phase approximation, are playing an important role in account of the exchange and short-range interactions that lead to essential corrections at minimal carrier densities in graphene. Further, distinct electronic instabilities upsurge in bilayer graphene depending on microscopic information of the electron−electron interactions. Although graphene has significant properties, this material possesses an inherent limitation for device applications due to its zero bandgap and comparative high toxicity [5, 14].

Along with graphene, other widely investigated 2D materials are topological insulators (TIs), layered transition metal dichalcogenides (TMDs), and black phosphorus (BP). These materials span the entire range of electronic structures and display analogous structure features and have distinct compositions as compared to that of graphene. These materials are strongly correlated as they are intensely influenced by the repellent nature of the electron–electron interactions [17]. Correlated materials frequently have their noteworthy properties in addition to the conversions between distinct and competing phases that show entirely different electronic and magnetic orders. The layered TMDs are versatile examples of correlated materials revealing diversified, challenging ordered states having superior phases together with charge density waves, superconductivity, and difficult-to-demagnetize ferromagnetism that may be altered by changing their composition, pressure, and applied magnetic fields. Electron–electron interactions are special in 2D TMD materials, as the electronic alignment of 2D materials merges the problem of competing many-body interactions with drastic sensitivity to atomistic structural details [18]. Thus, controlling the 2D material properties, albeit electronic and optical, by a coalesced "standard model" of their electronic structure remains unresolved. With the possibility to regulate interactions externally through screening made feasible by metallic gates, it is clear that many-body effects can be widely tunable in 2D materials. The TMDs provide abundant opportunities for one to explore the emerging competition between various ground states as well. Known for decades are more than 60 TMD systems, which, although they have varied and unexpected properties, still remain inexplicable.

The 2D TMDs have strong spin–orbit interactions that are promising in spintronics and valleytronic device applications [19–27]. For such applications, materials in which the polarization can be turned on and off are useful. In case of bilayers, the inversion symmetry is smashed by electric fields executed normal to the planes. This applied field strength creates spin–orbit splittings and a reduction of the electronic bandgap, linearly leading to a potential variation between distinct layers allowing valley polarization control resultant as spin and valley polarized semiconductors [27, 28]. The linear reduction of bandgap provides extra means to tune these systems from the semiconducting to the metallic phase, respectively. The applied electric field strength responses provide practical application possibilities for nanoelectronic devices.

As the 2D TMDs have distinct properties such as finite bandgap energies and large specific surface areas, 2D TMD materials are given attention by researchers. Especially, TMDs are semiconductor in nature, share resemblances of a bandgap present in the visible near-infrared (IR) range, prominent carrier mobility, and on/off ratio, and possess huge potential for ultra-small and flexible electronic and spintronic devices [30]. The 2Ds find vital importance for optoelectronics, spintronics, catalysts, chemical and biological sensors (TMDs are less toxic than that of graphene oxide) [14], supercapacitors, solar cells, lithium–ion batteries, and so on [2, 31–38]. It is evident that by simple surface modifications, the TMDs can be made biocompatible, although developing strategies to fabricate 2D TMD nanosheets that have uniform thickness, size, morphology, and high colloidal stability are needed. In addition, explicit and systematic studies must be performed according to their biodistribution, degradation, residence time, and clearance so as to promise TMDs for biosafety.

Observation of topological magnetic structures is a perspective for Skyrmion-based magnetic data storage (information can be gathered in a magnetic region between two domain barriers) in spintronics [39–42]. The mechanisms to evolve Skyrmions are different. They reveal themselves in thin films under perpendicular magnetic field as there is a struggle between the easy-axis anisotropy and dipolar interactions that favor out-of-plane and in-plane magnetizations. By dynamically controlling the intrinsic magnetic interactions by adjusting the high-frequency laser field strengths, the Skyrmionic features can be subsequently tuned and are reported [43]. These intrinsic magnetic interactions induce drastic switching from anti-ferromagnetic exchange interaction to ferromagnetism, making the direct exchange collaboration in equilibrium non-negligible. Additionally, the Dzyaloshinskii–Moriya interactions are renormalized by the high-frequency laser field, allowing for easy tuning of their radius and stabilization in the presence of perpendicular magnetic fields. The existence of Skyrmion crystal phases in decorated TMDs such as Y/MoS_2, for Y = Co, Ni, F, is reported [44]. To date, enormous efforts have been demanded to exploring the controlled fabrication of these 2D materials as well as their fascinating applications according to their properties. Therefore, this chapter describes the crystal structure and features of TIs, TMDs, and BPs. The chapter describes the building of an escalating foundation to synthesis and growth of the 2D materials through various approaches; through these methods, materials can be created, which have never before been realized. Further, we focus on the multiple application directions (mostly with TMDs) carried out so far.

3.2 Crystal Structure and Revealing Properties

3.2.1 Topological insulators

TIs are a series of Dirac materials with resemblance to that of graphene. The common TIs are HgTe quantum wells, BiSb alloys, Bi_2Se_3, and Bi_2Te_3

crystals. TIs have mostly stoichiometric layered structure, with five layers held in position and stacking together by relatively weak Van der Waals (vdW) forces [45]. TIs are materials used in electronics, with an intense spin–orbit coupling, “flipped” bandgap (edge or surface states that are “protected” by time-reversal symmetry), and have perpendicularly bounded spin as well as momentum. Lack of perfection in TIs such as point defects and line-edge roughness are not spin-splitting; thus, backscattering of carriers is intensively less probable, leading to exceptionally efficient (spin-polarized) exchange [46]. Doped TIs and several Bi_2X_3 intercalation compounds (X = Se and Te) have been predicted to be superconducting [47].

3.2.2 Transition metal dichalcogenides

The TMDs consist of hexagonal layers of metal atoms (*M* = Mo, W, Ti, Zr, or Hf) sandwiched between two chalcogen atoms (*X* = S, Se, or Te) layers. TMDs are layered materials having a structure similar to graphite with a layer thickness of about 6–7 Å. The interactions between the individual monolayers are of the vdW type. Typical crystal structures of monolayer TMDs include trigonal prismatic 2H for multilayer, octahedral 1T, and distorted 1T phases [48]. The electronic arrangement of TMDs is very responsive to that of the crystal phase, whereas their electronic properties are significantly sensitive to external factors such as temperature, pressure, and strain [49–54]. Through density functional theory (DFT) calculations that contain spin–orbit interactions, the main features of the electronic structure for TMDs are expressed [55]. The fundamental unit block of an MX_2 crystal structure is fabricated from an interior layer of *M* atoms on a triangular lattice placed between two layers of *X* atoms placed on the triangular net of alternate hollow sites. The MX_2 crystal almost forms a trigonal prism structure with a nearest neighbor *M*–*X* and a distance between the *M* and *X* planes very close to their ideal values. The valence and conduction bands are formed by hybridization of d_{xy}, $d_x{}^2{}_{-y}{}^2$, and $d_{3z}{}^2{}_{-r}{}^2$ orbitals of the transition metal *M* and the p_x, p_y, and p_z orbitals of the chalcogen atoms *X*. Single-layer TMDs are known to be direct gap semiconductors due to the gap lying at the two distinct *K* points of the hexagonal Brillouin zone. The orbital contribution at the *K* point (i.e., edge of the valence band) is due to a cooperation of d_{xy} and $d_x{}^2{}_{-y}{}^2$ of *M*, which hybridizes to the p_x and p_y orbitals of *X*. At the edge of the conduction band, the orbital contribution is due to $d_{3z}{}^2{}_{-r}{}^2$ of *M*, in addition to the minor contributions of the p_x and p_y orbitals of *X*. The spin–orbit coupling gives rise to the separation of the valence band and the conduction band in various types of formed TMDs. Furthermore, Coulomb interactions can modify the spin–orbit coupling split bands.

The TMDs are associated with a direct bandgap in the visible spectrum [56], located in K and $K\prime$ points of the hexagonal Brillouin zone along with a huge spin–orbit coupling. Relativistic effect in the spin–orbit coupling arises for honeycomb 2D lattices having smashed inversion symmetry, one example being the 2H TMD monolayers. For MoS_2, a 100 meV giant spin–orbit-induced band separation through Raman experiments [57] is reported. The TMDs ensure bandgaps in a range between 1 and 2 eV and a carrier mobility of 100 $cm^2\ V^{-1}\ s^{-1}$ at room temperature. Based on the chemical and structural configurations, the TMDs are metallic, semimetallic, semiconducting, or superconducting, allowing ways to tune the fluorescence and electrochemical performances [58].

In TMDs, various complexes of M are chemically and physically intercalated (i.e., insertion) at the vdW sites in between the layers [59]. Such insertions drive changes in the physical properties that are associated with changes in the structure, chemical composition, and nature of the intercalants.

Alkali metal ions, such as Li, Na, K, Rb, and Cs, can be introduced at the M sites, which are introduced in between the consecutive layers [34,60–62]. Through the insertion, conduction electrons are donated to the system, which becomes metallic. Thus, based on the changes in electron population, the Fermi surface-driven transitions are investigated and the electronic properties are controllably tuned. Superconductivity is demonstrated in MoS_2 by intercalation of potassium K. Intercalation induced the 2H phase of MoS_2 to metallic polytypes with 1T and 1T$\prime$ phase changes in MoS_2, and all three phases demonstrated a superconducting transition temperature of T_c = 6.9, 2.8, and 4.6 K, as shown in Figure 3.1, respectively [63]. Although $TiSe_2$ is exceptional, intercalation with Cu [64, 65] or Pd [66] unveiled an increase in the layer spacing corresponding to the structural change, thereby providing grounds for a novel superconducting state at finite layer spacing. However, here, the charge doping alone cannot be the only explanation for superconducting temperature dependence. With intercalation of organic molecules at the M site, interlayer spacing increases by one order of magnitude [67]. With intercalation of 3d transition metals, the magnetic properties greatly depend on the intercalant and the transition metal used [68]; the quantity of intercalant also determines the structural and electronic properties [69]. Another unpredictable behavior in the magnetic properties was observed with the intercalation of Fe in TaS_2, showing a sharp shift of the magnetic domain alignment at applied strong magnetic fields ($\sim$ 4 T)

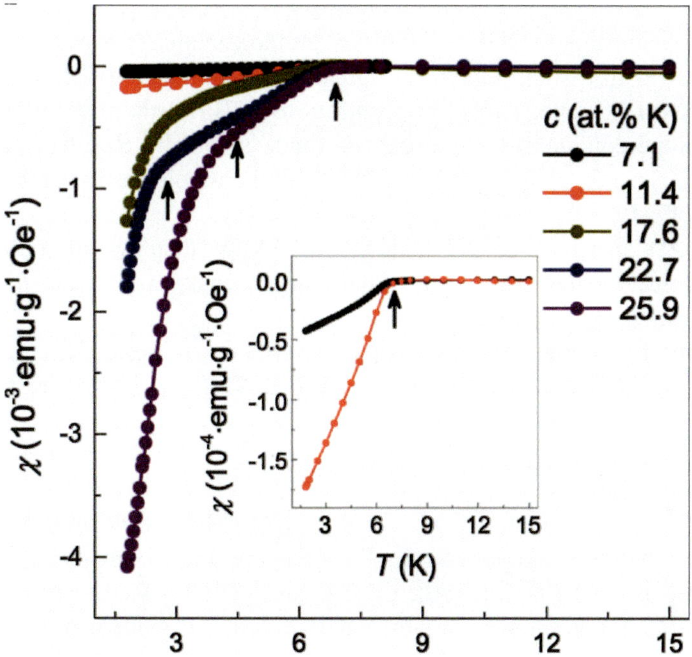

Figure 3.1 Superconductivity in potassium ion (K) intercalated MoS_2. The susceptibility $\chi = M/H$ increases with an average concentration of potassium, when direct current magnetic moment (M) is measured in the ZFC mode with applied field (H) = 10 Oe. The three arrows indicate superconducting transitions with respect to temperature ($d\chi/dT$). The inset shows a zoomed-in view of superconducting transition K concentration $c \approx 11$ atom % and $c \approx 7$ atom % [63]. Reprinted with permission from [63].

and temperatures (up to 100 K) [66]. The research work on the observed drastic shift in structural, electronic, or magnetic properties due to these intercalations are under process. The system can be made metallic by means of electric field effects as well [55]. An interesting aspect of intercalated and metallic MoS_2 is the observation of the superconducting transition depending on charge carrier density n and high-κ dielectrics. With desirable doping of $n \sim 1.2 \times 10^{14}$ cm^{-2}, superconducting transition with critical temperature of $T_c \sim 10.8$ K, which drops to zero with higher n [70, 71], is reported. According to theory, superconductivity is most likely caused by the induced electron–electron interactions, and it is feasible for a superconducting phase to have a nontrivial gap structure [55].

Thus, unlike graphene, a combination of spin order coupling and a non-trivial superconducting order parameter can result in additional fascinating features. According to DFT calculations, phonon-mediated superconductivity is possible with a small electron doping range, and wave of charge density might exist for higher carrier concentrations [72]. Furthermore, spin triplet p-wave superconducting phase can be made stable by Rashba spin−orbit coupling, and this phase is expected to be in the topologically trivial regime in MoS_2 [73].

The 2D structure of the majority of TMDs displays more than one charge density wave transitions. The charge density wave transitions and superconductivity compete in TMDs at low temperatures [72, 74]. This is understood by tuning TMDs with electron doping (in TaS_2) [75], changes in pressure (in $NbSe_2$ [76], TaS_3, $(TaSe_4)I$, and $NbSe_3$ [77, 78]), or to-gather (in $Lu_5Ir_4Si_{10}$ [79]). Through this tuning, at low temperatures, the charge density wave transitions are suppressed and conversely superconductivity critical temperature is enhanced.

TMD-based metal atoms (M = V, Nb, Ta, Cr, Mo, and W) and chalcogens (i.e., S, Se, and Te) except $CrS_2/Se_2/Te_2/VS_2$ are stable compounds and have diverse electronic behaviors [80]. 1T TaS_2 and 1T ReS_2 TMDs are studied for gate-tunable charge density waves and electrical and optical properties that are linearly anisotropic [81, 82]. Chalcogenides and halides such as pyrites and marcasites have both layer and nonlayer in their structures with show diverse electronic properties.

In the 2D limit, semiconducting TMDs, precisely MoS_2, WS_2, and WSe_2, are suitable for electronic devices as they have high charge carrier mobilities and substantial bandgaps enabling large switching ratios in field-effect transistors (FETs) [83–86]. 2D TMDs are capable of offering a broad range of unusual properties such as valley polarization, lattice symmetry-induced valley Hall effect, and superconductivity [25, 71, 87]. Furthermore, successful fabrication of 2D TMDs at the wafer scale improves their prospects as atomically thin integrated electronics [88].

3.2.3 Black phosphorus

A rare allotrope of phosphorus, BP, is another 2D material escalating as an alternate 2D semiconducting TMD. The bulk BP crystal structure is a layered orthorhombic, adjacent layer distanced by 5.3 A° and stacked together by vdWs force. A BP monolayer termed as phosphorene comprises an "armchair" vertically staggered hexagonal lattice structure. In this structure, there are four P atoms; three of them are situated within the same plane (bond length of 0.2224 nm) and the fourth P atom is found parallel in an adjoining plane (bond length is 0.2244 nm). BP in the z-direction is formed as a hexagonal shape having bond angles equal to 96.3° and 102.1° [89]. The example of BP is phosphorene, which is an anisotropic semiconductor. In phosphorene, the topologically stable chiral Dirac charge carriers (2D, 1D in bulk or along boundary) are achieved by tuning the polarization and intensity of a single external field. Monolayer phosphorene may go through Lifshitz transitions (shows that nontrivial phases may be induced in phosphorene) going from band insulating phase to semi-metallic and topological insulating phases if instantly driven by in-plane and time-periodic laser fields [90]. These linear polarized fields thus mimic electronic properties of graphene as massless Dirac quasiparticles. It can be merged into a thin-film form and it is really suitable for nanoelectronic devices. With respect to the stack of layers, BP's bandgap from 0.3 eV for bulk can be tuned to 2 eV or above. With extremely high hole mobility, it is observed that there is a high field-effect mobility up to $\sim$10,000 $cm^2\ V^{-1}s^{-1}$ for monolayer and wide in-plane anisotropy caused by its puckered atomic structure [91, 92]. Due to BP's intrinsic direct bandgap and carrier mobility, they provide a platform to bridge between graphene and TMDs.

3.3 Synthesis Methods

Generally, top-down and bottom-up methods are used to prepare the 2D materials; the former involves exfoliation of bulk to monolayer or few-layer 2D nanosheets by removing the vdWs force between the layers, and the latter method involves synthesizing nanomaterials directly on molecular level. According to the desired composition, size, thickness, crystal phase, and surface features of 2D materials, methods such as mechanical, liquid-phase, shear, electrochemical, and oxidation-assisted exfoliations as well as chemical vapor deposition (CVD) or wet chemical synthesis have been established. Figure 3.2 indicates the schematic of the methods, and the detailed procedure is as follows.

3.3.1 Mechanical exfoliation

Mechanical exfoliation technique is the most common method used. In this method, a scotch tape is utilized over to the desired material surface and then cleaved after exerting the normal force [93, 94]. This cleavage is repeated several times with regular force being used; the desired material layer is cleaved, and the materials become thinner and finally form a single-layer material.

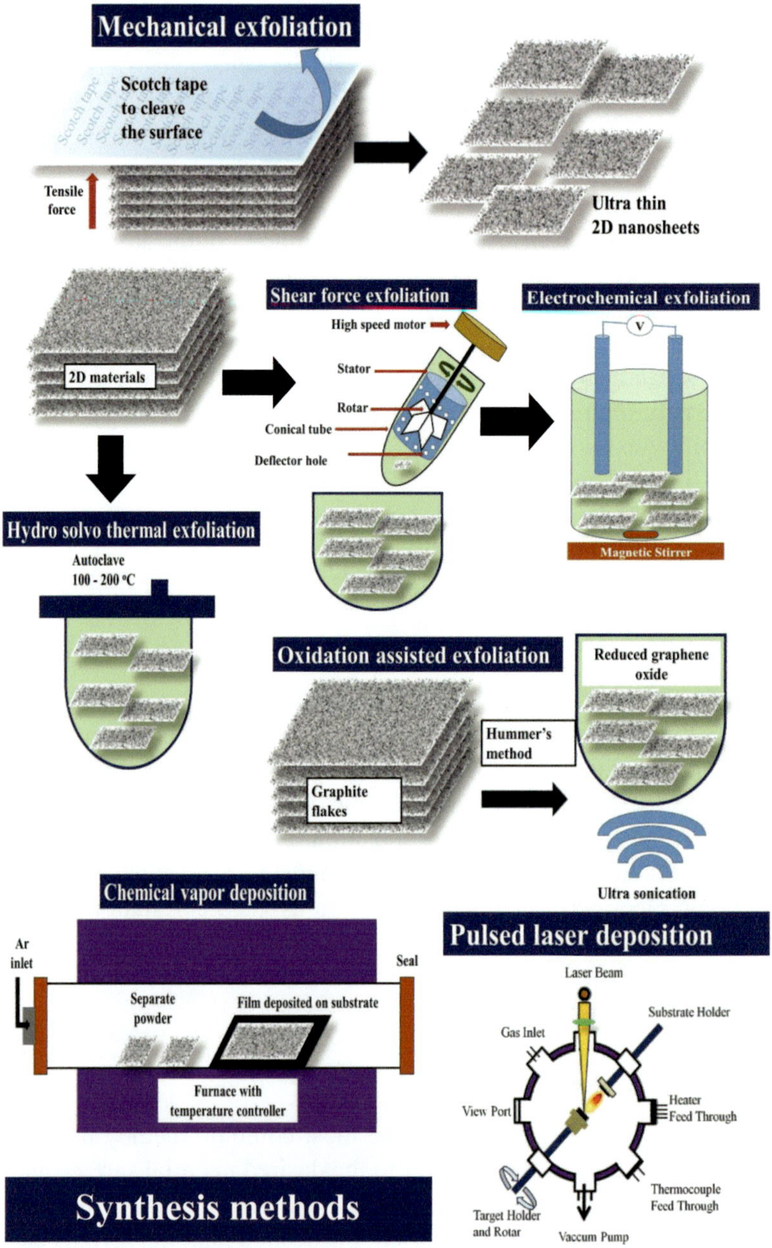

Figure 3.2 Schematic representation of the synthesis methods employed.

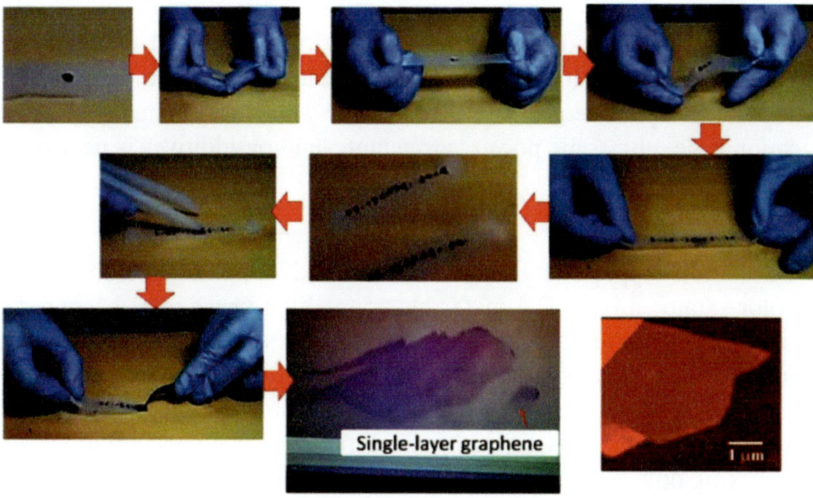

Figure 3.3 Micromechanical cleavage using scotch tape of highly oriented pyrolytic graphite [93]. Reprinted with permission from [93].

Ultrathin 2D nanomaterials are cleaved from tiny crystals of highly aligned pyrolytic graphite to form graphene as shown in Figure 3.3. Later, numerous ranges of TMD and TI materials have been cleaved. The size produced with mechanical exfoliation of 2D crystals can be few to tens of micrometers with clean chemical-free surfaces. The disadvantages of this production method are that the yield is quite low, thick flakes are present, there is a lack of precision, control, and reproduction, freestanding possibilities, and dispersions in solutions.

3.3.2 Liquid phase exfoliation

For the sonication-assisted liquid-phase exfoliation, the materials are dispersed in a solvent, exfoliated, and purified [95, 96]. Exfoliation is done by chemical wet dispersion by ultrasonication both in aqueous and non-aqueous solvents. Through sonication, an intense tensile stress is generated on the layers of the bulk crystals, which leads to the exfoliation of layered materials into thin-layer sheets. The essential aspect is that the surface energies of the solvent and layered material are matched, which are required to exfoliate. Besides graphene, other materials have been exfoliated, including TMDs, TIs, and BP. The solvent is key to stabilize the exfoliated nanosheets and to further prohibit from restacking and/or aggregation. Mostly exfoliation is done in pure water. This method ensures the fabrication of thin 2D nanomaterials for practical purposes. In this method, to procure the desired concentration and lateral size parameters, control over the solvent system, temperature, sonication time, ultrasonic power, surfactants, and/or additives is necessary.

3.3.3 Shear exfoliation

The exfoliation supported by shear force is based on tremendous shear added to a rotor–stator mixer. The mixer generates high shear rates triggered by the shear force in a liquid containing stacked bulk crystals under high-speed rotation. This method can produce a large-scale exfoliation of graphene flakes from dispersions from graphite by applied shear force [97]. By optimizing the parameters of the mixer along with the starting material concentration, rotor speed, mixing time, and liquid volume, one can achieve the order of milligram per minute scale productions. *h*-BN, MoS_2, and WS_2 have been exfoliated by means of this method.

3.3.4 Electrochemical exfoliation

Electrochemical reactions that occur on the electrode having layered structure result in intercalation and/or exfoliations [98]. The electrochemical process is either of cathodic or anodic type. The high quality of the exfoliated nanosheets extremely depends on and is impacted by the applied potential and the type of electrolyte used. Electrochemical exfoliation is simple, requires least processing time, has easy operation control, and has the capacity to scale up.

3.3.5 Oxidation-assisted exfoliation

This process converts graphite into graphene oxide by using potent oxidizing agents. Ultrasonication is frequently used in the Hummers method to generate graphene oxide and for delamination in an appropriate solvent [99].

3.3.6 Hydro/solvothermal synthesis

This method involves a wet-chemical synthetic approach that uses water or organic solvents. The prepared homogeneous reaction solution is enclosed in an autoclave at a reaction temperature usually between 100 and 240 °C, i.e., beyond the boiling point of the solvent, to generate high pressure promoting the formation of a desired crystalline material [100]. For synthesizing inorganic materials, 2D nanosheets of metal oxides, and metal di-chalcogenides, this method is implemented. Experimental circumstances such as concentrations of precursors, solvents, polymers, surfactants, and temperature are sensitive parameters, making this method difficult to properly govern the structure for repeatability.

3.3.7 Chemical vapor deposition

CVD works at high temperatures using chemically synthesized materials that are to be deposited onto substrates. Thin films are formed using the CVD process of metals, semiconductors, and insulators. CVD is a scalable method with high-quality/purity/control to synthesize 2D nanomaterials that can be obtained in large scales. Typically, gaseous or powder reactants are placed in the reaction chamber, through specific chemical reactions at appropriate reaction parameters. The 2D materials deposited onto desired substrates can have control over morphology, crystallinity, and defects using this method. By controlling reaction parameters, the number of layers of 2D materials with good accuracy [101] can be obtained. Alloyed TMD nanosheets can also be prepared by controlling the instrument parameters. By altering the ratio of constituent components, the bandgap of alloyed TMDs can be precisely controlled [102, 103]. Epitaxial 2D TMD heterostructures, for example, WSe_2-WS_2, $MoSe_2$-MoS_2, and $MoSe_2$-WSe_2, can be synthesized using CVD [104–106]. The epitaxial mode, i.e., lateral and vertical development, can be achieved by controlling the experimental parameters. Additionally, by merging a variety of solid, liquid, and gaseous precursors, it is possible to produce a broad range of 2D nanosheets and their hybrids, for which doped and functionalized 2D starting precursors are introduced with an additional precursor. In contrast, using the CVD method, tailoring the properties at atomic level, and controlling the stoichiometry and defects are still challenging and need several trials by optimizing the parameters.

Few other fabrication methods put forward by researchers for specific 2D nanomaterials are: pulsed laser deposition where the pulsed laser is utilized to ablate the initial material and to evaporate inside a vacuum chamber, where the molecules deposit on suitable substrate [107–109]. This method is widely known for depositions of complex compositions and for the production of stoichiometric as well as epitaxial films. Examples are a variety of complex oxides, perovskites, and bilayers. Other methods are molecular beam epitaxy and pulsed magnetron-sputtering technology that are reported to grow high-quality TI nanomaterials [110–112]. Methods such as vapor phase sulfurization take advantage of sulfur vapor to sulfurize W and Mo thin metal films synthesized at a high temperature and to synthesize few-layer TMDs [113]. Laser thinning approaches sublimates/cleaves upper layers of solid bulk material with heat caused by light absorption, thereby leaving single bottom layer onto the substrate [114].

Based on the prerequisite studies, the developed samples are mostly characterized with the help of optical microscopy, scanning probe microscopy, atomic force microscopy, conductive atomic force microscopy, electrostatic force microscopy and Kelvin probe force microscopy, scanning tunneling transmission electron microscopy/scanning, transmission electron microscopy, X-ray absorption fine structure spectroscopy, X-ray photoelectron spectroscopy, and Raman spectroscopy.

3.4 Electronic and Spintronic Devices using 2D Materials

With the possibility of creating artificial heterostructures and the intrinsic material properties with physical occurrences that are exotic in a wide range of novel 2D materials, it is conceivable to design novel 2D-based applications and devices. In this section, we present various application directions based on these emerging materials.

3.4.1 Field-effect transistors

Graphene with ultrahigh electron carrier mobilities between 3000 and 27,000 cmš/Vs is considered to show great promise for digital nanoelectronic device applications. However, graphene-based practical applications are hindered due to their zero bandgap. The zero bandgap in graphene yields at most a few tens of a current on/off ratio and a huge current off, and, as a consequence, graphene-based FETs cannot comply with international technology roadmap for semiconductors (ITRS) requirements [115]. Alternatives to graphene are the upsurging 2D semiconducting materials, such as TIs, TMDs, and BP nanosheets, which provide a remedy relative to the materials that strengthened carrier mobility and configurable band structures [2, 116, 117]. The main specific advantage of TMDs above to that of graphene in most of them is their $\sim 1-2$ eV bandgap, which enables their use as switches having low current off.

MoS_2-channel FETs assembled with a large current on/off ratio and a small subthreshold swing that approaches the desired measurements use FETs for the replacement of silicon in complementary metal oxide semiconductor (CMOS) like digital logic devices. TMD MoS_2 became highlighted after the realization that a single-layer MoS_2 FET fabricated using the micromechanical exfoliation demonstrated a current on/off ratio measured at room temperature that exceeded 10^8 and 200 $cm^2\ V^{-1}s^{-1}$ mobility values [118]. The MoS_2 FET, which is atomically thin, is considered to be promising for future nanoelectronic devices as it offers large bandgap, excellent stability chemically as well as thermally, and superior resistance fabricated in the short channel. A transistor that has a 1 nm physical gate utilizing a bilayer of MoS_2 as a channel with a single-walled carbon nanotube as a gate electrode has been constructed [119]. The current on/off ratio was reported to be $\sim 10^6$. This transistor, which has an ultrashort channel, excellently exhibited switching characteristics having 65 mV/dec of near-ideal subthreshold swing, which approaches the theoretical limit. Furthermore, CVD for preparing an entire integrated multistage circuit assembled using few-layers of MoS_2 has been developed. It is reported that the developed circuits indicated the integration of an inverter, in addition to a NAND gate along with a static random access memory with a five-stage ring oscillator developed on bilayers of MoS_2 flakes and is as shown in Figure 3.4 [120]. Several initiatives have been dedicated to enhance and optimize the performances of MoS_2-based transistors. Versatile methods such as chemical doping [121], channel suspending [122, 123], strain engineering [124], and chemically repairing the sulfur vacancies [125] have been demonstrated.

Molecular doping of the channel is an efficient approach for tuning and enhancing the MoS_2-based transistors and their performances [126]. The fabrication of thin forms that are atomic in size of layered materials such as semiconducting MoS_2, insulating hexagonal boron nitride (*h*-BN), and conducting graphene allows to construct devices that are perfectly transparent and flexible. Such high-performance flexible and transparent electronic devices have been developed using ultrathin 2D MoS_2 [30,127–129]. In such devices, the constructed MoS_2-based thin-film transistors were prepared on a carrier template and later transferred potentially on any substrate irrespective of the type or material used. Figure 3.5 displays the schematic representation of the process to enable fabrication based on few-layer MoS_2 thin-film transistors and later transferring them from a silicon template to substrates that are arbitrarily organic, inorganic, flexible, or rigid. In this work, it is showing that the fabricated transistor has 19 cm^2/Vs mobility, the I_{on}/I_{off} ratio is greater than 10^6, and the threshold voltage is about -2 V. Bending the radius to 5 mm, the performance of this transistor remains unchanged [129].

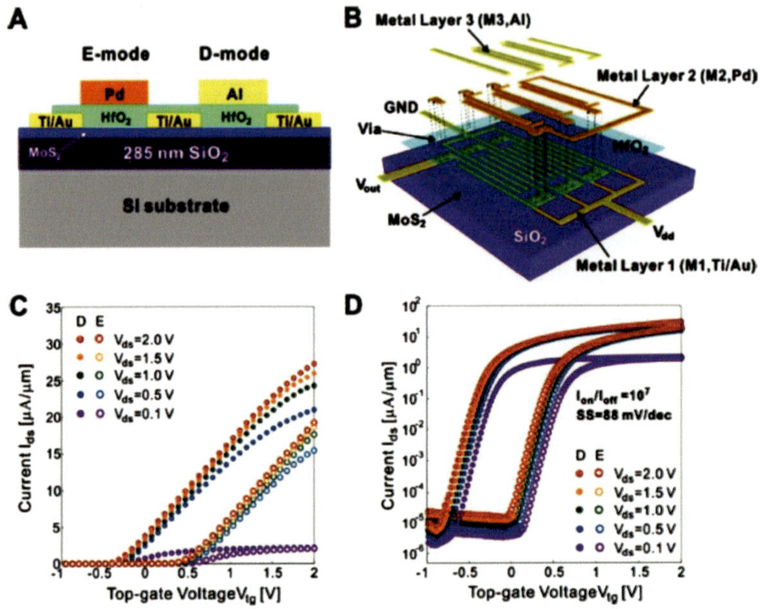

Figure 3.4 (a) Schematic representation of an *E*-mode and a *D*-mode device. (b) Schematic representation to show integrated five-stage ring oscillator on MoS_2 thin films. (c) Current–voltage transfer characteristics of the depletion (*D*) mode and the enhancement (*E*) mode for MoS_2 FETs. (d) The log-scale presentation of (c) [120]. Reprinted with permission from [120].

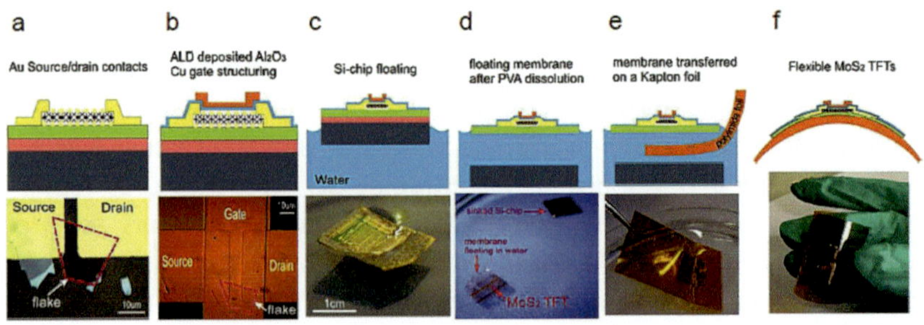

Figure 3.5 Step-wise (a–f) schematic representation for the fabrication of flexible MoS_2 thin film transistors [129]. Reprinted with permission from [129].

Devices using MoS_2 as channels built on *h*-BN dielectric, with graphene as gate electrodes demonstrating flexible FETs showed I_{on}/I_{off} ratio above 10^4, 45 $cm^2\ V^{-1}s^{-1}$ values of field-effect mobility, lower operation gate voltage (less than 10 V), and minimal strain influence [127]. With dielectric engineering, one can improve the performance of transfer of carriers in 2D semiconducting TMD transistors by encapsulating MoS_2 with HfO_2 or by using Al_2O_3, which gives an increase in the MoS_2 mobilities [118, 130]. By covering the transistors with insulating layer, one can achieve stable operation of these devices. 2D-based transistors find applications in chemical sensing [131], memory devices [132], and MoS_2-based FET biosensors used to identify proteins with superior sensitivity of more than 74-fold by surpassing graphene's sensitivity [35], gate-tunable memristors formed by grain boundaries in monolayer MoS_2 devices [133].

BP transistor based on a nanosheet using few BP layers, which are micromechanically exfoliated and show thickness-dependent charge-carrier mobility, was the obtained values of $\sim$1000 $cm^2\ V^{-1}s^{-1}$ for thickness of $\sim$10 nm [2]. It is reported that BP has exceptional electronic features, such as expected high-theoretical carrier mobility (10,000 $cm^2\ V^{-1}s^{-1}$), adjustable direct bandgap, and anisotropic transport [134]. BP is particularly useful for 2D-based electronics as it is a p-type 2D semiconductor as opposed to TMDs, which are usually n-type semiconductors [135]. Electronic effects of BP transistors are explored for gate-dielectric tuning [136], doped contact [137], and ultrashort channel [138].

3.4.2 Optoelectronics

The 2D TMD materials are suitable for optoelectronics due to their special property, i.e., the bandgap of MoS_2. A multilayer MoS_2 has a 1.2 eV indirect bandgap, which can be varied to a direct bandgap for a single layer (1.9 eV). The MoS_2 with single layer is a direct bandgap semiconductor due to quantum-mechanical confinement having large absorption coefficient and is thus popular in optoelectronics. $MoSe_2$, WSe_2, and WS_2 show similar sizable bandgap dependence to that of MoS_2 [84]. Mechanically exfoliated MoS_2-based single-layered phototransistor showed much higher photosensitivity compared to graphene. Due to incident-light control, quick photo switching, and excellent photo-responsivity of the mono-layer MoS_2 phototransistor, one can manufacture the optoelectronic-based device [139]. A photodetector grounded on MoS_2 that has been mechanically exfoliated was capable of a maximum exterior photo responsivity of 880 AW^{-1} at a wavelength of 561 nm and a photo-response in the 400–680 nm range [140]. Encapsulated with atomic hafnium oxide thin-layers based on MoS_2 photodetectors (monolayer and bilayer) are highly sensitive and show enhanced electronic performance up to $10-10^4$ AW^{-1} and photosensitive-response time from

~10 ms to 10 s [141], due to the devices being protected with hafnium oxide. Figure 3.6 shows the optoelectronic response and charge carrier dynamics of the photodetectors.

Flexible organic light emitting diode (OLED) displays devices using MoS_2 FETs on 6-mm-thick ultrathin polyethylene terephthalate flexible substrates are fabricated using the CVD method. The MoS_2 OLED devices exhibited a driving circuit of 10 μA and a I_{on}/I_{off} ratio of 1×10^9. OLEDs have future perspectives as wearable electronic displays. Choi et al. prepared an ultra-flexible active matrix-OLED-based wearable display utilizing MoS_2 thin-film transistors, which includes a high-κ dielectric Al_2O_3 as a passivation layer [142]. This MoS_2-based ultra-flexible OLED was used as a wrist display on the human skin, presenting the ultra-flexibility of substrates using polymers. In this work, the MoS_2 thin-film transistor as a gate electrode is sandwiched between two high-κ dielectric Al_2O_3 as passive layers. The OLED showed a rise in carrier mobility due to n-type doping of MoS_2 and the source/drain electrodes as the contact regions, enabling the function of flexible OLED displays.

BP in optoelectronics has a direct bandgap for almost all thicknesses and is thus useful for optoelectronics. For bulk BP, the bandgap is 0.33 eV, which increases between 1.7 and 2.1 eV for single layers [143]. High photoresponsivity was demonstrated by a BP-based transistor with a transparent

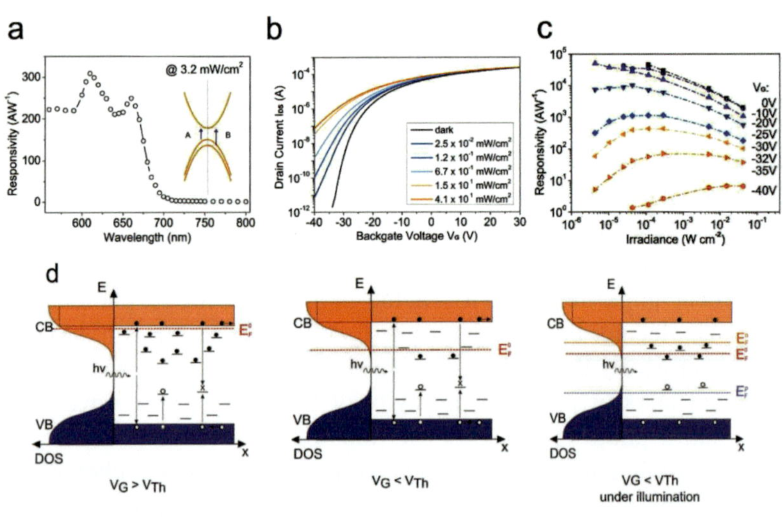

Figure 3.6 Optoelectronic response and carrier excitation and trapping dynamics of fabricated monolayer MoS_2 flakes [141]. Reprinted with permission from [141].

top-gate made of graphene. It was demonstrated that the graphene-based BP transistor function under bias with a minimal dark current can have an intrinsic responsivity up to 135 and 657 mAW^{-1} in 11.5- and 100-nm-thick devices at external temperature, along with a broad response bandwidth of more than 3 GHz [144]. Using BP/MoS_2 heterostructure to develop a p-n junction diode, a maximum photo-responsivity of 418 mAW^{-1} with an external quantum efficiency of 0.3% was achieved [145].

3.4.3 Van der Walls Heterostructures

vdW heterostructures are 2D building blocks that are stacked vertically by vdW force. By dry transfer method, atomically sharp interface and contamination-free vdW gaps are possible for ultrahigh performances. The first noticeable works are stacking vertically the double-layer graphene heterostructures [146], barrister based on graphene/silicon vertical junctions [147], and field-effect tunneling transistor based on formed graphene/*h*-BN/graphene heterojunction layers [148]. These studies help advance vertical FET technologies, such as graphene/WS_2 heterojunction on the h-BN support. These highly flexible and transparent hybrid devices demonstrated the combined effect of tunneling and thermal contribution to a I_{on}/I_{off} ratio of 10^6 at external temperature [149].

The first publication on the preparation of lateral p-n junctions of MoS_2/WSe_2 and WS_2/WSe_2 used home-built CVD and *in-situ* modifications of vapor-phase reactants by thermal evaporation [104]. The more complicating vertical heterostructures from WS_2/MoS_2 monolayer were fabricated by monitoring the system growth temperature. The formed WS_2/MoS_2 artificial heterostructures formed a monolayer that has an atomically clean and sharp interface forming the p-n junction, which is challenging. In addition, the grown junctions even showed a strong enhancement in the localized photoluminescence and the p-n junction diode behavior [106]. There are several efforts dedicated to perceive the synthesis of vdW heterostructures in large scales and the electronic performances are still lagging in success by the use of the CVD method.

In addition, vdW heterostructures have an extensive future in optoelectronics due to special intense excitonic effects and valley- and spin-dependent characters. These properties determine the strength of light interactions with matter, radiative recombination lifetime of electron–hole, and the material's optoelectronic response [31, 150]. Specifically, the p-n junctions that are formed as vdW heterostructures have exhibited remarkable accomplishments as photodiode and photodetector [145, 153–153], electroluminescence [154–156], and photovoltaics [157].

For instance, the very first use of electrically tunable BP (p-type) along with monolayer MoS_2 forming an atomically clean and sharp interface, i.e., type-II heterointerface, vdW heterojunction p-n diode was demonstrated [145]. The formed diode based on MoS_2 exhibited a gate that can be tuned along with current-rectifying characteristics and are as revealed in Figure 3.7. Upon illumination, the p-n diode highlighted as photodetector shows a 418-mA/W response, which is maximum and 100 times higher than the reported BP phototransistors. The photovoltaic power generation had 0.3% quantum efficiency. In addition, through mapping the photocurrent that is uniform over large areas, it was shown that the formed diode can be implemented in solar cells and photodetectors.

The tunable graphene/silicon heterojunction, which has similar structures to the vertical transistors, reported highly efficient photodiodes and photodetectors [158]. Further integrating graphene into a silicon optical waveguide reached extreme photo-responsivity with significantly enhanced graphene—light interaction [151]. More complex vdW heterostructures, such as high-quality graphene/WSe_2/graphene encapsulated in *h*-BN, are reported to reach a photo-response time as short as 5.5 pS, having quantum efficiency (internal) as high as 70% [152].

The optoelectronic device based on vdW heterostructure, on graphene/WS_2 stacks, opened up opportunities for the formation of flexible photovoltaic devices with quantum efficiency (external) as high as 30% [159], which has strong light—matter interaction, enhanced photon absorption, and electron—hole creation. The effect is improved in MoS_2/graphene heterojunction where the quantum efficiency was above 50% [160]. Atomic level p-n junctions formed by MoS_2/WSe_2 monolayers sandwiched in-between graphene layers enhanced photoexcited carrier collection and showed gate-tunable current rectification and photovoltaic response across the whole interface [161]. Light generating and harvesting of bulk semiconductor/TMD semiconductor p-n junction was reported. In this case, the silicon/MoS_2 p-n heterojunction works as both a light emitting diode and a solar cell [162]. A vertical vdW heterostructure utilizing bilayer MoS_2 (n-type) and thickness-controlled $MoTe_2$ (p-type) shows exceptional performances as forward/Zener/backward diodes, transistors, and multivalued logic depending on the thickness of layers, which contributes to changes in Fermi level and band structures [163]. The schematic and basic characterization of this structure is shown in Figure 3.8. Electroluminescence is observed in GaN-Al_2O_3-MoS_2

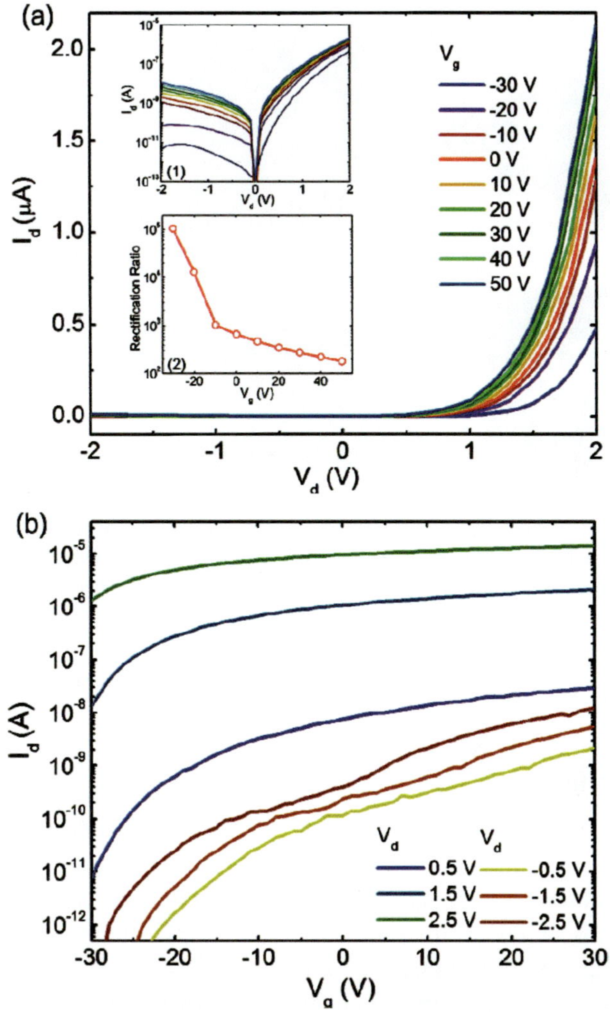

Figure 3.7 (a and b) Gate tunable current–voltage (*I-V*) characteristics and transfer curves of the 2D p-n diode. Inset in (a) shows *I-V* in semi-log scale and rectification ratio as a function of the back-gate voltage [145]. Reprinted with permission from [145].

and GaN-Al_2O_3-MoS_2-Al_2O_3-graphene vertical heterojunctions. The electric field induced direct bandgap emission in multilayer MoS_2 as a result of redistribution of bandgaps from indirect to direct [155]. Light emitting diodes on the basis of 2D (TMDs, h-BN, and graphene) materials use a prepared and desired band structure that is one atomic layer thin, i.e., quantum well

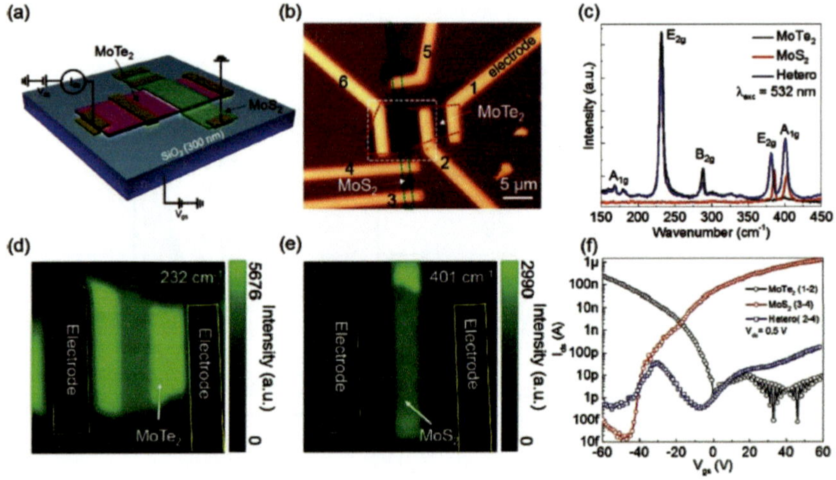

Figure 3.8 (a–f) Schematic view, optical image, Raman spectra, and mode-selective Raman mapping along with the current–voltage characteristics for $MoS_2/MoTe_2$ heterostructure prepared on a silicon/silicon dioxide substrate [163]. Reprinted with permission from [163].

heterostructures. These wells show quantum efficiency up to 8.4%, which can be tuned due to the atomistic precise layers [156].

3.4.4 Batteries

Enormous efforts are put forward to discover various nanostructured anodes and cathodes to improvise the electrochemical responses for developing lithium-ion batteries. Exfoliated testing of the restacked layered MoS_2 with enlarged *c*-axis as anode material for lithium-ion batteries exhibits a large reversible capacity of 800 $mAhg^{-1}$ after 50 cycles and extraordinarily good cycling stability at huge current density of 50 mA/g [164]. The high cycling performance was due to MoS_2 layers (0.635 nm), which had more spacing and surface area. Wang and co-workers reported that the MoS_2 nanotube as anode exhibited improved electrochemical properties in a lithium-ion battery, with a reversible capacity of ~839 $mAhg^{-1}$ after 50 current density cycles [165].

For improving the electrochemical performances, creating pores on ultrathin 2D nanomaterials is an efficient path to prevent the destruction of materials used for electrodes and create short Li^+ ion diffusion lengths for achieving a reversible capacity of 1040 $mAhg^{-1}$ at a current of 100 mAg^{-1}, although 833 $mAhg^{-1}$ can be retained after numerous current density cycles [166].

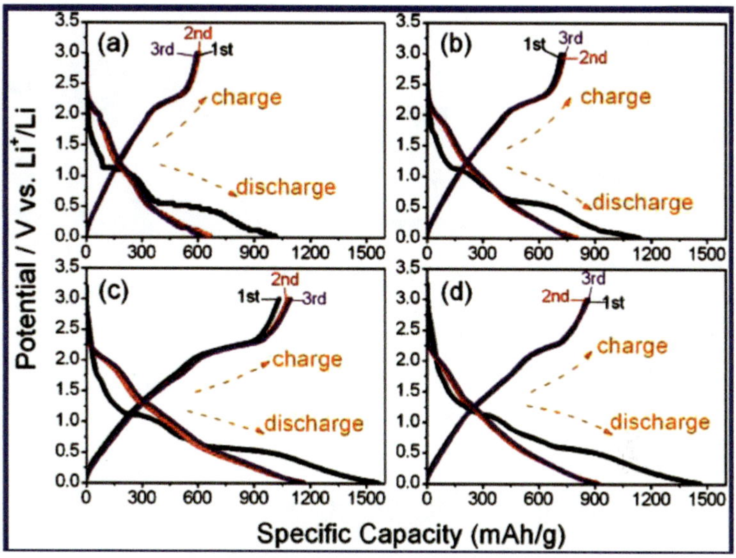

Figure 3.9 First three charge and discharge curves of (a) MoS_2, (b) MoS_2/graphene (1:1), (c) MoS_2/graphene (1:2), and (d) MoS_2/graphene (1:4) [171]. Reprinted with permission from [171].

In addition, ultrathin 2D nanomaterials have been studied as cathodes for Li^+ ion batteries [167–170]. A composite of MoS_2-graphene 3D network as anodes showed reversible capacity of 877 $mAhg^{-1}$ at a current of 100 $mAhg^{-1}$ during 50 cycles and excellent high-current-density performance [168]. Inclusion of graphene prevented the aggregation of 2D nanomaterials, thereby achieving thin and small grain size layers, which reduced the energy barrier for Li^+ ion intercalation and improved the surface/interface Li storage [170, 171]. The MoS_2/graphene composite with a Mo:C molar ratio of 1:2 displays a reversible capacity of 1100 $mAhg^{-1}$ and a current of 100 mAg^{-1} with excellent cycle stability and high-rate capability. Figure 3.9 shows that the discharge capacities of the MoS_2, MoS_2/graphene (1:1), MoS_2/graphene (1:2), and MoS_2/graphene (1:4) electrodes are 1012, 1142, 1571, and 1462 $mAhg^{-1}$, correspondingly, while the charge capacities are 600, 718, 1031, and 855 $mAhg^{-1}$, correspondingly.

The layered chalcogenide MoS_2 was used as a host material in zinc-based ion batteries. In zinc-based ion batteries, the bivalent Zn involves a two-electron transfer during charging–discharging cycle, enabling plenty of storage capacity [172]. Oxygen incorporation in small amounts like 5% into

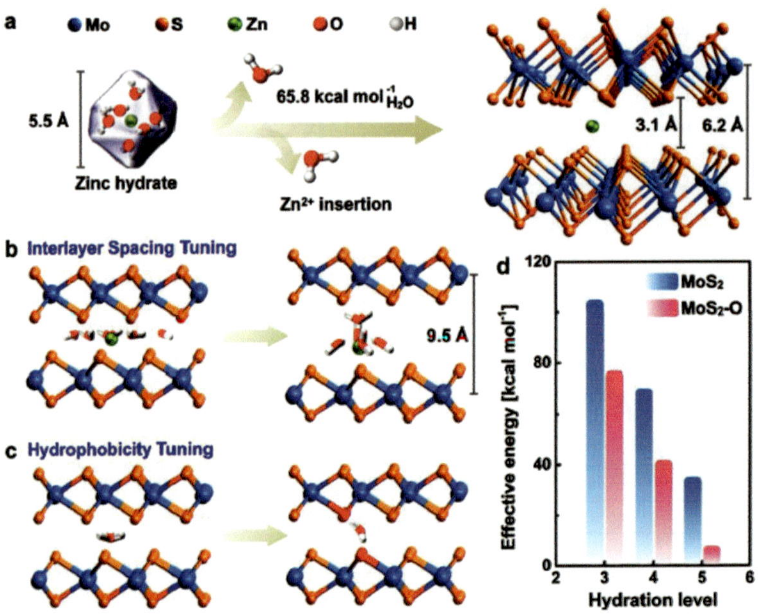

Figure 3.10 Schematic illustration of MoS_2 intercalated with Zn hydrate and their structure [172]. Reprinted with permission from [172].

MoS_2 increased the interlayer spacing to 9.5 Å and increased hydrophiles by significantly minimizing the Zn^{2+} intercalation energy, as shown in Figure 3.10. As a result of oxygen incorporation, three times enhanced Zn^{2+} diffusivity to 232 $mAhg^{-1}$ in Zn^{2+} storage capacity is reported, which is tens of magnitude higher than MoS_2 alone. The promoted Zn^{2+} intercalation to produce metallic 1T-phase MoS_2 is promising but requires safety procedures using the common Li^+ intercalation method.

3.4.5 Supercapacitors

High power density, rapid charging–discharging rates, cycling life, cost, and being safer make supercapacitors a promising energy storage device. As graphene has high surface area (theoretical $\sim$ 2630 m^2g^{-1}), mobility of charge transporters (200,000 $cm^2V^{-1}s^{-1}$), and a specific structure, capacitance by theory up to 550 F/g precedes them as supercapacitors [173–175]. However, fast charging–discharging rate along with high power density deteriorate due to the aggregates that form graphite. Thus, efforts have been dedicated to solve restacking issues, study performances using graphene electrodes, and search of new 2D materials for supercapacitors. Materials

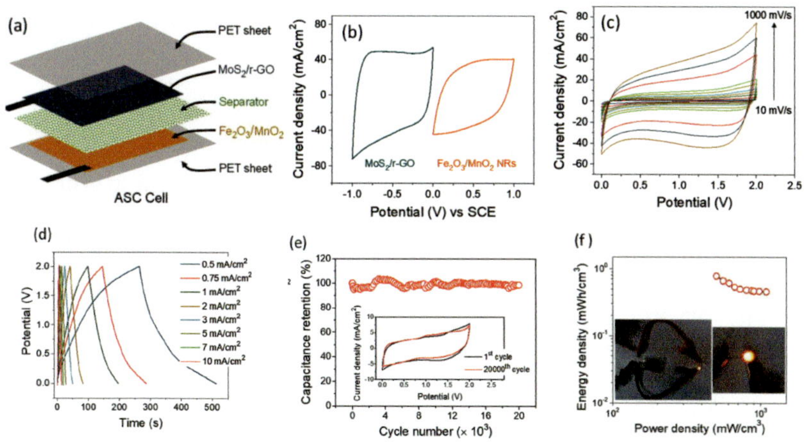

Figure 3.11 Schematics and performance of MoS_2/r-GO//Fe_2O_3/MnO_2 asymmetric supercapacitors [176]. Reprinted with permission from [176].

in 2D are 2D nanosheets of metal oxides/hydroxides, TMDs, MXenes, and so on. Charge-storage in free-standing MoS_2/r-GO (reduced graphene oxide) nanoflakes deposited on molybdenum foil in Na_2SO_4 solution is reported as asymmetric supercapacitor with a desired cycling performance. The 2 V MoS_2/r-GO//Fe_2O_3/MnO_2 asymmetric supercapacitor exhibits a cycling performance over 20,000 cycles in an aqueous electrolyte as shown in Figure 3.11. The asymmetric supercapacitor displays a volumetric capacitance of 1.54 F/cm^3 at a scan rate of 10 mV/s, maintains initial capacitance, and ensures a good electronic conductivity of individual electrode materials [176]. An electrode constituted by 1T-phase MoS_2 nanosheet in thin-film form worked as a supercapacitor with enhanced performance [177]. The reported capacitance is 650 F/cm^3 in an aqueous electrolyte solution of H2SO4, which is the highest using TMD as supercapacitor. The 1T-phase MoS_2 electrode has high power and volumetric energy densities. Contrarily, 2H-phase MoS_2 is inapt due to low specific capacitance (e.g., 129.2 Fg^{-1} at 1 Ag^{-1}) [178].

3.4.6 Solar cells

Graphene is suitable for use as a transparent conductive oxide electrode due to higher optical transparency and electrical conductivity [179]. Furthermore, the flexible form and chemical stability of graphene and graphene-based composites make them appropriate for flexible photovoltaic devices and as solar cells. Absorption features of monolayer TMDs are used to manufacture ultrathin photovoltaic devices. TMDs can act as interlayers or active layers in photovoltaic devices [83].

Solid-state solar cell photovoltaic properties of Schottky barriers based on WSe_2/Al and WSe_2/ZnO p-n heterojunctions exposed conversion efficiencies of 5.3% and 6.1% [180, 181]. Ultrafast charge transfer in photoexcited carriers was reported for atomically thin MoS_2/WS_2 p-n heterojunctions, although with low conversion efficiency when using WSe_2 instead of WS_2 [182]. In solar cells, 2D TMDs serve as light-absorbing layers. The photoactive layer of the Schottky barrier solar cell is made of 2D TMDs that can be either a p-type or an n-type semiconductor [183]. 2D TMDs can be apt as charge transport layers in organic photovoltaic cells [184]. Nowadays, 2D TMDs with potential for photovoltaic performance, such as MoS_2, WS_2, TaS_2, and $NbSe_2$, have been successfully utilized in organic solar cells [185–187]. An n-type MoS_2 single layer deposited on p-type Si substrate reported power conversion efficiency of 5.23 % among all TMD single-layer-based solar cells fabricated so far [37]. The MoS_2 single layer creates a built-in electric field adjacent to the interface sandwiched between MoS_2 and p-Si to promote photogeneration and carrier distinction [129]. Figure 3.12 (top) shows the band diagram of n-MoS_2/p-Si heterojunction attained from UPS measurements; Φ is a work function and EG is an energy bandgap. Figure 3.12 (bottom) reveals the J-V characteristics of n-MoS_2/p-Si heterojunction solar cells. Dye-sensitized solar cells with TiO_2/MoS_2 (few-layer) heterostructure showed enhanced photovoltaic performance. The results showed low power conversion efficiency (1.08%) due to weak bonding of MoS_2 nanosheet and TiO_2 matrix [188]. Due to their many edge sites that produce excellent electro-catalytic activities, 2D TMDs hold promise as electro-catalysts on counter electrodes in dye-sensitized solar cells [189].

3.4.7 Photocatalysis and sensors

Graphene-based photocatalysts are a viable alternative to increase photocatalytic hydrogen and oxygen production performance in converting solar energy to chemical energy and result in improved photocatalytic activity [190–192].For electro-catalytic water splitting, MoS_2 and WS_2 are being developed [193, 194]. Nanosheets made of MoS_2 and WS_2 in conjunction with CdS and TiO_2 photosensitizers are highly effective co-catalysts for photocatalytic H_2 production. The TMDs not only activate the locations for H2 production but also enhance the stability and separation of the charges.

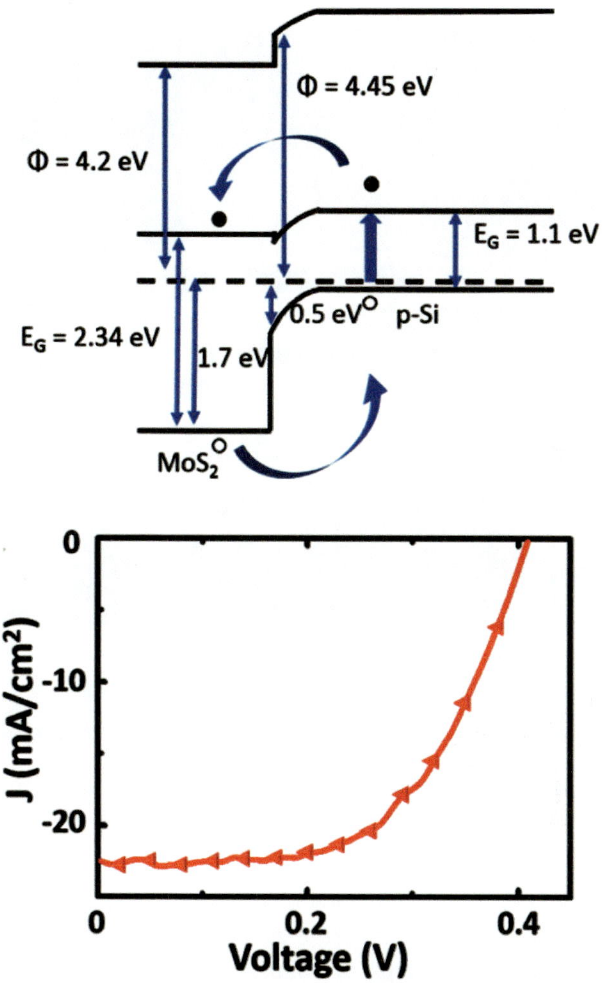

Figure 3.12 (Top) Band diagram obtained from UPS measurements. Φ is a work function and EG is an energy bandgap. (Bottom) Current–voltage characteristics of MoS_2/p-Si heterojunction solar cells [129]. Reprinted with permission from [129].

Many TMD (MoS_2/WS_2/NiS_2) nanosheets are excellent co-catalysts [195–200]. Photoelectrochemical activities of lateral MoS_2 heterostructures consist of improved photocatalytic activities [201]. Introduction of CO_2 in the 2H-phase MoS_2 host promoted the phase alteration of partial 2H-phase MoS_2 lattices into metallic 1T-phase MoS_2. Figure 3.13 indicates the electron microscopy images of the metallic 1T-phase MoS_2 nanosheets. By using

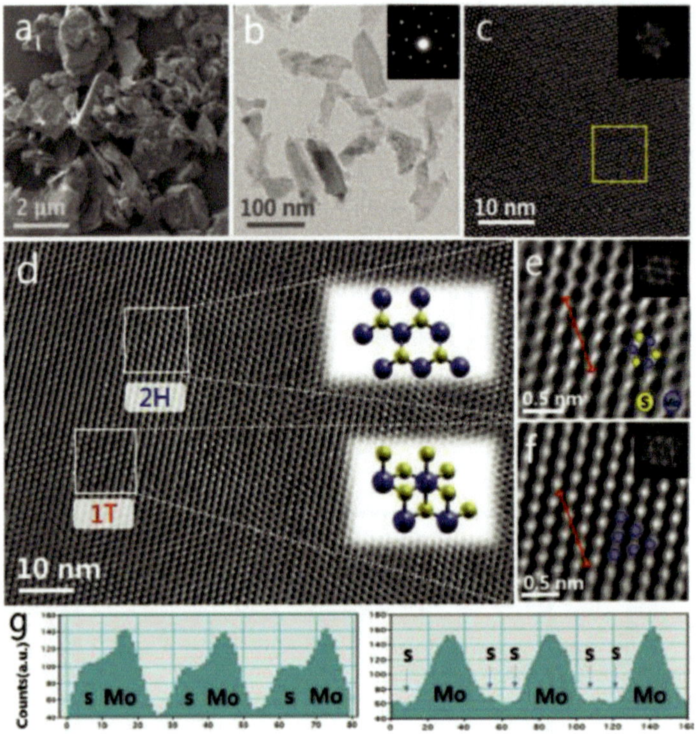

Figure 3.13 SEM morphology and HRTEM images of MoS_2 and lateral heterostructures of 1T-phase and 2H-phase MoS_2 [201]. Reprinted with permission from [201].

external bias, the enhanced efficiency of charge transfer was monitored. The existence of 1T-phase MoS_2 can serve as an excellent electron acceptor and transport material, which helps to control the recombination processes of photogenerated charges, boosting the photocatalytic activity in visible light as a result of this. The BP nanosheets have even attracted growing enthusiasm due to their high carrier mobilities and unique optoelectronic characteristics. The bandgap of BP can be tuned depending on the layers from 0.3 (bulk) to 2.1 eV (monolayer), which is useful for water splitting. To date, only BP-TiO2 hybrids have demonstrated improved photocatalytic activity for the breakdown of dye when exposed to light for the dye degradation [202].

Graphene and TMDs have outstanding conductivity, excellent specific surface area, high sensitivity to exterior stimuli, and fluorescence quenching abilities, which allow them to be useful for the detection of guest species. Electronic sensors fabricated using MoS_2 nanosheets can detect biomolecules and prostate-specific antigen [35, 203, 204]. Ultrasensitive CVD built up monolayer MoS_2 FET-based DNA sensors that match the requirements for down-syndrome screening, including sensitivity DNA with a detection limit of 0.1 fM and good specificity for 3-nucleotide polymorphism discrimination [205]. Gold nanoparticles with optimized size and density dropped on MoS_2 can probe DNA. Monolayer MoS_2-based FET was used to utilize the high performance as the sensing base, optimize DNA pattern, and probe immobilizing approach to collect target DNA. In Figure 3.14, the material characterization and device fabrication are presented. Ultrathin 2D nanomaterial fluorescent sensors exhibit superior sensitivity due to the presence of a 2D structure that features superior fluorescence quenching efficiency.

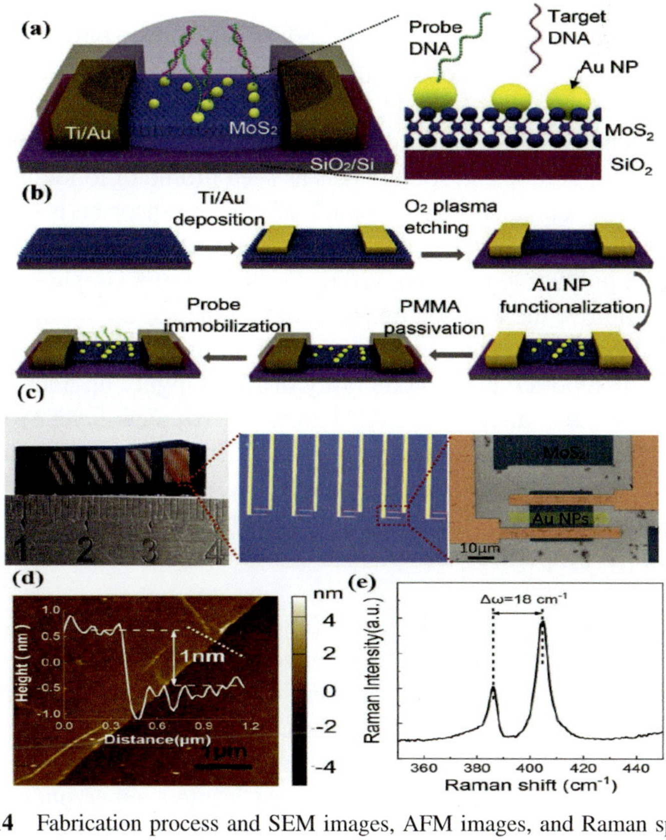

Figure 3.14 Fabrication process and SEM images, AFM images, and Raman spectrum of MoS_2 FET-based biosensor [205]. Reprinted with permission from [205].

3.5 Recent Advances and Outlook

High performance to human friendly electronic is looked for today's research. Till date, an n-type $MoTe_2$ unipolar transistor has high saturation and relatively low contact resistance with the use of silver as contacts and AlO_x as encapsulation [206]. Such unipolar transistors with scandium as contacts combined with h-BN show pinning factors. Chen et al. fabricated a monolayer MoS_2 nanoribbon transistor creating ultranarrow channel transistors. The nanoribbon devices have an I_{on}/I_{off} ratio more than 10^4 and an extrinsic field-mobility of 8.53 $cm^2\ V^{-1}s^{-1}$. With a 30-nm-wide device, fabrication using an MoS_2 flake has a small value of mobility, low I_{on}/I_{off} ratio, and a large subthreshold [207]. Figure 3.15 shows the electrical performance of monolayer MoS_2 FET microribbon and nanoribbon coated with an atomic layer deposition passivation layer. Based on the sulfurization growth and sodium molybdate dehydrate as dual-purpose precursors, MoS_2 FET electron migration up to ~90 $cm^2\ V^{-1}s^{-1}$ according to the sulfurization growth and sodium molybdate dehydrate as dual-purpose precursors are demonstrated. Here Mo acts as source and sodium acts as seed promotor to facilitate lateral growth of MoS_2 crystals [208]. Flexible Ebola biosensor built on semiconducting 2H phase-MoS_2 flakes that are stacked using a cathodic exfoliation method on a polyimide substrate [209]. Furthermore, a simple yet effective method to improve the electron transport of BP, such as the n-doping method for few-layer BP FET via *in situ* aluminum surface functionalization, was achieved. The reported electron mobility of BP is highly upgraded from 1.6 to 10.6 $cm^2\ V^{-1}s^{-1}$ after 4.8 nm Al alteration [210]. Few-layer MoS_2/WSe_2 heterostructure transistor devices report that local sheet conductance varies monotonically as a function of applied back-gate voltage [211]. Currently, research efforts envisaged to improve the methodology and enhance the electronic properties and performances using the 2D materials are underway.

Fabrication of twistronic devices through the weak vdW forces between the atomic planes in 2D materials is reported just by arbitrary rotational order [212, 213] and by placing a single sheet of graphene above another sheet with a rotated twist at the "magic" angle of 1.1°. This twist firstly tunes the metallic property of the material to that of an insulator, and, secondly, by applying a strong electric field, it converts it into a superconductor. Until now, only studies on graphene and boron nitride are reported and several 2D materials find these possibilities and remain unexplored. It has even been demonstrated that the energy gap can be altered in graphene by varying

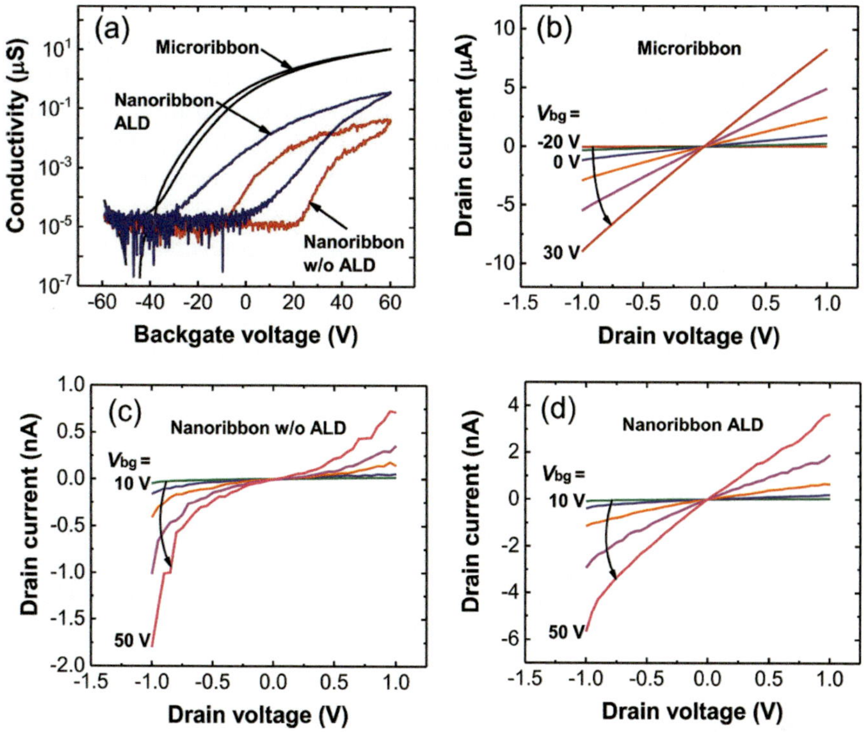

Figure 3.15 Electrical performance of fabricated monolayer MoS_2 FET microribbons and nanoribbons [207]. Reprinted with permission from [207].

the layer orientations. For instance, a 2D graphene sheet with no bandgap develops one when placed in contact with hexagonal boron nitride having nearly similar lattice constant forming "Moiré superlattice." If these layers are twisted, the bandgap re-diminishes. Thus, graphene can be made metallic, insulating, semiconducting, magnetic, as well as superconducting.

These materials achieved wide attention due to their specific optical and electronic performances. Despite this enormous potential, very few findings have occurred to date on the integration of 2D materials apart from graphene into spintronic devices. The use of graphene is limited for logical circuits for low-power electronic switching (i.e., this material flops in switch control) due to its overlapped band structure. With bandgaps of semiconducting TMDs, large on/off current ratios can be attained. The bandwidth for MoS_2 (TMD) can be tuned although its electron mobility is peerless to that of graphene, which prevents this substance from serving as transparent electrodes. Thus, manufacture of hybrid structures grounded on 2D materials is gaining interests. The major challenge that remains is controlled fabrication of 2D TMDs with huge-area uniformity. Even a minute percent of substitute can lower carrier mobility or vary the band structure in atomically thin TMDs, and, thus, methods to have controlled doping strategies are needed. Purely 2D p-n heterojunctions are not realized due to the absence of supervised doping methods. Degradation due to adsorbate interaction is reversible in TMDs contrasting in organic semiconductors where the degradation is irreversible. Despite the TMDs showing high response, the phototransistors using TMDs suffer long response times when evaluated to photodiodes. TMDs induce inverters and logic gates to consume a lot of electricity. Also, large mobility differences are reported for a variety of TMDs in comparison with encapsulated TMD devices [214–217]. Their carrier mobilities are sensitive to local dielectric environment, charged impurities, and sub-optimal sample quality. Improving the responses lacks industrial standard production of TMDs due to impossibility to achieve high yield and reproducibility. The studies of TMD device applications are still in their initial investigation stages. Newer methods and platforms to design and fabricate TMDs for devices are being explored.

3.6 Future Perspective

Recent theoretical work and experiments suggest that in semiconducting TMDs, which are ultrathin, the optical transitions are led by excitons instead of direct interband transitions. This opens up the way for using semiconducting TMDs with unique symmetry and an additional valley degree of freedom, thus promising valleytronic devices. The additional variable in non-TMDs can be retrieved by applying stain or magnetic fields, whereas for TMDs by circularly polarized light. Starting works using MoS_2 and WSe_2 monolayers by applying circular polarized light to achieve valley population are demonstrated. Semiconducting TMD monolayers provide an alternative approach for light emission sensing studied through changing resistances.

Although semiconducting TMDs are looked forward to in future FETs, deep understanding and optimizations are necessary for their high performances. Their band alignment, carrier distinction, depletion regions, and charge flowing mechanisms need to be entangled for p-n atomistic thin heterojunctions. For improving light absorption capabilities and their quantum yields for optoelectronics, new methods need to be developed. Solar cells based on semiconducting TMD ultrathin layers tackle performance challenges such as high efficiency, plasmon absorption, strain, and vertical stacking. Surface engineering, encapsulation, doping, and design of TMD devices are underway to approach traditional light emitters. Doping and work function control of TMDs is looked forward for viable tuning of threshold voltages.

In the near future, as presented earlier, semiconducting TMDs show promise for the formation of flexible, stretchable, and transparent electronic devices. Studies related to thickness-dependent lateral TMD junction, flexible optoelectronic devices, and energy storage devices exist, although feasible methods to approach comparison with commercial working device are still underway. Feasible methods are necessary for highly improving the TMD performances in electronics and for optical electronics. However, as the intrinsic structure is specific to a particular TMD, it is difficult for one sample to satisfy all parameters for practical functioning. Pushing devices into real practical applications, achieved by combining various structures, achieving a desired bandgap, and studying their electronic-optical physical property performance, will be a prominent topic in the near future.

References

[1] Nagaosa N. 2007 PHYSICS: A New State of Quantum Matter. Science 318, 758–759. (doi:10.1126/science.1150199)

[2] Li L, Yu Y, Ye GJ, Ge Q, Ou X, Wu H, Feng D, Chen XH, Zhang Y. 2014 Black phosphorus field-effect transistors. Nat. Nanotechnol. 9, 372–377. (doi:10.1038/nnano.2014.35)

[3] Bhimanapati GR et al. 2015 Recent Advances in Two-Dimensional Materials beyond Graphene. ACS Nano 9, 11509–11539. (doi:10.1021/acsnano.5b05556)

[4] He M, Sun H, He QL. 2019 Topological insulator: Spintronics and quantum computations. Front. Phys. 14, 43401. (doi:10.1007/s11467-019-0893-4)

[5] Novoselov KS, Fal/ko VI, Colombo L, Gellert PR, Schwab MG, Kim K. 2012 A roadmap for graphene. Nature 490, 192–200. (doi:10.1038/nature11458)

[6] Novoselov KS, Geim AK, Morozov SV, Jiang D, Katsnelson MI, Grigorieva IV, Dubonos SV, Firsov AA. 2005 Two-dimensional gas of massless Dirac fermions in graphene. Nature 438, 197–200. (doi:10.1038/nature04233)

[7] Vozmediano MAH, Guinea F. 2012 Effect of Coulomb interactions on the physical observables of graphene. Phys. Scr. T146, 014015. (doi:10.1088/0031-8949/2012/T146/014015)

[8] Lin Y-M, Dimitrakopoulos C, Jenkins KA, Farmer DB, Chiu H-Y, Grill A, Avouris P. 2010 100-GHz Transistors from Wafer-Scale Epitaxial Graphene. Science 327, 662–662. (doi:10.1126/science.1184289)

[9] Liu M, Yin X, Ulin-Avila E, Geng B, Zentgraf T, Ju L, Wang F, Zhang X. 2011 A graphene-based broadband optical modulator. Nature 474, 64–67. (doi:10.1038/nature10067)

[10] Wu Z-S, Parvez K, Feng X, Müllen K. 2013 Graphene-based in-plane micro-supercapacitors with high power and energy densities. Nat. Commun. 4, 1–8. (doi:10.1038/ncomms3487)

[11] Artiles MS, Rout CS, Fisher TS. 2011 Graphene-based hybrid materials and devices for biosensing. Adv. Drug Deliv. Rev. 63, 1352–1360. (doi:10.1016/j.addr.2011.07.005)

[12] Yang G, Lee C, Kim J, Ren F, Pearton SJ. 2013 Flexible graphene-based chemical sensors on paper substrates. Phys. Chem. Chem. Phys. 15, 1798–1801. (doi:10.1039/C2CP43717A)

[13] Garaj S, Hubbard W, Reina A, Kong J, Branton D, Golovchenko JA. 2010 Graphene as a subnanometre trans-electrode membrane. Nature 467, 190–193. (doi:10.1038/nature09379)

[14] Chng ELK, Pumera M. 2014 Toxicity of graphene related materials and transition metal dichalcogenides. RSC Adv. 5, 3074–3080. (doi:10.1039/C4RA12624F)

[15] Asgari R, Davoudi B, Polini M, Giuliani GF, Tosi MP, Vignale G. 2005 Quasiparticle self-energy and many-body effective mass enhancement in a two-dimensional electron liquid. Phys. Rev. B 71, 045323. (doi:10.1103/PhysRevB.71.045323)

[16] Asgari R, Tanatar B. 2006 Many-body effective mass and spin susceptibility in a quasi-two-dimensional electron liquid. Phys. Rev. B 74, 075301. (doi:10.1103/PhysRevB.74.075301)

[17] Morosan E, Natelson D, Nevidomskyy AH, Si Q. 2012 Strongly Correlated Materials. Adv. Mater. 24, 4896–4923. (doi:10.1002/adma.201202018)

[18] Guinea F, Katsnelson MI, Wehling TO. 2014 Two-dimensional materials: Electronic structure and many-body effects. Ann. Phys. 526, A81–A82. (doi:10.1002/andp.201470096)

[19] Mak KF, He K, Shan J, Heinz TF. 2012 Control of valley polarization in monolayer MoS2 by optical helicity. Nat. Nanotechnol. 7, 494–498. (doi:10.1038/nnano.2012.96)

[20] Mak KF, He K, Lee C, Lee GH, Hone J, Heinz TF, Shan J. 2013 Tightly bound trions in monolayer MoS2. Nat. Mater. 12, 207–211. (doi:10.1038/nmat3505)

[21] Cao T et al. 2012 Valley-selective circular dichroism of monolayer molybdenum disulphide. Nat. Commun. 3, 1–5. (doi:10.1038/ncomms1882)

[22] Sallen G et al. 2012 Robust optical emission polarization in MoS2 monolayers through selective valley excitation. Phys. Rev. B 86, 081301. (doi:10.1103/PhysRevB.86.081301)

[23] Xiao D, Liu G-B, Feng W, Xu X, Yao W. 2012 Coupled Spin and Valley Physics in Monolayers of MoS2 and Other Group-VI Dichalcogenides. Phys. Rev. Lett. 108, 196802. (doi:10.1103/PhysRevLett.108.196802)

[24] Wu S et al. 2013 Electrical tuning of valley magnetic moment through symmetry control in bilayer MoS2. Nat. Phys. 9, 149–153. (doi:10.1038/nphys2524)

[25] Zeng H, Dai J, Yao W, Xiao D, Cui X. 2012 Valley polarization in MoS2 monolayers by optical pumping. Nat. Nanotechnol. 7, 490–493. (doi:10.1038/nnano.2012.95)

[26] Ochoa H, Roldán R. 2013 Spin-orbit-mediated spin relaxation in monolayer MoS2. Phys. Rev. B 87, 245421. (doi:10.1103/PhysRevB.87.245421)

[27] Zibouche N, Philipsen P, Kuc A, Heine T. 2014 Transition-metal dichalcogenide bilayers: Switching materials for spintronic and valleytronic applications. Phys. Rev. B 90, 125440. (doi:10.1103/PhysRevB.90.125440)

[28] Uchihashi T. 2016 Two-dimensional superconductors with atomic-scale thickness. Supercond. Sci. Technol. 30, 013002. (doi:10.1088/0953-2048/30/1/013002)

[29] Roldán R, Silva-Guillén JA, López-Sancho MP, Guinea F, Cappelluti E, Ordejón P. 2014 Electronic properties of single-layer and multilayer transition metal dichalcogenides MX2 (M = Mo, W and X = S, Se). Ann. Phys. 526, 347–357. (doi:10.1002/andp.201400128)

[30] Cheng R, Jiang S, Chen Y, Liu Y, Weiss N, Cheng H-C, Wu H, Huang Y, Duan X. 2014 Few-layer molybdenum disulfide transistors and circuits for high-speed flexible electronics. Nat. Commun. 5, 1–9. (doi:10.1038/ncomms6143)

[31] Mak KF, Shan J. 2016 Photonics and optoelectronics of 2D semiconductor transition metal dichalcogenides. Nat. Photonics 10, 216–226. (doi:10.1038/nphoton.2015.282)

[32] Pesin D, MacDonald AH. 2012 Spintronics and pseudospintronics in graphene and topological insulators. Nat. Mater. 11, 409–416. (doi:10.1038/nmat3305)

[33] Kong D, Cha JJ, Wang H, Lee HR, Cui Y. 2013 First-row transition metal dichalcogenide catalysts for hydrogen evolution reaction. Energy Environ. Sci. 6, 3553–3558. (doi:10.1039/C3EE42413H)

[34] Tan C et al. 2015 High-Yield Exfoliation of Ultrathin Two-Dimensional Ternary Chalcogenide Nanosheets for Highly Sensitive and Selective Fluorescence DNA Sensors. J. Am. Chem. Soc. 137, 10430–10436. (doi:10.1021/jacs.5b06982)

[35] Sarkar D, Liu W, Xie X, Anselmo AC, Mitragotri S, Banerjee K. 2014 MoS2 Field-Effect Transistor for Next-Generation Label-Free Biosensors. ACS Nano 8, 3992–4003. (doi:10.1021/nn5009148)

[36] Peng X, Peng L, Wu C, Xie Y. 2014 Two dimensional nanomaterials for flexible supercapacitors. Chem. Soc. Rev. 43, 3303–3323. (doi:10.1039/C3CS60407A)

[37] Tsai M-L, Su S-H, Chang J-K, Tsai D-S, Chen C-H, Wu C-I, Li L-J, Chen L-J, He J-H. 2014 Monolayer MoS2 Heterojunction Solar Cells. ACS Nano 8, 8317–8322. (doi:10.1021/nn502776h)

[38] Zhao Y, Li X, Yan B, Xiong D, Li D, Lawes S, Sun X. 2016 Recent Developments and Understanding of Novel Mixed Transition-Metal Oxides as Anodes in Lithium Ion Batteries. Adv. Energy Mater. 6, 1502175. (doi:10.1002/aenm.201502175)

[39] Jonietz F et al. 2010 Spin Transfer Torques in MnSi at Ultralow Current Densities. Science 330, 1648–1651. (doi:10.1126/science.1195709)

[40] Fert A, Cros V, Sampaio J. 2013 Skyrmions on the track. Nat. Nanotechnol. 8, 152–156. (doi:10.1038/nnano.2013.29)

[41] Romming N, Hanneken C, Menzel M, Bickel JE, Wolter B, Bergmann K von, Kubetzka A, Wiesendanger R. 2013 Writing and Deleting Single Magnetic Skyrmions. Science 341, 636–639. (doi:10.1126/science.1240573)

[42] Tomasello R, Martinez E, Zivieri R, Torres L, Carpentieri M, Finocchio G. 2014 A strategy for the design of skyrmion racetrack memories. Sci. Rep. 4, 1–7. (doi:10.1038/srep06784)

[43] Stepanov EA, Dutreix C, Katsnelson MI. 2017 Dynamical and Reversible Control of Topological Spin Textures. Phys. Rev. Lett. 118, 157201. (doi:10.1103/PhysRevLett.118.157201)

[44] Raeliarijaona A, Belashchenko K, Kovalev A. 2019 Skyrmion Crystals in Decorated Transition Metal Dichalcogenides., H41.008.

[45] Lind H, Lidin S, Häussermann U. 2005 Structure and bonding properties of (Bi2Se3)m(Bi2)n stacks by first-principles density functional theory. Phys. Rev. B 72, 184101. (doi:10.1103/PhysRevB.72.184101)

[46] Vandenberghe WG, Fischetti MV. 2017 Imperfect two-dimensional topological insulator field-effect transistors. Nat. Commun. 8, 1–8. (doi:10.1038/ncomms14184)

[47] Sasaki S, Mizushima T. 2015 Superconducting doped topological materials. Phys. C Supercond. Its Appl. 514, 206–217. (doi:10.1016/j.physc.2015.02.018)

[48] Xu M, Liang T, Shi M, Chen H. 2013 Graphene-Like Two-Dimensional Materials. Chem. Rev. 113, 3766–3798. (doi:10.1021/cr300263a)

[49] Lu P, Wu X, Guo W, Zeng XC. 2012 Strain-dependent electronic and magnetic properties of MoS2 monolayer, bilayer, nanoribbons and nanotubes. Phys. Chem. Chem. Phys. 14, 13035–13040. (doi:10.1039/C2CP42181J)

[50] Peelaers H, Van de Walle CG. 2012 Effects of strain on band structure and effective masses in MoS2. Phys. Rev. B 86, 241401. (doi:10.1103/PhysRevB.86.241401)

[51] Yun WS, Han SW, Hong SC, Kim IG, Lee JD. 2012 Thickness and strain effects on electronic structures of transition metal dichalcogenides: 2H-MX2 semiconductors (M= Mo, W; X = S, Se, Te). Phys. Rev. B 85, 033305. (doi:10.1103/PhysRevB.85.033305)

[52] mGhorbani-Asl M, Borini S, Kuc A, Heine T. 2013 Strain-dependent modulation of conductivity in single-layer transition-metal dichalcogenides. Phys. Rev. B 87, 235434. (doi:10.1103/PhysRevB.87.235434)

[53] Shi H, Pan H, Zhang Y-W, Yakobson BI. 2013 Quasiparticle band structures and optical properties of strained monolayer MoS2 and WS2. Phys. Rev. B 87, 155304. (doi:10.1103/PhysRevB.87.155304)

[54] Hromadová L, Martoňák R, Tosatti E. 2013 Structure change, layer sliding, and metallization in high-pressure MoS2. Phys. Rev. B 87, 144105. (doi:10.1103/PhysRevB.87.144105)

[55] Roldán R, Silva-Guillén JA, López-Sancho MP, Guinea F, Cappelluti E, Ordejón P. 2014 Electronic properties of single-layer and multilayer transition metal dichalcogenides MX2 (M = Mo, W and X = S, Se). Ann. Phys. 526, 347–357. (doi:10.1002/andp.201400128)
[56] Cappelluti E, Roldán R, Silva-Guillén JA, Ordejón P, Guinea F. 2013 Tight-binding model and direct-gap/indirect-gap transition in single-layer and multilayer MoS2. Phys. Rev. B 88, 075409. (doi:10.1103/PhysRevB.88.075409)
[57] Sun L et al. 2013 Spin-Orbit Splitting in Single-Layer MoS2 Revealed by Triply Resonant Raman Scattering. Phys. Rev. Lett. 111, 126801. (doi:10.1103/PhysRevLett.111.126801)
[58] Wilson JA, Di Salvo FJ, Mahajan S. 1974 Charge-Density Waves in Metallic, Layered, Transition-Metal Dichalcogenides. Phys. Rev. Lett. 32, 882–885. (doi:10.1103/PhysRevLett.32.882)
[59] Friend RH, Yoffe AD. 1987 Electronic properties of intercalation complexes of the transition metal dichalcogenides. Adv. Phys. 36, 1–94. (doi:10.1080/00018738700101951)
[60] Hwang H, Kim H, Cho J. 2011 MoS2 Nanoplates Consisting of Disordered Graphene-like Layers for High Rate Lithium Battery Anode Materials. Nano Lett. 11, 4826–4830. (doi:10.1021/nl202675f)
[61] Hu Z, Wang L, Zhang K, Wang J, Cheng F, Tao Z, Chen J. 2014 MoS2 nanoflowers with expanded interlayers as high-performance anodes for sodium-ion batteries. Angew. Chem. Int. Ed Engl. 53, 12794–12798. (doi:10.1002/anie.201407898)
[62] Liu H, Su D, Zhou R, Sun B, Wang G, Qiao SZ. 2012 Highly Ordered Mesoporous MoS2 with Expanded Spacing of the (002) Crystal Plane for Ultrafast Lithium Ion Storage. Adv. Energy Mater. 2, 970–975. (doi:10.1002/aenm.201200087)
[63] Zhang R, Tsai I-L, Chapman J, Khestanova E, Waters J, Grigorieva IV. 2016 Superconductivity in Potassium-Doped Metallic Polymorphs of MoS2. Nano Lett. 16, 629–636. (doi:10.1021/acs.nanolett.5b04361)
[64] Morosan E, Zandbergen HW, Dennis BS, Bos JWG, Onose Y, Klimczuk T, Ramirez AP, Ong NP, Cava RJ. 2006 Superconductivity in Cu x TiSe 2. Nat. Phys. 2, 544–550. (doi:10.1038/nphys360)
[65] Barath H, Kim M, Karpus JF, Cooper SL, Abbamonte P, Fradkin E, Morosan E, Cava RJ. 2008 Quantum and Classical Mode Softening Near the Charge-Density-Wave-Superconductor Transition of CuxTiSe2. Phys. Rev. Lett. 100, 106402. (doi:10.1103/PhysRevLett.100.106402)

[66] Morosan E, Wagner KE, Zhao LL, Hor Y, Williams AJ, Tao J, Zhu Y, Cava RJ. 2010 Multiple electronic transitions and superconductivity in PdxTiSe2. Phys. Rev. B 81, 094524. (doi:10.1103/PhysRevB.81.094524)

[67] Hannay N, editor. 1976 Treatise on Solid State Chemistry: Volume 3 Crystalline and Noncrystalline Solids. Springer US. (doi:10.1007/978-1-4684-2664-9)

[68] Parkin SSP, Friend RH. 1980 3d transition-metal intercalates of the niobium and tantalum dichalcogenides. I. Magnetic properties. Philos. Mag. B 41, 65–93. (doi:10.1080/13642818008245370)

[69] Parkin SSP, Friend RH. 1980 3d transition-metal intercalates of the niobium and tantalum dichalcogenides. II. Transport properties. Philos. Mag. B 41, 95–112. (doi:10.1080/13642818008245371)

[70] Taniguchi K, Matsumoto A, Shimotani H, Takagi H. 2012 Erratum: "Electric-field-induced superconductivity at 9.4 K in a layered transition metal disulphide MoS2" [Appl. Phys. Lett. 101, 042603 (2012)]. Appl. Phys. Lett. 101, 109902. (doi:10.1063/1.4750245)

[71] Ye JT, Zhang YJ, Akashi R, Bahramy MS, Arita R, Iwasa Y. 2012 Superconducting Dome in a Gate-Tuned Band Insulator. Science 338, 1193–1196. (doi:10.1126/science.1228006)

[72] Rösner M, Haas S, Wehling TO. 2014 Phase diagram of electron-doped dichalcogenides. Phys. Rev. B 90, 245105. (doi:10.1103/PhysRevB.90.245105)

[73] Yuan NFQ, Mak KF, Law KT. 2014 Possible Topological Superconducting Phases of MoS2. Phys. Rev. Lett. 113, 097001. (doi:10.1103/PhysRevLett.113.097001)

[74] Roldán R, Cappelluti E, Guinea F. 2013 Interactions and superconductivity in heavily doped MoS2. Phys. Rev. B 88, 054515. (doi:10.1103/PhysRevB.88.054515)

[75] Fang L, Wang Y, Zou PY, Tang L, Xu Z, Chen H, Dong C, Shan L, Wen HH. 2005 Fabrication and superconductivity of NaxTaS2 crystals. Phys. Rev. B 72, 014534. (doi:10.1103/PhysRevB.72.014534)

[76] Morris RC. 1975 Connection between Charge-Density Waves and Superconductivity in NbSe2. Phys. Rev. Lett. 34, 1164–1166. (doi:10.1103/PhysRevLett.34.1164)

[77] Escudero R, Briggs A, Monceau P. 2001 Point contact characteristics of NbSe3 in the superconducting state. J. Phys. Condens. Matter 13, 6285–6295. (doi:10.1088/0953-8984/13/29/302)

[78] Núñez-Regueiro M, Mignot J-M, Jaime M, Castello D, Monceau P. 1993 Superconductivity under pressure in linear chalcogenides. Synth. Met. 56, 2653–2659. (doi:10.1016/0379-6779(93)90013-M)

[79] Singh Y, Nirmala R, Ramakrishnan S, Malik SK. 2005 Competition between superconductivity and charge-density-wave ordering in the Lu5Ir4(Si1-xGex)10 alloy system. Phys. Rev. B 72, 045106. (doi:10.1103/PhysRevB.72.045106)

[80] Wilson JA, Yoffe AD. 1969 The transition metal dichalcogenides discussion and interpretation of the observed optical, electrical and structural properties. Adv. Phys. 18, 193–335. (doi:10.1080/00018736900101307)

[81] Tongay S et al. 2014 Monolayer behaviour in bulk ReS2 due to electronic and vibrational decoupling. Nat. Commun. 5, 1–6. (doi:10.1038/ncomms4252)

[82] Yu Y et al. 2015 Gate-tunable phase transitions in thin flakes of 1T-TaS2. Nat. Nanotechnol. 10, 270–276. (doi:10.1038/nnano.2014.323)

[83] Jariwala D, Sangwan VK, Lauhon LJ, Marks TJ, Hersam MC. 2014 Emerging Device Applications for Semiconducting Two-Dimensional Transition Metal Dichalcogenides. ACS Nano 8, 1102–1120. (doi:10.1021/nn500064s)

[84] Wang QH, Kalantar-Zadeh K, Kis A, Coleman JN, Strano MS. 2012 Electronics and optoelectronics of two-dimensional transition metal dichalcogenides. Nat. Nanotechnol. 7, 699–712. (doi:10.1038/nnano.2012.193)

[85] Fiori G, Bonaccorso F, Iannaccone G, Palacios T, Neumaier D, Seabaugh A, Banerjee SK, Colombo L. 2014 Electronics based on two-dimensional materials. Nat. Nanotechnol. 9, 768–779. (doi:10.1038/nnano.2014.207)

[86] Bhimanapati GR et al. 2015 Recent Advances in Two-Dimensional Materials beyond Graphene. ACS Nano 9, 11509–11539. (doi:10.1021/acsnano.5b05556)

[87] Mak KF, McGill KL, Park J, McEuen PL. 2014 The valley Hall effect in MoS2 transistors. Science 344, 1489–1492. (doi:10.1126/science.1250140)

[88] Kang K, Xie S, Huang L, Han Y, Huang PY, Mak KF, Kim C-J, Muller D, Park J. 2015 High-mobility three-atom-thick semiconducting films with wafer-scale homogeneity. Nature 520, 656–660. (doi:10.1038/nature14417)

[89] Ling X, Wang H, Huang S, Xia F, Dresselhaus MS. 2015 The renaissance of black phosphorus. Proc. Natl. Acad. Sci. 112, 4523–4530. (doi:10.1073/pnas.1416581112)

[90] Dutreix C, Stepanov EA, Katsnelson MI. 2016 Laser-induced topological transitions in phosphorene with inversion symmetry. Phys. Rev. B 93, 241404. (doi:10.1103/PhysRevB.93.241404)

[91] Qiao J, Kong X, Hu Z-X, Yang F, Ji W. 2014 High-mobility transport anisotropy and linear dichroism in few-layer black phosphorus. Nat. Commun. 5, 1–7. (doi:10.1038/ncomms5475)

[92] Liu H, Neal AT, Zhu Z, Luo Z, Xu X, Tománek D, Ye PD. 2014 Phosphorene: An Unexplored 2D Semiconductor with a High Hole Mobility. ACS Nano 8, 4033–4041. (doi:10.1021/nn501226z)

[93] Yi M, Shen Z. 2015 A review on mechanical exfoliation for the scalable production of graphene. J. Mater. Chem. A 3, 11700–11715. (doi:10.1039/C5TA00252D)

[94] Li H, Wu J, Yin Z, Zhang H. 2014 Preparation and Applications of Mechanically Exfoliated Single-Layer and Multilayer MoS2 and WSe2 Nanosheets. Acc. Chem. Res. 47, 1067–1075. (doi:10.1021/ar4002312)

[95] Hernandez Y et al. 2008 High-yield production of graphene by liquid-phase exfoliation of graphite. Nat. Nanotechnol. 3, 563–568. (doi:10.1038/nnano.2008.215)

[96] Coleman JN et al. 2011 Two-Dimensional Nanosheets Produced by Liquid Exfoliation of Layered Materials. Science 331, 568–571. (doi:10.1126/science.1194975)

[97] Paton KR et al. 2014 Scalable production of large quantities of defect-free few-layer graphene by shear exfoliation in liquids. Nat. Mater. 13, 624–630. (doi:10.1038/nmat3944)

[98] Parvez K, Wu Z-S, Li R, Liu X, Graf R, Feng X, Müllen K. 2014 Exfoliation of Graphite into Graphene in Aqueous Solutions of Inorganic Salts. J. Am. Chem. Soc. 136, 6083–6091. (doi:10.1021/ja5017156)

[99] Marcano DC, Kosynkin DV, Berlin JM, Sinitskii A, Sun Z, Slesarev A, Alemany LB, Lu W, Tour JM. 2010 Improved Synthesis of Graphene Oxide. ACS Nano 4, 4806–4814. (doi:10.1021/nn1006368)

[100] Xie J, Zhang H, Li S, Wang R, Sun X, Zhou M, Zhou J, Lou XW (David), Xie Y. 2013 Defect-Rich MoS2 Ultrathin Nanosheets with Additional Active Edge Sites for Enhanced Electrocatalytic Hydrogen Evolution. Adv. Mater. 25, 5807–5813. (doi:10.1002/adma.201302685)

[101] Xia H, Li H, Lan C, Li C, Zhang X, Zhang S, Liu Y. 2014 Ultrafast erbium-doped fiber laser mode-locked by a CVD-grown molybdenum disulfide (MoS2) saturable absorber. Opt. Express 22, 17341–17348. (doi:10.1364/OE.22.017341)

[102] Xie LM. 2015 Two-dimensional transition metal dichalcogenide alloys: preparation, characterization and applications. Nanoscale 7, 18392–18401. (doi:10.1039/C5NR05712D)

[103] Li H et al. 2014 Growth of Alloy MoS2xSe2(1–x) Nanosheets with Fully Tunable Chemical Compositions and Optical Properties. J. Am. Chem. Soc. 136, 3756–3759. (doi:10.1021/ja500069b)

[104] Duan X et al. 2014 Lateral epitaxial growth of two-dimensional layered semiconductor heterojunctions. Nat. Nanotechnol. 9, 1024–1030. (doi:10.1038/nnano.2014.222)

[105] Huang C et al. 2014 Lateral heterojunctions within monolayer MoSe2–WSe2 semiconductors. Nat. Mater. 13, 1096–1101. (doi:10.1038/nmat4064)

[106] Gong Y et al. 2014 Vertical and in-plane heterostructures from WS2/MoS2 monolayers. Nat. Mater. 13, 1135–1142. (doi:10.1038/nmat4091)

[107] Mona J, Kale SN, Choudhary RJ, Phase DM. 2008 Role of substrate on the electrical properties of SnO2–La0.7Sr0.3MnO3 bilayers. Appl. Phys. Lett. 92, 142109. (doi:10.1063/1.2907502)

[108] Mona J, Mamgain H, Jejurikar S, Rawat RR, Ganesan V, Choudhary RJ, Phase DM, Kale SN. 2008 Surface and transport studies on La0.7Ba0.3MnO3:SnO2 bilayer. Appl. Surf. Sci. 254, 4808–4812. (doi:10.1016/j.apsusc.2008.01.158)

[109] Mona J, Bertram F, Shukla D, Choudhary RJ, Kale SN, Seeck O. 2015 Role of interface on La0.7Sr0.3MnO3 - Si junction properties. Appl. Sci. Lett. 1, 54–58. (doi:10.17571/appslett.2015.01014)

[110] Yan P, Lin R, Ruan S, Liu A, Chen H, Zheng Y, Chen S, Guo C, Hu J. 2015 A practical topological insulator saturable absorber for mode-locked fiber laser. Sci. Rep. 5, 1–5. (doi:10.1038/srep08690)

[111] Boguslawski J, Sobon G, Zybala R, Sotor J. 2015 Dissipative soliton generation in Er-doped fiber laser mode-locked by Sb2Te3 topological insulator. Opt. Lett. 40, 2786–2789. (doi:10.1364/OL.40.002786)

[112] Kowalczyk M et al. 2016 All-normal dispersion Yb-doped fiber laser mode-locked by Sb2Te3 topological insulator. In Laser Sources and Applications III, p. 98930T. International Society for Optics and Photonics. (doi:10.1117/12.2225893)

[113] Zhang S et al. 2015 Direct Observation of Degenerate Two-Photon Absorption and Its Saturation in WS2 and MoS2 Monolayer and Few-Layer Films. ACS Nano 9, 7142–7150. (doi:10.1021/acsnano.5b03480)

[114] Castellanos-Gomez A, Barkelid M, Goossens AM, Calado VE, van der Zant HSJ, Steele GA. 2012 Laser-Thinning of MoS2: On Demand Generation of a Single-Layer Semiconductor. Nano Lett. 12, 3187–3192. (doi:10.1021/nl301164v)

[115] Lemme MC, Echtermeyer TJ, Baus M, Kurz H. 2007 A Graphene Field-Effect Device. IEEE Electron Device Lett. 28, 282–284. (doi:10.1109/LED.2007.891668)

[116] Butch NP, Kirshenbaum K, Syers P, Sushkov AB, Jenkins GS, Drew HD, Paglione J. 2010 Strong surface scattering in ultrahigh-mobility Bi2Se3 topological insulator crystals. Phys. Rev. B 81, 241301. (doi:10.1103/PhysRevB.81.241301)

[117] Bao W, Cai X, Kim D, Sridhara K, Fuhrer MS. 2013 High mobility ambipolar MoS2 field-effect transistors: Substrate and dielectric effects. Appl. Phys. Lett. 102, 042104. (doi:10.1063/1.4789365)

[118] Radisavljevic B, Radenovic A, Brivio J, Giacometti V, Kis A. 2011 Single-layer MoS 2 transistors. Nat. Nanotechnol. 6, 147–150. (doi:10.1038/nnano.2010.279)

[119] Desai SB et al. 2016 MoS2 transistors with 1-nanometer gate lengths. Science 354, 99–102. (doi:10.1126/science.aah4698)

[120] Wang H et al. 2012 Integrated Circuits Based on Bilayer MoS2 Transistors. Nano Lett. 12, 4674–4680. (doi:10.1021/nl302015v)

[121] Yang L et al. 2014 Chloride Molecular Doping Technique on 2D Materials: WS2 and MoS2. Nano Lett. 14, 6275–6280. (doi:10.1021/nl502603d)

[122] Jin T, Kang J, Su Kim E, Lee S, Lee C. 2013 Suspended single-layer MoS2 devices. J. Appl. Phys. 114, 164509. (doi:10.1063/1.4827477)

[123] Wang F, Stepanov P, Gray M, Lau CN, Itkis ME, Haddon RC. 2015 Ionic Liquid Gating of Suspended MoS2 Field Effect Transistor Devices. Nano Lett. 15, 5284–5288. (doi:10.1021/acs.nanolett.5b01610)

[124] Liu Z et al. 2014 Strain and structure heterogeneity in MoS2 atomic layers grown by chemical vapour deposition. Nat. Commun. 5, 1–9. (doi:10.1038/ncomms6246)

[125] Yu Z et al. 2014 Towards intrinsic charge transport in monolayer molybdenum disulfide by defect and interface engineering. Nat. Commun. 5, 1–7. (doi:10.1038/ncomms6290)

[126] Du Y, Liu H, Neal AT, Si M, Ye PD. 2013 Molecular Doping of Multilayer MoS2 Field-Effect Transistors: Reduction in Sheet and Contact Resistances. IEEE Electron Device Lett. 34, 1328–1330. (doi:10.1109/LED.2013.2277311)

[127] Lee G-H et al. 2013 Flexible and Transparent MoS2 Field-Effect Transistors on Hexagonal Boron Nitride-Graphene Heterostructures. ACS Nano 7, 7931–7936. (doi:10.1021/nn402954e)

[128] Chang H-Y, Yang S, Lee J, Tao L, Hwang W-S, Jena D, Lu N, Akinwande D. 2013 High-Performance, Highly Bendable MoS2 Transistors with High-K Dielectrics for Flexible Low-Power Systems. ACS Nano 7, 5446–5452. (doi:10.1021/nn401429w)

[129] Salvatore GA, Münzenrieder N, Barraud C, Petti L, Zysset C, Büthe L, Ensslin K, Tröster G. 2013 Fabrication and Transfer of Flexible Few-Layers MoS2 Thin Film Transistors to Any Arbitrary Substrate. ACS Nano 7, 8809–8815. (doi:10.1021/nn403248y)

[130] Liu H, Ye P. 2011 MoS2 Dual-Gate MOSFET with Atomic-Layer-Deposited Al2O3 as Top-Gate Dielectric. (doi:10.1109/LED.2012.218452)

[131] Liu B, Chen L, Liu G, Abbas AN, Fathi M, Zhou C. 2014 High-Performance Chemical Sensing Using Schottky-Contacted Chemical Vapor Deposition Grown Monolayer MoS2 Transistors. ACS Nano 8, 5304–5314. (doi:10.1021/nn5015215)

[132] Bertolazzi S, Krasnozhon D, Kis A. 2013 Nonvolatile memory cells based on MoS2/graphene heterostructures. ACS Nano 7, 3246–3252. (doi:10.1021/nn3059136)

[133] Sangwan VK, Jariwala D, Kim IS, Chen K-S, Marks TJ, Lauhon LJ, Hersam MC. 2015 Gate-tunable memristive phenomena mediated by grain boundaries in single-layer MoS2. Nat. Nanotechnol. 10, 403–406. (doi:10.1038/nnano.2015.56)

[134] Liu H, Du Y, Deng Y, Ye PD. 2015 Semiconducting black phosphorus: synthesis, transport properties and electronic applications. Chem. Soc. Rev. 44, 2732–2743. (doi:10.1039/C4CS00257A)

[135] Chhowalla M, Jena D, Zhang H. 2016 Two-dimensional semiconductors for transistors. Nat. Rev. Mater. 1, 1–15. (doi:10.1038/natrevmats.2016.52)

[136] Kamalakar MV, Madhushankar BN, Dankert A, Dash SP. 2015 Effect of high-k dielectric and ionic liquid gate on nanolayer black-phosphorus field effect transistors. Appl. Phys. Lett. 107, 113103. (doi:10.1063/1.4930236)

[137] Du Y, Liu H, Deng Y, Ye PD. 2014 Device Perspective for Black Phosphorus Field-Effect Transistors: Contact Resistance, Ambipolar Behavior, and Scaling. ACS Nano 8, 10035–10042. (doi:10.1021/nn502553m)

[138] Miao J, Zhang S, Cai L, Scherr M, Wang C. 2015 Ultrashort Channel Length Black Phosphorus Field-Effect Transistors. ACS Nano 9, 9236–9243. (doi:10.1021/acsnano.5b04036)

[139] Yin Z et al. 2012 Single-Layer MoS2 Phototransistors. ACS Nano 6, 74–80. (doi:10.1021/nn2024557)

[140] Lopez-Sanchez O, Lembke D, Kayci M, Radenovic A, Kis A. 2013 Ultrasensitive photodetectors based on monolayer MoS 2. Nat. Nanotechnol. 8, 497–501. (doi:10.1038/nnano.2013.100)

[141] Kufer D, Konstantatos G. 2015 Highly Sensitive, Encapsulated MoS2 Photodetector with Gate Controllable Gain and Speed. Nano Lett. 15, 7307–7313. (doi:10.1021/acs.nanolett.5b02559)

[142] Choi M, Park YJ, Sharma BK, Bae S-R, Kim SY, Ahn J-H. 2018 Flexible active-matrix organic light-emitting diode display enabled by MoS2 thin-film transistor. Sci. Adv. 4. (doi:10.1126/sciadv.aas8721)

[143] Woomer AH, Farnsworth TW, Hu J, Wells RA, Donley CL, Warren SC. 2015 Phosphorene: Synthesis, Scale-Up, and Quantitative Optical Spectroscopy. ACS Nano 9, 8869–8884. (doi:10.1021/acsnano.5b02599)

[144] Youngblood N, Chen C, Koester SJ, Li M. 2015 Waveguide-integrated black phosphorus photodetector with high responsivity and low dark current. Nat. Photonics 9, 247–252. (doi:10.1038/nphoton.2015.23)

[145] Deng Y et al. 2014 Black Phosphorus–Monolayer MoS2 van der Waals Heterojunction p–n Diode. ACS Nano 8, 8292–8299. (doi:10.1021/nn5027388)

[146] Ponomarenko LA et al. 2011 Tunable metal–insulator transition in double-layer graphene heterostructures. Nat. Phys. 7, 958–961. (doi:10.1038/nphys2114)

[147] Yang H et al. 2012 Graphene Barristor, a Triode Device with a Gate-Controlled Schottky Barrier. Science 336, 1140–1143. (doi:10.1126/science.1220527)

[148] Britnell L et al. 2012 Field-Effect Tunneling Transistor Based on Vertical Graphene Heterostructures. Science 335, 947–950. (doi:10.1126/science.1218461)

[149] Georgiou T et al. 2013 Vertical field-effect transistor based on graphene–WS2 heterostructures for flexible and transparent electronics. Nat. Nanotechnol. 8, 100–103. (doi:10.1038/nnano.2012.224)

[150] Britnell L et al. 2013 Strong Light-Matter Interactions in Heterostructures of Atomically Thin Films. Science 340, 1311–1314. (doi:10.1126/science.1235547)

[151] Wang X, Cheng Z, Xu K, Tsang HK, Xu J-B. 2013 High-responsivity graphene/silicon-heterostructure waveguide photodetectors. Nat. Photonics 7, 888–891. (doi:10.1038/nphoton.2013.241)

[152] Massicotte M, Schmidt P, Vialla F, Schädler KG, Reserbat-Plantey A, Watanabe K, Taniguchi T, Tielrooij KJ, Koppens FHL. 2016 Picosecond photoresponse in van der Waals heterostructures. Nat. Nanotechnol. 11, 42–46. (doi:10.1038/nnano.2015.227)

[153] Yu WJ, Liu Y, Zhou H, Yin A, Li Z, Huang Y, Duan X. 2013 Highly efficient gate-tunable photocurrent generation in vertical heterostructures of layered materials. Nat. Nanotechnol. 8, 952–958. (doi:10.1038/nnano.2013.219)

[154] Cheng R et al. 2014 Electroluminescence and Photocurrent Generation from Atomically Sharp WSe2/MoS2 Heterojunction p–n Diodes. Nano Lett. 14, 5590–5597. (doi:10.1021/nl502075n)

[155] Li D, Cheng R, Zhou H, Wang C, Yin A, Chen Y, Weiss NO, Huang Y, Duan X. 2015 Electric-field-induced strong enhancement of electroluminescence in multilayer molybdenum disulfide. Nat. Commun. 6, 1–8. (doi:10.1038/ncomms8509)

[156] Withers F et al. 2015 Light-emitting diodes by band-structure engineering in van der Waals heterostructures. Nat. Mater. 14, 301–306. (doi:10.1038/nmat4205)

[157] Furchi MM, Pospischil A, Libisch F, Burgdörfer J, Mueller T. 2014 Photovoltaic Effect in an Electrically Tunable van der Waals Heterojunction. Nano Lett. 14, 4785–4791. (doi:10.1021/nl501962c)

[158] An X, Liu F, Jung YJ, Kar S. 2013 Tunable Graphene–Silicon Heterojunctions for Ultrasensitive Photodetection. Nano Lett. 13, 909–916. (doi:10.1021/nl303682j)

[159] Zhang W, Wang Q, Chen Y, Wang Z, Wee ATS. 2016 Van der Waals stacked 2D layered materials for optoelectronics. 2D Mater. 3, 022001. (doi:10.1088/2053-1583/3/2/022001)

[160] Yu WJ, Li Z, Zhou H, Chen Y, Wang Y, Huang Y, Duan X. 2013 Vertically stacked multi-heterostructures of layered materials for logic transistors and complementary inverters. Nat. Mater. 12, 246–252. (doi:10.1038/nmat3518)

[161] Lee C-H et al. 2014 Atomically thin p–n junctions with van der Waals heterointerfaces. Nat. Nanotechnol. 9, 676–681. (doi:10.1038/nnano.2014.150)

[162] Lopez-Sanchez O, Alarcon Llado E, Koman V, Fontcuberta i Morral A, Radenovic A, Kis A. 2014 Light Generation and Harvesting in a van der Waals Heterostructure. ACS Nano 8, 3042–3048. (doi:10.1021/nn500480u)

[163] Duong NT, Lee J, Bang S, Park C, Lim SC, Jeong MS. 2019 Modulating the Functions of MoS2/MoTe2 van der Waals Heterostructure via Thickness Variation. ACS Nano 13, 4478–4485. (doi:10.1021/acsnano.9b00014)

[164] Du G, Guo Z, Wang S, Zeng R, Chen Z, Liu H. 2010 Superior stability and high capacity of restacked molybdenum disulfide as anode material for lithium ion batteries. Chem. Commun. 46, 1106–1108. (doi:10.1039/B920277C)

[165] Wang P, Sun H, Ji Y, Li W, Wang X. 2014 Three-Dimensional Assembly of Single-Layered MoS2. Adv. Mater. 26, 964–969. (doi:10.1002/adma.201304120)

[166] Fang Y, Lv Y, Che R, Wu H, Zhang X, Gu D, Zheng G, Zhao D. 2013 Two-Dimensional Mesoporous Carbon Nanosheets and Their Derived Graphene Nanosheets: Synthesis and Efficient Lithium Ion Storage. J. Am. Chem. Soc. 135, 1524–1530. (doi:10.1021/ja310849c)

[167] Sun W, Cao F, Liu Y, Zhao X, Liu X, Yuan J. 2012 Nanoporous LiMn2O4 nanosheets with exposed {111} facets as cathodes for highly reversible lithium-ion batteries. J. Mater. Chem. 22, 20952–20957. (doi:10.1039/C2JM32658B)

[168] Cao X, Shi Y, Shi W, Rui X, Yan Q, Kong J, Zhang H. 2013 Preparation of MoS2-Coated Three-Dimensional Graphene Networks for High-Performance Anode Material in Lithium-Ion Batteries. Small 9, 3433–3438. (doi:10.1002/smll.201202697)

[169] Xiong P, Liu B, Teran V, Zhao Y, Peng L, Wang X, Yu G. 2014 Chemically Integrated Two-Dimensional Hybrid Zinc Manganate/Graphene Nanosheets with Enhanced Lithium Storage Capability. ACS Nano 8, 8610–8616. (doi:10.1021/nn5041203)

[170] Chang K, Chen W. 2011 Single-layer MoS2/graphene dispersed in amorphous carbon: towards high electrochemical performances in rechargeable lithium ion batteries. J. Mater. Chem. 21, 17175–17184. (doi:10.1039/C1JM12942B)

[171] Chang K, Chen W. 2011 l-Cysteine-Assisted Synthesis of Layered MoS2/Graphene Composites with Excellent Electrochemical Performances for Lithium Ion Batteries. ACS Nano 5, 4720–4728. (doi:10.1021/nn200659w)

[172] Liang H, Cao Z, Ming F, Zhang W, Anjum DH, Cui Y, Cavallo L, Alshareef HN. 2019 Aqueous Zinc-Ion Storage in MoS2 by Tuning the Intercalation Energy. Nano Lett. 19, 3199–3206. (doi:10.1021/acs.nanolett.9b00697)

[173] Zhao G, Wen T, Chen C, Wang X. 2012 Synthesis of graphene-based nanomaterials and their application in energy-related and environmental-related areas. RSC Adv. 2, 9286–9303. (doi:10.1039/C2RA20990J)

[174] Xia J, Chen F, Li J, Tao N. 2009 Measurement of the quantum capacitance of graphene. Nat. Nanotechnol. 4, 505–509. (doi:10.1038/nnano.2009.177)

[175] Wang Y, Wu Y, Huang Y, Zhang F, Yang X, Ma Y, Chen Y. 2011 Preventing Graphene Sheets from Restacking for High-Capacitance Performance. J. Phys. Chem. C 115, 23192–23197. (doi:10.1021/jp206444e)

[176] Sarkar D, Das D, Das S, Kumar A, Patil S, Nanda KK, Sarma DD, Shukla A. 2019 Expanding Interlayer Spacing in MoS2 for Realizing an Advanced Supercapacitor. ACS Energy Lett. 4, 1602–1609. (doi:10.1021/acsenergylett.9b00983)

[177] Acerce M, Voiry D, Chhowalla M. 2015 Metallic 1T phase MoS 2 nanosheets as supercapacitor electrode materials. Nat. Nanotechnol. 10, 313–318. (doi:10.1038/nnano.2015.40)

[178] Huang K-J, Zhang J-Z, Shi G-W, Liu Y-M. 2014 Hydrothermal synthesis of molybdenum disulfide nanosheets as supercapacitors electrode material. Electrochimica Acta 132, 397–403. (doi:10.1016/j.electacta.2014.04.007)

[179] Du J, Pei S, Ma L, Cheng H-M. 2014 25th Anniversary Article: Carbon Nanotube- and Graphene-Based Transparent Conductive Films for Optoelectronic Devices. Adv. Mater. 26, 1958–1991. (doi:10.1002/adma.201304135)

[180] Clemen C, Saldaña XI, Munz P, Bucher E. 1978 Photovoltaic properties of some semiconducting layer structures. Phys. Status Solidi A 49, 437–443. (doi:10.1002/pssa.2210490203)

[181] Paranthaman MP, Wong-Ng W, Bhattacharya RN, editors. 2016 Semiconductor Materials for Solar Photovoltaic Cells. Springer International Publishing. (doi:10.1007/978-3-319-20331-7)

[182] Hong X et al. 2014 Ultrafast charge transfer in atomically thin MoS 2 /WS 2 heterostructures. Nat. Nanotechnol. 9, 682–686. (doi:10.1038/nnano.2014.167)

[183] McOuat RF, Pulfrey DL. 1976 A model for Schottky-barrier solar cell analysis. J. Appl. Phys. 47, 2113–2119. (doi:10.1063/1.322857)

[184] Liu Z, Lau SP, Yan F. 2015 Functionalized graphene and other two-dimensional materials for photovoltaic devices: device design and processing. Chem. Soc. Rev. 44, 5638–5679. (doi:10.1039/C4CS00455H)

[185] Gu X, Cui W, Li H, Wu Z, Zeng Z, Lee S-T, Zhang H, Sun B. 2013 A Solution-Processed Hole Extraction Layer Made from Ultrathin MoS2 Nanosheets for Efficient Organic Solar Cells. Adv. Energy Mater. 3, 1262–1268. (doi:10.1002/aenm201300549)

[186] Le QV, Nguyen TP, Kim SY. 2014 UV/ozone-treated WS2 hole-extraction layer in organic photovoltaic cells. Phys. Status Solidi RRL – Rapid Res. Lett. 8, 390–394. (doi:10.1002/pssr.201409165)

[187] Liu W, Yang X, Zhang Y, Xu M, Chen H. 2014 Ultra-stable two-dimensional MoS2 solution for highly efficient organic solar cells. RSC Adv. 4, 32744–32748. (doi:10.1039/C4RA04116J)

[188] Du T, Wang N, Chen H, He H, Lin H, Liu K. 2015 TiO2-based solar cells sensitized by chemical-bath-deposited few-layer MoS2. J. Power Sources 275, 943–949. (doi:10.1016/j.jpowsour.2014.11.048)

[189] Meng K, Chen G, Thampi KR. 2015 Metal chalcogenides as counter electrode materials in quantum dot sensitized solar cells: a perspective. J. Mater. Chem. A 3, 23074–23089. (doi:10.1039/C5TA05071E)

[190] Yeh T-F, Teng C-Y, Chen L-C, Chen S-J, Teng H. 2016 Graphene oxide-based nanomaterials for efficient photoenergy conversion. J. Mater. Chem. A 4, 2014–2048. (doi:10.1039/C5TA07780J)

[191] Xiang Q, Yu J. 2013 Graphene-Based Photocatalysts for Hydrogen Generation. J. Phys. Chem. Lett. 4, 753–759. (doi:10.1021/jz302048d)

[192] Xiang Q, Yu J, Jaroniec M. 2012 Graphene-based semiconductor photocatalysts. Chem. Soc. Rev. 41, 782–796. (doi:10.1039/C1CS15172J)

[193] Laursen AB, Kegnæs S, Dahl S, Chorkendorff I. 2012 Molybdenum sulfides—efficient and viable materials for electro - and photoelectrocatalytic hydrogen evolution. Energy Environ. Sci. 5, 5577–5591. (doi:10.1039/C2EE02618J)

[194] Benck JD, Hellstern TR, Kibsgaard J, Chakthranont P, Jaramillo TF. 2014 Catalyzing the Hydrogen Evolution Reaction (HER) with Molybdenum Sulfide Nanomaterials. ACS Catal. 4, 3957–3971. (doi:10.1021/cs500923c)

[195] Zong X, Han J, Ma G, Yan H, Wu G, Li C. 2011 Photocatalytic H2 Evolution on CdS Loaded with WS2 as Cocatalyst under Visible Light Irradiation. J. Phys. Chem. C 115, 12202–12208. (doi:10.1021/jp2006777)

[196] Zong X, Yan H, Wu G, Ma G, Wen F, Wang L, Li C. 2008 Enhancement of Photocatalytic H2 Evolution on CdS by Loading MoS2 as Cocatalyst under Visible Light Irradiation. J. Am. Chem. Soc. 130, 7176–7177. (doi:10.1021/ja8007825)

[197] Zong X, Wu G, Yan H, Ma G, Shi J, Wen F, Wang L, Li C. 2010 Photocatalytic H2 Evolution on MoS2/CdS Catalysts under Visible Light Irradiation. J. Phys. Chem. C 114, 1963–1968. (doi:10.1021/jp904350e)

[198] Zhou W, Yin Z, Du Y, Huang X, Zeng Z, Fan Z, Liu H, Wang J, Zhang H. 2013 Synthesis of Few-Layer MoS2 Nanosheet-Coated TiO2 Nanobelt Heterostructures for Enhanced Photocatalytic Activities. Small 9, 140–147. (doi:10.1002/smll.201201161)

[199] Chen J, Wu X-J, Yin L, Li B, Hong X, Fan Z, Chen B, Xue C, Zhang H. 2015 One-pot Synthesis of CdS Nanocrystals Hybridized with Single-Layer Transition-Metal Dichalcogenide Nanosheets for Efficient Photocatalytic Hydrogen Evolution. Angew. Chem. Int. Ed. 54, 1210–1214. (doi:10.1002/anie.201410172)

[200] Chang K, Mei Z, Wang T, Kang Q, Ouyang S, Ye J. 2014 MoS2/Graphene Cocatalyst for Efficient Photocatalytic H2 Evolution under Visible Light Irradiation. ACS Nano 8, 7078–7087. (doi:10.1021/nn5019945)

[201] Qi Y, Xu Q, Wang Y, Yan B, Ren Y, Chen Z. 2016 CO2-Induced Phase Engineering: Protocol for Enhanced Photoelectrocatalytic Performance of 2D MoS2 Nanosheets. ACS Nano 10, 2903–2909. (doi:10.1021/acsnano.6b00001)

[202] Lee HU et al. 2015 Stable semiconductor black phosphorus (BP)@titanium dioxide (TiO 2) hybrid photocatalysts. Sci. Rep. 5, 1–6. (doi:10.1038/srep08691)

[203] Lee J, Dak P, Lee Y, Park H, Choi W, Alam MA, Kim S. 2014 Two-dimensional Layered MoS 2 Biosensors Enable Highly Sensitive Detection of Biomolecules. Sci. Rep. 4, 1–7. (doi:10.1038/srep07352)

[204] Wang L, Wang Y, Wong JI, Palacios T, Kong J, Yang HY. 2014 Functionalized MoS2 Nanosheet-Based Field-Effect Biosensor for Label-Free Sensitive Detection of Cancer Marker Proteins in Solution. Small 10, 1101–1105. (doi:10.1002/smll.201302081)

[205] Liu J, Chen X, Wang Q, Xiao M, Zhong D, Sun W, Zhang G, Zhang Z. 2019 Ultrasensitive Monolayer MoS2 Field-Effect Transistor Based DNA Sensors for Screening of Down Syndrome. Nano Lett. 19, 1437–1444. (doi:10.1021/acs.nanolett.8b03818)

[206] Mleczko MJ et al. 2019 Contact Engineering High-Performance n-Type MoTe2 Transistors. Nano Lett. 19, 6352–6362. (doi:10.1021/acs.nanolett.9b02497)

[207] Chen S, Kim S, Chen W, Yuan J, Bashir R, Lou J, van der Zande AM, King WP. 2019 Monolayer MoS2 Nanoribbon Transistors Fabricated by Scanning Probe Lithography. Nano Lett. 19, 2092–2098. (doi:10.1021/acs.nanolett.9b00271)

[208] Tong SW et al. 2019 Employing a Bifunctional Molybdate Precursor To Grow the Highly Crystalline MoS2 for High-Performance Field-Effect Transistors. ACS Appl. Mater. Interfaces 11, 14239–14248. (doi:10.1021/acsami.9b01444)

[209] Zhang P et al. 2019 Electrochemically Exfoliated High-Quality 2H-MoS2 for Multiflake Thin Film Flexible Biosensors. Small 15, 1901265. (doi:10.1002/smll.201901265)

[210] Zheng Y et al. 2019 Black phosphorus inverter devices enabled by in-situ aluminum surface modification. Nano Res. 12, 531–536. (doi:10.1007/s12274-018-2246-y)

[211] Wu D et al. 2019 Visualization of Local Conductance in MoS2/WSe2 Heterostructure Transistors. Nano Lett. 19, 1976–1981. (doi:10.1021/acs.nanolett.8b05159)

[212] Rice M. 2006 Superconductivity with a Twist. Science 314, 1248–1249. (doi:10.1126/science.1134921)

[213] Gibney E. 2019 How 'magic angle' graphene is stirring up physics. Nature 565, 15–18. (doi:10.1038/d41586-018-07848-2)

[214] Radisavljevic B, Radenovic A, Brivio J, Giacometti V, Kis A. 2011 Single-layer MoS 2 transistors. Nat. Nanotechnol. 6, 147–150. (doi:10.1038/nnano.2010.279)

[215] Radisavljevic B, Kis A. 2013 Mobility engineering and a metal-insulator transition in monolayer MoS2. Nat. Mater. 12, 815–820. (doi:10.1038/nmat3687)

[216] Perera MM, Lin M-W, Chuang H-J, Chamlagain BP, Wang C, Tan X, Cheng MM-C, Tománek D, Zhou Z. 2013 Improved Carrier Mobility in Few-Layer MoS2 Field-Effect Transistors with Ionic-Liquid Gating. ACS Nano 7, 4449–4458. (doi:10.1021/nn401053g)

[217] Cho K, Park W, Park J, Jeong H, Jang J, Kim T-Y, Hong W-K, Hong S, Lee T. 2013 Electric Stress-Induced Threshold Voltage Instability of Multilayer MoS2 Field Effect Transistors. ACS Nano 7, 7751–7758. (doi:10.1021/nn402348r)

4

Highly Correlated Systems Associated with Magnetoelectricity in Multiferroics

Jayalakshmy M. S.

International and Inter University Centre for Nanoscience and Nanotechnology, Mahatma Gandhi University, India

Abstract

According to solid-state physics, it is the spin of electrons that results the second-order relativistic quantum mechanical phenomena such as magnetism. The fundamental matter responsible for the magnetoelectric effect is the electromagnetic interaction. But, the interaction or coupling amongst the spin and charge is very feeble. The reason behind the already existing coupling is the relativistic spin–orbit coupling, which correlates the spin of the electron to the field developed as a consequence of the motion of electron. This is why most of the materials that order either magnetically or electrically show very weak coupling between electrical and magnetical properties. However, there are materials in which the magnetoelectric coupling is significantly high. Such kinds of materials that exhibit a simultaneous occurrence of, or an interaction amongst, the magnetic and electric order parameters are known as magnetoelectric multiferroics, which provide magnetic and electric properties in response to electric and magnetic fields, respectively [1, 2]. The area of multiferroics, especially the subdivision of magnetoelectricity in it, has greatly flourished especially due to the inventions of various kinds of multiferroic materials.

So, the main objective of this chapter is to give some ideas and basic understandings of the characteristics of the systems having strong interactions and correlations between its constituent particles or characteristics, and how we can anticipate finding the same kind of effect in systems that we have been examining all along. Here, the author mainly focuses on the

DOI: 10.1201/9781003337102-4

magnetoelectricity in single-phase and multiphase compounds of multiferroic systems. Moreover, some applications and technology, in addition to future perspectives, are also included to shed light into the scope for future investigations in this field.

Keywords: Magnetoelectricity, Multiferroic, Electron spin, Highly correlated system, Correlation in multiferroics, Type-I multiferroics, Type-II multiferroics

4.1 Introduction to Highly Correlated Systems

Almost all of the physical systems familiar to us are similar to ensembles of free particles. For instance, most of the known metals and semiconductors could be depicted as having electrons that are non-interacting. This description is valid since the energy of interaction of electrons (i.e., the Coulomb energy) is far smaller than their kinetic energy. Another case is alkali atoms. They can be considered as non-interacting bosons since the length at which they interact with each other (i.e., the scattering length) is far smaller than the mean separation among the particles. At the same time, there are lots of other systems that can be treated as very important since the interactions (correlations) between their particles are not weak as in the previous cases. In such systems, these particle–particle interactions play important roles in deciding their characteristics. These kinds of systems are known as highly correlated systems. Highly correlated materials are a division of heavy fermion compounds, which encompass insulators and electronic (semiconductor) materials and exhibit phenomenal electronic and magnetic characteristics, viz. metal insulator transitions, half-metallicity, spin charge separation, etc. [3]. Different cases for highly correlated systems are provided below for a better understanding of the topic.

- Considering *conventional superconducting materials*, the Coulomb interaction amongst their ions and electrons will be in a state that can support an electrical current flow without much dissipation.
- The transition temperature for *high-temperature super conductors* is astonishingly high. Even though the origin of superconductivity is not clear, it is considered that the Coulomb interaction between electrons is the main reason for this. Whereas, for conventional super conductors, it is the electron–ion interaction.
- Coulomb interactions between electrons will contribute to different orders of spins in *magnetic systems*. Examples include ferromagnetism

and anti-ferromagnetism in which spin of the adjacent particles will be aligned in parallel and anti-parallel directions, respectively.

- Charge and spin are the two significant quantum numbers carried by electrons. When an electron travels through an insulator or vacuum, propagation of charge and spin together can be observed. Whereas, in systems that are one-dimensional, propagation of disintegrated charges and spins at varying velocities can be observed.

4.2 Ferroics and Multiferroics

The term *ferroics* is a common title that deals with the study of materials exhibiting ferroic properties such as ferromagnetism, ferroelectricity, ferroelasticity, and, rarely, ferrotoroidicity. Ferrotoroidicity is not as common as the other three properties and much research outcomes are not available on this characteristic phenomenon. The category of ferroics can also be broadened to anti-ferroics, such as anti-ferromagnetic as well as anti-ferroelectric. The prefix "*ferro*" stands historically for iron (the Latin word for iron is *ferrum*) that exhibits a spontaneous magnetization, which is controllable under a magnetic field applied externally. Materials that exhibit coupling between ferroelectric and ferroelastic orders have been explored elaborately and they are now at the edge of countless applications. An example is $Pb(ZrTi)O_3$ (PZT).

Multiferroics are multifunctional materials, of course, to a great extent because of the reason that they simultaneously showcase some of the above-mentioned fundamental *ferroic* properties. It means that there are materials in which any two different kinds of ordering, for example, (ferro)magnetism (in which instinctive ordering of orbitals as well as magnetic moments of spins takes place) and (ferro)electricity (in which instinctive ordering of electric dipole moments takes place) coexist even if no magnetic or electric fields are provided. In principle, in the multiferroic group of materials, there is one more category, the above-mentioned ferroelasticity, in which spontaneous deformation takes place. But, in the present scenario, we mean by multiferroics, preponderantly applicable to the simultaneous existence of ferromagnetism and electricity (not necessarily ferromagnetism). Figure 4.1 represents a schematic diagram for different kinds of couplings that exist in various materials.

The origin of magnetism is the subsistence of partly filled *d* or *f* subshells of electrons of rare earth or transition metals. These electrons in the partly filled subshells inhibit hybridization with the electrons of *p* subshells of the nearby oxygen anions and thus develop a cation displacement, which is

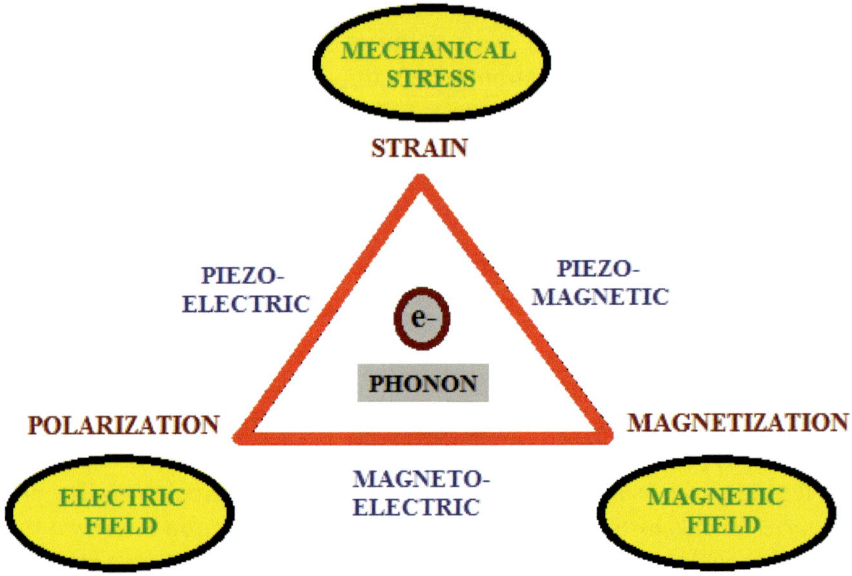

Figure 4.1 Schematic representation of various kinds of couplings that occur in materials.

essential for ferroelectricity [1]. Also, electrons in the partly filled subshells possess respective spin or magnetic moments. The transfer processes that take place among these magnetic moments guide to magnetic order. The ferroelectric materials should be good insulators so that the ferroelectric polarization will not get neutralized by free charges. Most of the ferromagnetic materials available are quite conducting.

There are various microscopic sources of multiferroicity. Depending upon these, multiferroics are classified as type-I and type-II. Type-I shows diverse sources for ferromagnetism and electricity that are highly independent of one another, but there exists a coupling among them, which is very weak. For these kinds of multiferroics, temperatures at which ferroelectricity (T_E) and magnetism (T_M) occur differ. Such materials show ferroelectricity (larger polarization $\sim 10 - 100\,\mu C/cm^2$) at elevated temperatures than magnetism (i.e., $T_E \neq T_M$). Whereas, in type-II multiferroics, magnetism induces ferroelectricity (smaller polarization $\sim 10^{-2}$ $\mu C/cm^2$), implying a very strong coupling among them. So, for these kinds of materials, the temperatures that correspond to the generation of ferroelectricity and magnetism are almost the same (i.e., $T_E = T_M$). That means type-II materials have a common origin of multiferroicity.

Type-I multiferroelectricity includes the effects in multiferroic perovskites (e.g., $BaTiO_3$, PZT, etc.), ferroelectricity as the resultant of charge ordering and lone pairs, *geometric* ferroelectricity, etc. Type-II multiferroics include spiral type-II multiferroics, type-II multiferroics with collinear magnetic structures, etc. [4]. These subdivisions are explained in the following subsections.

4.2.1 Multiferroic effects in type-I materials

4.2.1.1 Multiferroicity in perovskites

The perovskite family includes a lot of magnetic as well as ferroelectric materials and exhibits mutual exclusion of ferroelectricity and magnetism. This mutual exclusion is a matter of the number of electrons filled in the d shells. In most of the cases, magnetic d^n ions will be balanced in the center of their O_6 octahedra; still, there are chances for cases in which it is not like this. So, a solution to this problem is to make mixed perovskites having d^0 as well as d^n ions. Generally, the coupling among magnetic as well as ferroelectric parts in mixed perovskites will be quite weak.

4.2.1.2 Multiferroicity due to charge ordering

This kind of mechanism, which leads to ferroelectricity and to type-I multiferroicity, is usually exhibited by transition metal compounds, particularly by those having transition metal ions with different valences. This can be noticed in systems such as nickelates. Another chance is when the bonds are not equivalent due to material structure.

4.2.1.3 Multiferroicity due to lone pairs

Some materials contain ions for which there will be two exterior 6s electrons that do not take part in chemical bonding. These are known as *lone pairs* or *dangling bonds*. They possess very high polarizability, necessary for ferroelectricity. In such compounds, the generation of ferroelectricity is because of the ordering of such lone pairs in a single direction. Seemingly, it is the same thing that happens in the case of some kind of multiferroics like $BiFeO_3$.

4.2.1.4 Geometric multiferroicity

In some kind of materials similar to that shown in Figure 4.2, ferroelectricity and thereby multiferroicity occur as an outcome of the closer packing that resulted from a geometrical tilting.

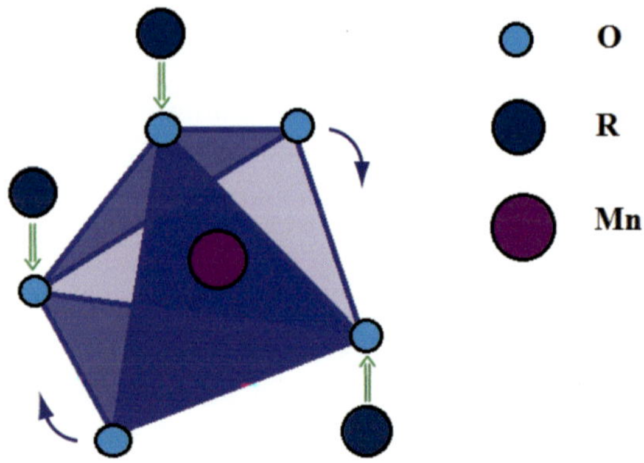

Figure 4.2 The *geometric* mechanism exhibited by rare earth manganite multiferroic materials during polarization. (Here, the production of ferroelectricity happens by the leaning of almost stiff MnO_5 block. As a result of this tilting, a tighter packing generates and the oxygen ions shift closer to the *R* ions.)

4.2.2 Multiferroic effects in type-II materials

4.2.2.1 Spiral type-II multiferroics

Majority of the type-II multiferroics belong to this class. Here, ferroelectricity occurs in coincidence with a spiraling magnetic phase (generally of cycloid type). Various kinds of electron spin patterns relevant to type-II multiferroics are shown in Figure 4.3. Perhaps, this can be a kind of anti-ferromagnetic

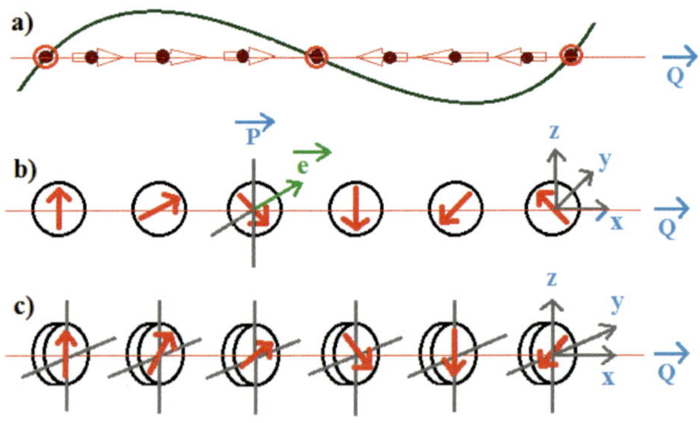

Figure 4.3 Different kinds of spin patterns that can be observed in type-II multiferroics. (a) Sinusoidal spin wave with spins oriented along a single direction and differing in magnitude. (b) Cycloid spiral wave with spins twirling in the $x-z$ plane (this is the case for non-zero polarization). (c) "P*roper screw*" propagation where spins rotate in the $y-z$ plane.

phase as the total moment for the magnet becomes zero. In the case of insulators, the sources of spiral magnetic ordering are magnetic frustrations. So, these kinds of multiferroics are normally found in frustrated systems.

4.2.2.2 Collinear magnetic structured type-II multiferroics

Multiferroic materials have attracted tremendous attention of scientists and engineers, and the field is considered a hot topic in correlated electronics, especially due to the excellent physics and magnetoelectric functionalities behind them. But, it is not mandatory that the property of magnetoelectricity is related to multiferroicity and vice versa. For example, magnetic materials like Cr_2O_3 exhibit linear magnetoelectric behavior and do not possess spontaneous ferroelectric polarizations. In this way, they are not multiferroics [5]. A lot of comprehensive research and review articles on systems based on multiferroicity and magnetoelectricity are available in the literature [1, 6–15].

4.3 Correlation in Multiferroics (General Point of View)

In the context of multiferroics, mainly there are two types of correlations: one that contributes to the statistical nature of particles (either bosons or fermions) and another that contributes to the interaction among them.

Consider an N-particle system in which the number of particles is always greater than 1. Regarding type-I correlation, based on if it is a boson or a fermion, the eigenstates should be symmetric and anti-symmetric, respectively. Whereas, concerning type-II correlation, the higher the coupling co-efficient of the inter-particle interaction, this kind of correlation will be the strongest. The terminology "highly correlated system" has its basis in the considerateness if the ground state as well as the low-lying excited states of a system could be precisely found out with the help of a low-order perturbation expansion about the ground state of a mean-field approximation of the full Hamiltonian of a system. While on the contrary, a mean-field approximation already satisfies for a "weakly correlated system."

The trait of a mean-field approximation is that for an N-particle system, it is possible to express the whole particle mean-field Hamiltonian as the sum of N single-particle Hermitian Hamiltonians, so that the N-particle problem can be minimized to a one-particle problem. This creates the foundation of any band-structure computations. A "band-structure" program can also be reached out by going beyond mean-field, but the single-particle Hamiltonian gets energy-dependent and non-Hermitian, indicating that Hamiltonian of the single-particle problem systematically relies on energy of the single-particle "eigenstate." The pertinent formalism is that of Green functions (both single as well as multi-particle), and the previously stated energy-dependent non-Hermitian single-particle Hamiltonian demands the energy-dependent self-energy. Considering the low-lying eigenstates, "strong correlation" demonstrates itself in the way the self-energy varies as a function of its energy argument in the vicinity of the chemical potential. If the variation is strong, on the measure of the band width (t) (and U the interaction strength), we have to deal with a strongly correlated system; else, the mean-field theory about which perturbation expansion is performed corresponds to a fairly correct depiction of the ground and low-lying excited states of the system. A trait of mean-field theories is that the previously stated single-particle Hamiltonian is not energy-dependent, so that the extent of the variation of the self-energy as a function of its energy parameter corresponds to a quantitative measure for the deviation of the exact theory from the adopted theory.

Concerning a cluster of atoms, for determinacy, take the electron interaction potential to be the Coulomb potential. In an N-body Hamiltonian, one has *kinetic energy* (KE) and *interaction* terms proportional to δ and $1/|k-k'|$ respectively. Here, δ and $1/|k-k'|$ have the dimensions of *the inverse of length-squared* and *the inverse length*, respectively. So, the higher the density of electrons, the KE dominates the interaction energy and vice versa if the lower the density is. Or, if it is a *dense* cluster of atoms, the less correlated its ground and low-lying states, and if it is a dilute assembly of atoms, the more strongly correlated its ground and low-lying excited states. That is, in strongly correlated systems, atoms preserve some of their traits as in isolation, and one of them is magnetic moment. Magnetism is thus an attribute side of nearly every highly correlated systems (although the ground state of a highly correlated system might not be magnetically aligned, the magnetic variations are frequently of great impact on its low-lying excited states. Intriguingly, magnetism feeds back on the type-I correlation, as magnetic order demands utilization of several electronic configurations, including heterogeneity of spin configurations; type-I correlation is commonly known as "exchange," but this terminology frequently overly conceals the intricacy hidden satisfaction of the anti-symmetry needed for the N-particle eigenstates of Fermions). This prospect and the previously stated disparity among the *dense* as well as *dilute* cluster of atoms is well demonstrated by the two-atom problem and the characteristics of the singlet and triplet states as functions of the inter-atomic distance.

Correlated electron systems can also be explained in respect of Hubbard Hamiltonian (single-band or multi-band). Consider the single-band

Hubbard Hamiltonian explaining nearest-neighbor hoppings and on-site interactions. The hopping amplitude t and the on-site U originate from the previously stated δ and electron−electron interaction potential respectively. The quantity U/t helps to analyze the nature of the correlation. The higher (lesser) the value of this ratio, the stronger (weaker) the correlation. For adequately big U/t value, the Hubbard model predicts an insulating ground state at a fractional filling of the band *without* symmetry breaking, a probability that is discarded in the band theory or mean-field theory. This means that in employing perturbation theory, it is essential to know beforehand if the system is nearer to an independent-electron limit or to the atomic limit. In the latter case, we have to apply the strong-coupling perturbation theory.

Double perovskites and multiferroics are systems with strong atomic trait. The terminology "multiferroics" expresses the relevance of magnetism in these materials [16, 17].

4.3.1 Correlation in single-phase compounds (intrinsic multiferroics)

In the area of multiferroics, single-phase multiferroic materials that are also known as intrinsic multiferroics, in which ferroelectricity and magnetism exist simultaneously, are very rare; and if it coexists, the magnetoelectric coupling will be, generally, very weak. This hinders the developments in the field of such materials. Initial studies of magnetoelectric coupling in single-phase compounds were based on materials such as Cr_2O_3 and because of the weak coupling found in such compounds failed to achieve the initial expectations of high coupling. However, later on, studies that focused on single-phase intrinsic multiferroic compounds like $BiFeO_3$ (the perovskites), $BiMnO_3$, $RMnO_3$ (where R is rare earth materials) (the hexagonal manganites), Fe_3O_4 (the spinels), etc., shed light into the developments of a relatively good number of multiferroic materials that exhibit good magnetic and electric orders simultaneously below their critical temperatures [9, 18–22]. Subsequent research works proved that their magnetoelectric properties and coupling effect can be modified drastically at the nanometer scale.

Possibly, the most widely studied multiferroic material would be the perovskite BiFeO3. It was for thin films that a large residual polarization for this material was firstly published; however, it has also been manifested for its ceramic materials as well as highly pure single crystals [23–25]. The magnetic structure of this material is a complicated one. It has a G-type anti-ferromagnetic ordering having each Fe^{3+} ions encompassed by six anti-parallel spins. Its weak coupling to the electrical polarization directs to minor tilting of the magnetic moment and the average value comes to nil as a result of long-range unequal spin cycloid [26–29].

Another example, the magnetite or Fe3O4, is a highly correlated single-phase oxide multiferroic that exhibits a complex interaction among its magnetic, electrical, and structural properties [30]. While comparing the Fe3O4 with the ideal spinal structure AB2O4 (i.e., $Fe^{3+}(Fe^{3+}Fe^{2+})O^{2-}4$), where the Fe^{3+} and Fe^{2+} atoms engage in the octahedral sites, the ferromagnetism happens due to the anti-ferromagnetic coupling among spins in the A and B sublattices and the ferromagnetic coupling among Fe^{3+} and Fe^{2+} atoms in the B sublattice [31].

In the case of rare earth multiferroic manganites also, the polarization takes place as a consequence of electron–electron interactions. Thus, the effects of electronic correlation are of primary relevance and it is the dominant reason behind the traits of such compounds of multiferroic transition metals.

In spite of many limitations that restrict the count of multiferroics that are single-phase, the invention of novel compounds and the excellent physical properties that they bring to the forefront keep on increasing. Since the mechanism takes place as a result of a strong correlation between electrons, understanding of the fundamental physical phenomena is demanding [1, 4, 10, 12, 32–34]. However, single-phase multiferroic materials can be used as component parts in many composite multiferroics and they still persist as important ingredients to enhance the performance of multiferroic composites.

4.3.2 Correlation in multiphase compounds (artificial multiferroics)

In the previous section, we have discussed the single-phase multiferroics. As a remedy to the weakened electromagnetic coupling in single-phase materials, alternative methods have been formulated to investigate proximity and interfacial effects among different materials (ferromagnetic as well as electric) to develop composite materials having improved coupling among the magnetic and electric orders. These kinds of fabricated composites are known as *artificial* multiferroics. Since the 1970s, various strategies have been established to attain this goal [35–37]. Different kinds of composite multiferroics include strain-mediated composite multiferroics, charge-mediated ones, composite heterostructures mediated by exchange-bias, multiferroic heterostructures based on electron transport, etc. The strain-mediated composite multiferroics themselves include epitaxial thin film nano/hetero structures (such as controlled growth process via self-assembly), other nanostructured composites, laminated composites, and particulate composites.

In the exchange-bias to couple anti-ferromagnetic to a ferromagnetic layer, exchange coupling among the uncompensated interfacial spins of the anti-ferromagnetic layer with the spins of ferromagnetic layer takes place, and this leads to unidirectional anisotropy in the ferromagnetic hysteresis loop. Thus, variations in the anti-ferromagnetic order through electric fields guide to corresponding modifications in the exchange-bias and thus in the characteristics of the ferromagnetic element [38, 39].

It is proved that (artificial) multiferroic composites that integrate both ferromagnetic as well as electric order materials are able to produce huge electromagnetic coupling. Sometimes neither the ferromagnetic nor the electric phase exhibits magnetoelectric coupling, but their composite will develop a remarkable coupling. For example, in the case of a piezo-magnetoelectric material, the coupling mechanism can be a consequence of the magnetostriction in the piezomagnetic order material and piezoelectricity in the piezoelectric phase. Here, coupling among the constituents is mediated by strain [40].

4.3.2.1 Correlation in complex oxide multiferroic structures

In order to enhance the coupling, a regulated development of films of (crystalline) complex oxides and heterostructures has also been developed [41–44]. Such correlated oxide systems are very fascinating and, at the same time, very challenging in the subject area of condensed matter physics. The charge, spin, orbital, and lattice degrees of freedom of such oxide networks create new magnetic and electric phases [30, 48]. Nanostructure developments by lithography and self-assembly have been keys to the developments [49–52]. This probed the charge sensitivities of magnetic orders in complex oxides [53–58] and dilute magnetic semiconductors [59–63] to attain control over the state of the spins by regulating charge employing a ferroelectric field effect approach [64–72].

In general, composite systems provide a very fascinating and productive strategy concerning the formation of novel materials having characteristics customized to take the benefits of particular coupling among distinct constituents [35, 73]. This kind of approach permits us to deviate the rivet from intrinsic to extrinsic characteristics.

4.4 Applications and Technology

Some anticipating technology-based applications of multiferroics involve energy transformation, signal generation, signal processing, etc. High-temperature superconductors find their applications in the fields of energy transmission and are used for manufacturing different kinds of electronic devices.

Materials exhibiting magnetic properties are widely used for the fabrication of memory devices. For instance, a large amount of information/data brought forth from consumer electronic products are usually stored in ferromagnets as portions, which are polarized magnetically in opposite directions. Moreover, the progress of research and technology in the concept of magnetoelectric effect succeeded in the development of totally fresh device prototypes like magnetic data storage that can be controlled by electric field [74, 75], a kind of four-stage memory. This kind of a memory does not demand coupling among ferromagnetism and electricity. If there exists a coupling, application like writing data magnetically and storing it in the electric polarization shall be materialized. This leads to non-volatile memory. The multiferroic bits can be utilized to enhance the magnetic anisotropy to extend the decay time for magnetic data storage.

Additional applications involve capacitors that are tunable with magnetic field. Using them, an electronic circuit's frequency dependence can be tuned with magnetic field. The industry of sensors and transducers intemperately trusts on switchable ferroelectric materials that can be electrically polarized using an externally applied electric field [74].

The complex oxide multiferroic materials, which are controllable by employing stress, optical excitations, electric field, and magnetic field, find their major uses in the areas of spintronics, information storage, and communications [76].

Some ferroelectric materials that exhibit ferroelasticity as well are employed in sonar detectors to transform acoustic waves to electrical pulses. Such materials also find their applications in the fields of actuators where they are used to transform electrical signals to movement [74].

4.5 Future Perspectives

It is expected that, in the near future, Quantum Hall systems as well as bosonic atoms in optical lattices can be used for quantum computations and communications. A general tendency toward the miniaturization of devices has resulted in an urge in merging magnetic as well as electronic characteristics into multifunctional materials. This enables an individual device component to execute many tasks. Also, the future generation electronic devices that probe the magnetoelectric coupling include spintronic devices, solid-state transformers, highly sensitive magnetic field sensors, electro-magneto-optic actuators, etc. [77].

By expanding the idea of multiferroic phenomena to other physical concepts, like electron transport, an array of developments can be probed, extending the scope of multifunctional characteristics, which is the behavior of highly correlated materials.

As a matter of fact, ferroelectric-magnets are very rare to be available in nature. Those which exhibit a magnetoelectric direct coupling are even rarer. So, the consequence is that research for the inventions of new materials and coupling mechanisms are the essential requirements for a clear comprehending of electromagnetic coupling mechanisms. In a nut shell, the area of multiferroics comprises deep and interesting physics in addition to lots of fascinating applications.

4.6 Conclusion

Research and technological interests on magnetoelectric materials have been energized by advancements in the regulated development of highly complex systems as well as by the requirement for materials with practicalities worth of next-generation electronic equipment and circuitries. By getting deep knowledge about different coupling mechanisms between electricity and magnetism, a variety of ways have been invented to explore such mechanisms at the atomic/molecular scale so that sizable couplings can be achieved in the solid state. Such couplings are preponderantly indirect, but the results will be rewarding.

We can think of various promising and challenging directions in this field. Inventions of new multiferroics keep on increasing, and each freshly invented multiferroic is a challenge in itself. Without doubt, the exploration of such kind of materials would be a very significant pillar for research based on multiferroics. So, it can be concluded that the areas of multiferroics and magnetoelectrics are already an exceedingly dynamic field and it is pretty sure that there is still a lot to come out.

Acknowledgements

The author is extremely grateful to University Grants Commission for a financial support under the scheme of Dr. D. S. Kothari Postdoctoral Fellowship (No.F.4-2/2006 (BSR)/PH/15-16/0111). The author would like to sincerely thank Mahatma Gandhi University, Kottayam, India, for providing access to its library, which was very helpful for preparing this chapter. All the friends and colleagues are greatly acknowledged for their inspirations and supports.

References

[1] N. A. Hill, "Why are there so few magnetic ferroelectrics?", *J. Phys. Chem. B*, **104** (29), pp. 6694-6709 (2000)

[2] H. Schmid, "Multi-ferroic magnetoelectrics", *Ferroelectrics*, 162 (1), pp. 317-338 (**1994**).

[3] https://en.wikipedia.org/wiki/Strongly_correlated_material

[4] D. Khomskii, "Classifying multiferroics: mechanisms and effects", *Physics*, **2** (20), (2009).

[5] S. Dong, J-M Liu, S-W Cheong, Z. Ren, "Multiferroic materials and magnetoelectric physics: Symmetry, entanglement, excitation, and topology", *Advances in Physics*, **64** (519-626), pp. 1-120, (2015).

[6] H. Schmid, "Multi-ferroic magnetoelectrics", *Ferroelectrics*, **162**, pp. 317-338, (1994).

[7] S. M. Skinner, *IEEE Trans. Parts Mater. Pack.* **1970**, *PMP-6*, 68.

[8] G. A. Smolenskii and I. E. Chupis, "Ferroelectromagnets", *Sov. Phys. Usp.*, ***25***, 475 (1982).

[9] M. Fiebig, "Revival of the magnetoelectric effect", *J. Phys. D: Appl. Phys.*, **38** (8), R123, (2005).

[10] C. Ederer and N. A. Spaldin, "Recent progress in first-principles studies of magnetoelectric multiferroics", *Current Opinion in Solid State and Materials Science*, **9**, 128 (2005).

[11] N. A. Spaldin , in *Physics of Ferroelectrics*, Topics in Applied Physics, (Eds: K. Rabe, C. H. Ahn, and J.-M. Triscone), Springer-Verlag, Vol. **105**, p. 175 (2007).

[12] D. I. Khomskii, "Multiferroics: Different ways to combine magnetism and ferroelectricity", *J. Magn. Magn. Mater.*, **306** (1), pp. 1-8, (2006).

[13] C. N. R. Rao and C. R. Serrao, "New routes to multiferroics", *J. Mater. Chem.*, **17** (47), pp. 4931-4938, (2007).

[14] H. Schmid, "Some symmetry aspects of ferroics and single phase multiferroics", *J. Phys.: Condens. Matter*, **20** (43), 434201 (2008).

[15] K. F. Wang, J. M. Liu, and Z. F. Ren, "Multiferroicity: the coupling between magnetic and polarization orders", *Adv. Phys.*, **58** (4), 321 (2009).

[16] *Electron Correlations in Molecules and Solids*, 3rd edition (Springer, Berlin) by Peter Fulde (2002).

[17] *Lecture Notes on Electron Correlation and Magnetism* (World Scientific, Singapore, by Patrik Fazekas (1999).

[18] D. N. Astrov, "The magnetoelectric effect in antiferromagnetics", *Sov. Phys. JETP*, **11** (3), pp. 708-709 (1960).

[19] D. N. Astrov, "Magnetoelectric effect in Chromium oxide", *Sov. Phys. JETP*, **13** (4), pp.1035-1041 (1961).

[20] V. J. Folen, G. T. Rado, and E. W. Stalder, "Anisotropy of the magnetoelectric effect in Cr_2O_3", *Phys. Rev. Lett.*, **6** (11), p. 607 (1961).

[21] S. M. Skinner, "Magnetically ordered ferroelectric materials", *IEEE Trans. Parts Mater. Pack.*, *PMP-6* , pp. 68-90 (1970).

[22] G. A. Smolenskii, and I. E. Chupis, "Ferroelectromagnets", *Sov. Phys. Usp.*, **25** (7), p. 475 (1982).

[23] D. Lebeugle, D. Colson, A. Forget, and M. Viret, "Very large spontaneous electric polarization in $BiFeO_3$ single crystals at room temperature and its evolution under cycling fields", *Appl. Phys. Lett.*, **91** (2), p. 022907 (2007).

[24] D. Lebeugle, D. Colson, A. Forget, M. Viret, P. Bonville, J. F. Marucco, and S. Fusil, "Room temperature coexistence of large electric polarization and magnetic order in $BiFeO_3$ single crystalls", *Phys. Rev. B*, **76** (2), p. 024116 (2007).

[25] V. V. Shvartsman, W. Kleemann, R. Haumont, and J. Kreisel, "Large bulk polarization and regular domain structure in ceramic $BiFeO_3$", *Appl. Phys. Lett.*, **90** (17), pp. 172115 (2007).

[26] A. M. Kadomtseva, Y. F. Popov, A. P. Pyatakov, G. P. Vorob"ev, A. K. Zvezdin and D. Viehland, "Phase transitions in multiferroic $BiFeO_3$ crystals, thin-layers, and ceramics: enduring potential for a single phase, room-temperature magnetoelectric "holy grail"", *Phase Transitions*, **79** (12), pp. 1019-1042 (2006).

[27] J. F. Scott, "Magnetic phases of bismuth ferrite", *J. Magn. Magn. Mater.*, **321**, pp. 1689-1691 (2009).

[28] I. Sosnowska, T. P. Neumaier, and E. Steichele, "Spiral magnetic ordering in bismuth ferrite", *J. Phys. C:Sol. State Phys.*, **15** (23), p. 4835 (1982).

[29] I. Sosnowska, M. Loewenhaupt, W. I. F. David, and R. M. Ibberson, "Investigation on the unusual magnetic spiral arrangement in $BiFeO_3$", *Physica B: Condensed Matter*, **180-181**, pp. 117-118 (1992).

[30] N. Tsuda, K. Nasu, A. Fujimori, and K. Siratori, *Electronic Conduction in Oxides*, 2nd ed., Springer-Verlag, Berlin, (2000).

[31] C. A. F. Vaz, C. H. Ahn, and V. E. Henrich, in *Epitaxial Ferromagnetic Films and Spintronic Applications* (Eds: A. Hirohata and Y. Otani), Research Signpost, p. 145 (2009).

[32] N. A. Spaldin, and W. E. Pickett, "Computational design of multifunctional materials", *J. Solid State Chem.*, **176** (2), pp. 615-632 (2003).

[33] S.-W. Cheong, and M. Mostovoy , "Multiferroics: a magnetic twist for ferroelectricity", *Nature Mater.*, **6**, pp. 13-20 (2007).

[34] T. Kimura, "Spiral magnets as magnetoelectrics", *Annu. Rev. Mater. Res.*, **37**, pp. 387-413 (2007).

[35] J. van Suchtelen, "Product properties: a new application of composite materials", *Philips Res. Repts*, **27**, pp. 28-37 (1972).

[36] L. Mitoseriu, and V. Buscaglia, "Intrinsic/extrinsic interplay contributions to the functional properties of ferroelectric-magnetic composites", *Phase Transitions*, **79** (12), pp. 1095-1121 (2006).

[37] C. -G. Stefanita, *From Bulk to Nano*, Springer Series in Materials Science No. **117**, Springer-Verlag, Berlin (2008).

[38] W. H. Meiklejohn, and C. P. Bean, "New magnetic anisotropy", *Phys. Rev.*, 102 (5), p. 1413 (1956).

[39] J. Nogués, and I. K. Schuller, "Exchange bias", *J. Magn. Magn. Mater.*, 192 (2), pp. 203-232 (1999).

[40] C. W. Nan, M.I. Bichurin, S. X. Dong, D. Viehland and G. Srinivasan, "Multiferroic magnetoelectric composites: historical perspective, status, and future directions", *J. Appl. Phys.* **103** (3), p. 031101 (2008).

[41] J. N. Eckstein and I. Bozovic, "High-temperature superconducting multilayers and heterostructures grown by atomic layer-by-layer molecular beam epitaxy", *Annu. Rev. Mater. Sci.*, **25**, pp. 679-709 (1995).

[42] R. A. McKee, F. J. Walker, and M. F. Chisholm, "Crystalline oxides on silicon: the first five monolayers", *Phys. Rev. Lett.*, **81** (14), pp. 3014 (1998).

[43] D. G. Schlom, J. Haeni, J. Lettieri, C. D. Theis, W. Tian, J. C. Jiang, and X. Q. Pan, "Oxide nano-engineering using MBE", *Mater. Sci. Eng. B*, **87**, p. 282 (1992).

[44] A.-B. Posadas, M. Lippmaa, F. J. Walker, M. Dawber, C. H. Ahn, and J.-M. Triscone, in *Physics of ferroelectrics*, Topics in Applied Physics, Vol. *105* (Eds: K. Rabe, C. H. Ahn, and J.-M. Triscone), p. 219, Springer-Verlag (2007).

[45] S. A. Chambers, "Molecular beam epitaxial growth of doped oxide semiconductors", *J. Phys.: Condens. Matter*, **20** (26), p. 264004 (2008).

[46] L. W. Martin, S. P. Crane, Y.-H. Chu, M. B. Holcomb, M. Gajek, M. Huijben, C.-H. Yang, N. Balke, and R. Ramesh, "Multiferroics and magnetoelectrics; thin films and nanostructures", *J. Phys.: Condens. Matter*, 20 (43), p. 434220 (2008).

[47] J. W. Reiner, F. J. Walker, and C. H. Ahn, "Atomically engineered oxide interfaces", *Science*, **323** (5917), pp. 1018-1019 (2009).

[48] M. Imada, A. Fujimori, and Y. Tokura, Metal_insulator transitions, Rev. Mod. Phys. **70** (4), p. 1039 (1998).

[49] H. Zheng, Q. Zhan, F. Zavaliche, M. Sherburne, F. Straub, M. P. Cruz, L.-Q. Chen, U. Dahmen, and R. Ramesh, "Controlling self-assembled perovskite-spinel nanostructures", *Nano Lett.*, **6** (7), pp. 1401-1407 (2006).

[50] I. Levin, J. Li, J. Slutsker, and A. L. Roytburd, "Design of self assembled multiferroic nanostructures in epitaxial films", *Adv. Mater.*, **18** (15), pp. 2044-2047, (2006).

[51] J. Slutsker, I. Levin, J. Li, A. Artemev, and A. L. Roytburd, "Effect of elastic interactions on the self assembly of multiferroic nanostructures in epitaxial films", *Phys. Rev. B*, **73** (18), p. 184127 (2006) .

[52] L. Yan, Y. Yang, Z. Wang, Z. Xing, J. Li, and D. Viehland, *J. Mater. Sci.* **2009** , *44* , 5080 .

[53] H. J. A. Molegraaf, J. Hoffman, C. A. F. Vaz, S. Gariglio, D. Van Der Marel, C. H. Ahn, and J.-M. Triscone, "Magneto electric effects in complex odides with competing ground states", *Adv. Mater.,* 21, p. 3470 (2009).

[54] H. Tanaka, J. Zhang, and T. Kawai, "Giant electric fied modulation of double exchange ferromagnetism at room temperature in the perovskite manganite/titanate p-n junction", *Phys. Rev. Lett.*, 88 (2), p. 027204 (2002).

[55] X. Hong, A. Posadas, A. Lin, and C. H. Ahn, "Ferroelectric-field-induced tuning of magnetism in the colossal magnetoresistive oxide $La_{1-x}Sr_xMnO_3$", *Phys. Rev. B*, 68 (13), p. 134415 (2003).
[56] X. Hong, J. Yau, J. D. Hoffman, C. H. Ahn, Y. Bason, and L. Klein, "Effect of electric field doping on the anisotropic magnetoresistance in doped manganites", *Phys. Rev. B*, **74** (17), p. 174406 (2006).
[57] T. Kanki, H. Tanaka, and T. Kawai, "Electric control of room temperature magnetism in a $Pb(Zr_{0.2}Ti_{0.8})O_3$ field-effect transistor/ $La_{0.85}Ba_{0.15}MnO_3$", *Appl. Phys. Lett.*, **89**, p. 242506 (2006).
[58] A. S. Dhoot, C. Israel, X. Moya, N. D. Mathur, and R. H. Friend, "Large electric field effect in Electrolyte-Gated Manganites", *Phys. Rev. Lett.*, **102** (13), p. 136402 (2009).
[59] H. Ohno, D. Chiba, F. Matsukura, T. Omiya, E. Abe, T. Dietl, Y. Ohno, and K. Ohtani, "Electric-field control of ferromagnetism", *Nature*, **408** (6815), pp. 944-946 (2000).
[60] H. Boukari, P. Kossacki, M. Bertolini, D. Ferrand, J. Cibert, S. Tatarenko, A. Wasiela, J. A. Gaj, and T. Dietl, "Light and electric field control of ferromagnetism in magnetic Quantum structures", *Phys. Rev. Lett.*, **88** (20), p. 207204 (2002).
[61] Y. D. Park, A. T. Hanbicki, S. C. Erwin, C. S. Hellberg, J. M. Sullivan, J. E. Mattson, T. F. Ambrose, A. Wilson, G. Spanos, and B. T. Jonker, "A group-IV ferromagnetic semiconductor: Mn_xGe_{1-x}", *Science*, **295** (5555), pp. 651-654 (2002).
[62] D. Chiba, M. Yamanouchi, F. Matsukura, and H. Ohno, "Electrical manipulation of magnetization reversal in a ferromagnetic semiconductor", *Science*, **301** (5635), pp. 943-945 (2003).
[63] I. Stolichnov, S. W. E. Riester, H. J. Trodahl, N. Setter, A. W. Rushforth, K. W. Edmonds, R. P. Campion, C. T. Foxon, B. L. Gallagher, and T. Jungwirth, "Non-volatile ferroelectric control of ferromagnetism in (Ga, Mn)As", *Nature Mater.*, **7**, pp. 464-467 (2008).
[64] C. H. Ahn, J. M. Triscone, N. Archibald, M. Decroux, R. H. Hammond, T. H. Geballe, O. Fischer, and M. R. Beasley, "Ferroelectric field effect in epitaxial thin film oxide $SrCuO_2/Pb(Zr_{0.52}Ti_{0.48})O_3$ heterostructures", *Science*, **269** (5222), pp. 373-376 (1995).
[65] C. H. Ahn, R. H. Hammond, T. H. Geballe, M. R. Beasley, J.-M. Triscone, M. Decroux, O. Fischer, L. Antognazza, and K. Char, "Ferroelectric field effect in ultra thin $SrRuO_3$ films", *Appl. Phys. Lett.*, **70**, p. 206 (1997).

[66] C. H. Ahn, S. Gariglio, P. Paruch, T. Tybell, L. Antognazza, and J.-M. Triscone, "Electrostatic modulation of superconductivity in ultra-thin $GdBa_2Cu_3O_{7-x}$ films", *Science*, **284** (5417), pp. 1152-1155 (1999).

[67] S. Gariglio, C. H. Ahn, D. Matthey, and J.-M. Triscone, "Electrostatic tuning of the Hole density in $NdBa_2Cu_3O_{7-x}$ films and its effect on the hall response", *Phys. Rev. Lett.*, **88** (6), p. 067002 (2002).

[68] C. H. Ahn, J.-M. Triscone, and J. Mannhart, "Electric field effect in correlated oxide systems", *Nature*, **424**, pp. 1015-1018 (2003).

[69] C. H. Ahn, A. Bhattacharya, M. D. Ventra, J. N. Eckstein, C. D. Frisbie, M. E. Gershenson, A. M. Goldman, I. H. Inoue, J. Mannhart, A. J. Millis, A. F. Morpurgo, D. Natelson, and J.-M. Triscone, *Rev. Mod. Phys.*, **78**, p. 1185 (2006).

[70] T. Venkatesan, D. C. Kundaliya, T. Wu, and S. B. Ogale, "Noval approaches to field modulation of electronic and magnetic properties of oxides", *Phil. Mag. Lett.*, **87** (3-4), pp. 279-292 (2007).

[71] M. Weisheit, S. Fähler, A. Marty, Y. Souche, C. Poinsignon, and D. Givord, "Electric field-induced modification of magnetism in thin film ferromagnets", *Science*, **315** (5810), pp. 349-351 (2007).

[72] X. Hong, A. Posadas, K. Zou, C. H. Ahn, and J. Zhu, "High-mobility few-layer Graphene Field Effect Transistors fabricated on epitaxial ferroelectric gate oxides", *Phys. Rev. Lett.*, **102** (13), p. 136808 (2009).

[73] R. E. Newnham, D. P. Skinner, and L. E. Cross, "Connectivity and piezoelectric-pyroelectric composites", *Mat. Res. Bull.*, **13** (5), pp. 525-536 (1978).

[74] N. A. Spaldin, M. Fiebig, "The Renaissance of magnetoelectric multi-ferroics", *Science*, **309** (5733), pp. 391-392, (2005).

[75] M. Gajek,M. CH4bibes, S. Fusil, K. Bouzehouane, J. Fontcuberta, A. Barthélémy and A. Fert, "Tunnel junctions with multiferroic barriers", *Nature Materials* **6** (4), pp. 296-302 (2007).

[76] J. Seidel, L. W. Martin, Q. He,......., and R. Ramesh, "Conduction at domain walls in oxide multiferroics", *Nature Materials*, **8**, pp. 229-234 (2009).

[77] C. A. F. Vaz, J. Hoffman, C. H. Ahn, and R. Ramesh, "Magnetoelectric coupling effects in multiferroic complex oxide composite structures", *Adv. Mater.*, **22**, pp. 2900-2918 (2010).

5

Topological Aspects of Strongly Correlated Systems

Gaurav Kumar Gupta

Technion-Israel Institute of Technology, Israel
E-mail: gkgupta.bhu@campus.technion.ac.il

Keywords: Topological Order, Symmetry breaking, Phase transition

5.1 Introduction

Condensed matter is the area of physics which deals with the phases of matter in condensed form with an aim to understand the properties of these phases and the transition between them based on the known laws of physics. This is usually done by figuring out some characteristic properties of the phase which sets it apart from the others, e.g., symmetry, topological order, entanglement, etc. The most common way of characterizing the transition between these phases is the broken symmetry. In this classification method, one finds the symmetry which phase breaks spontaneously during the transition and develops a local long-range order. This is the famous Landau's symmetry breaking paradigm [1]. Landau's symmetry breaking theory is a phenomenological theory which describes the transition from a symmetric phase (high-temperature phases) to a less symmetric/broken symmetry phase (phases at low temperature). A zero-temperature phase transition is also possible (by tuning the parameters of the Hamiltonian) and is known as quantum phase transition [2]. Although there were other paradigms of phase transition available, like Anderson localization [3], Berezinsky–Kosterlitz–Thouless transition [4] (broken symmetry but no long-range order), etc., Landau's spontaneous symmetry breaking dominated over the others.

However, in the last three decades or so, it became clear that there are phases which cannot be characterized (or completely characterized) by local order and instead they possess non-local orders (sometimes, highly

DOI: 10.1201/9781003337102-5

non-local); fractional quantum Hall (FQH) effect was one of the first experimentally verifiable examples with highly non-local order [5]. These phases are called "topological" phases and are characterized by topological invariants. A general definition of topological phase would be "a phase of matter characterized by a topological invariant (a non-local order parameter) rather than a local order parameter." Topological phase recently got a surge of attention because of the proposal of "topological quantum computation" using non-Abelian FQH state [6]. As being highly non-local, FQH states are robust to the decoherence caused by the local fluctuation [6, 7].

The area of topological insulators (TIs) started with the discovery of the quantum Hall effect (QHE) back in the 1980s by von Klitzing *et al.* [8]. They found plateaus in the Hall conductivity at the high magnetic field instead of the expected linear Hall conductivity as in the low magnetic field in a usual two-dimensional electron gas (2DEG). The plateaus were exactly quantized as

$$\sigma_{xy} = \nu \frac{e^2}{h}, \tag{5.1}$$

where e is the electronic charge, h is the Plank's constant, and is an integer. This was dubbed as the quantum version of the Hall effect or QHE. It was later shown by Thouless *et al.* [9] and Simon [10] that the topologies of the Bloch wave function and the QHE are related and the Hall conductivity is quantized as long as there exists an insulating bulk gap. was shown to be an integer which is nothing but the first Chern number. Later on, Tsui *et al.* [11] found that there are plateaus for non-integer values of ν as well; more specifically, they found plateau for $= 1/3$. It turns out that it is impossible to explain this fractional quantum Hall effect (FQHE) with the non-interacting theory and interaction plays a crucial role in stabilizing this state. More interestingly, the elementary excitation in this state carries a $e/3$ charge. Going back to the QHE, in 1988, Haldane showed that magnetic field is not a necessary ingredient for QHE, but breaking the time-reversal symmetry (TRS) is must [12].

As σ_{xy} is odd under time reversal (edge state changes its direction on flipping magnetic field), it vanishes in TRS system, and, hence, this scheme of classification is not applicable on TRS systems. In 2005, Kane and Mele [13] showed (almost 20 years after the discovery of QHE) that one can define a new topological invariant for TRS system by taking the difference of the Hall conductance of the two copies of the Hamiltonian which are time-reversed conjugate to each other. These two copies both break the TRS individually, but, together, TRS is preserved. Later on, a more general classification scheme for the TRS system was proposed based on the "TR polarization" giving rise to a $\mathcal{Z}_2$classification for TIs [14].

5.2 Non-Interacting Topological Phases

To start the chapter, we will first discuss the topology in non-interacting systems. Since the area of TIs figuratively exploded recently, there are some very good review articles on this area 1. This section is in no sense meant to replace them but rather to give the necessary background for TIs which will help in understanding the later section and to make this chapter self-contained. First, we will briefly discuss the mother of all topological phases, the integer QHE (IQHE) and then move on to discuss the lattice version of it in the presence/absence of an external magnetic field. Then we will introduce the concept of topology in the TRS system. The main advantages of these models are their direct generalization to the interacting system. We will discuss this in the later section. It is worthwhile to mention that the term TI is almost always synonymously used for TRS $\mathcal{Z}_2$ invariant TIs, but we will use the term in a more general fashion, i.e., any phase with non-trivial topology is a TI.

5.2.1 Integer Quantum Hall Effect

In 1879, Edwin Hall found the development of transverse voltage in a 2D sample when subjected to a perpendicular electric and magnetic field [18]. Using the classical equation of motion of electrons in the presence of the electro-magnetic field, it can be shown that the transverse resistivity takes the form

$$\rho_{xy}^{\text{classical}} = -\frac{B}{ne}. \tag{5.2}$$

When the applied magnetic field is high, electrons are strictly confined in 2D and the sample is pure enough; von Klitzing *et al.* [8] found the plateaus (see Figure 5.1) in the Hall resistivity ρ_{xy} as a function of the magnetic field.

$$\rho_{xy}^{\text{quantum}} = -\frac{h}{\nu e^2}, \tag{5.3}$$

where ν is an integer. This is contrary to the classical Hall effect [18] where $\rho_{xy}^{\text{quantum}}$ changes linearly with the applied magnetic field. Along with this, the longitudinal resistivity (ρ_{xx}) vanishes at these plateaus. The quantization of ν has been confirmed in the present start-of-the-art experiments with the accuracy of a part in a billion.

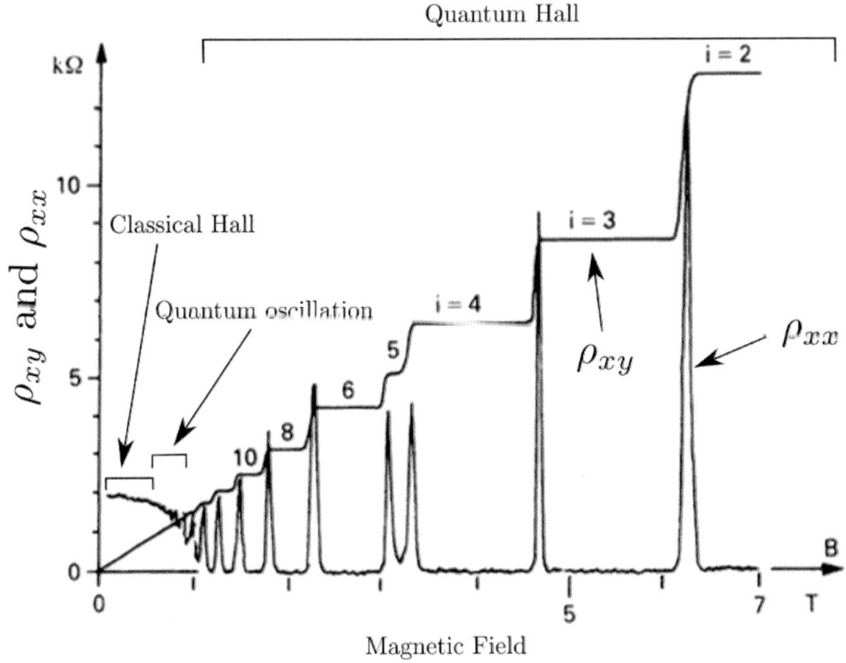

Figure 5.1 Experimental plot showing the plateaus in ρ_{xy} and zeros of ρ_{xx}. This figure can be broadly divided into three regimes: classical Hall regimes, where ρ_{xy} goes linearly with the magnetic field; quantum oscillation regime, where ρ_{xx} shows oscillation; and quantum Hall regime, where ρ_{xy} shows plateaus and $\rho_{xx} \to 0$ at those plateaus.

The relationship between the Bloch wave function and Hall conductivity/resistivity was famously demonstrated by Thouless-Kohmoto-Nightingale-den Nilis (TKNN) [9] along with Simon [10] using the linear response theory (Kubo formula)

$$\nu = \frac{1}{2\pi\, i} \sum_{\alpha} \left(\int_{\mathrm{BZ}} d\boldsymbol{k} \, (<\partial_y u^{\alpha}{}_k \mid \partial_x u^{\alpha}{}_k >) - (<\partial_x u^{\alpha}{}_k \mid \partial_y u_k >) \right), \tag{5.4}$$

where u_k^{α} are the Bloch wave function, and the summation of α runs over the filled bands and the integral is performed over the entire Brillouin Zone (BZ). It turns out that the above equation gives an integer as long as the bulk is gapped. The above equation is also interesting in a sense that it does not depend on the energy (except for the fact that summation runs over the occupied bands); so a small deformation in the Hamiltonian does not affect

v and hence plateaus of ρ_{xy} are robust against small deformation. The above equation can be conveniently written as

$$\nu = \frac{1}{2\pi}\int_{\mathrm{BZ}} dk \left(\frac{\partial A_y}{\partial k_x} - \frac{\partial A_x}{\partial k_y}\right)$$
$$A^{\alpha}{}_{\beta} = -i\sum_{\alpha} < u_k^{\alpha}\left|\partial_{k_\beta}\right| u_k^{\alpha} >, \tag{5.5}$$

where the vector A is known as the Berry connection which is equivalent to the vector potential in the momentum space. One can define the corresponding magnetic field ($\mathcal{F}$, also known as Berry connection) as

$$\mathcal{F}^{\alpha} = \nabla \times A^{\alpha}, \tag{5.6}$$

and the corresponding phase in a closed loop (γ^{α}) known as Berry phase [19] associated with this magnetic field as

$$\gamma^{\alpha} = \oint A^{\alpha}.dk. \tag{5.7}$$

Undoubtedly, Berry phase is one of the most fundamental measures of the topology, even in the gapless systems such as graphene, Wyle semi-metal, etc.

5.2.1.1 Chiral edge states

Using quantum mechanics, it can be shown that the bulk of the QH system is an insulator, and, hence, the quantized Hall conductance stems from the edge. One can get the basic intuition of edge state from the classical theory as well. In the presence of the magnetic field, charged particles follow a circular orbit known as cyclotron orbit. In the bulk, these orbits are localized as shown in Figure 5.2 (a), while on the edge, it collides with the boundary and moves along the edge (as shown in Figure 5.2 (a)). These orbits along the edge are known as skipping orbits and they form the edge states.

5.2.2 Topological Phase in 1D: SSH Model

In 1D, the simplest model for non-trivial TI would be the celebrated Su–Schrieffer–Hegger (SSH) model for polyacetylene [20]. Polyacetylene is a 1D chain of carbon atoms with staggered hopping t and t' as shown in Figure 5.3. The corresponding Hamiltonian can be written as

$$\mathcal{H}_{\mathrm{SSH}} = \sum_{i=1}^{N} t c_{A,i}^{\dagger} c_{B,i} + \sum_{i=1}^{N-1} t' c_{A,i+1}^{\dagger} c_{B,i} + h.c. \tag{5.8}$$

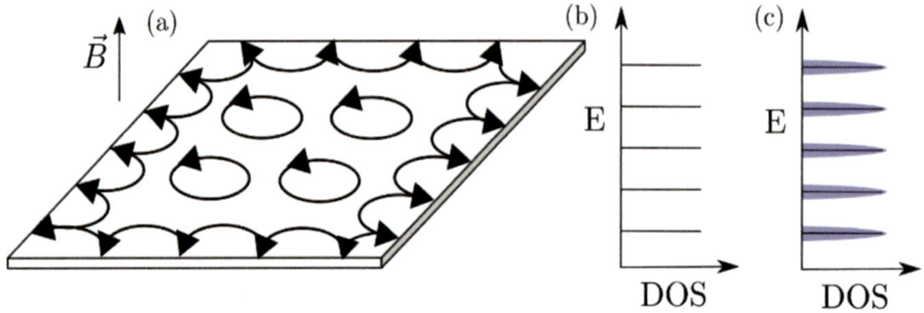

Figure 5.2 (a) Cartoon showing the cyclotron orbit of electrons in the presence of a perpendicular magnetic field as well as skipping orbit on the edge. The density of state in QH system in (b) the absence and (c) the presence of disorder.

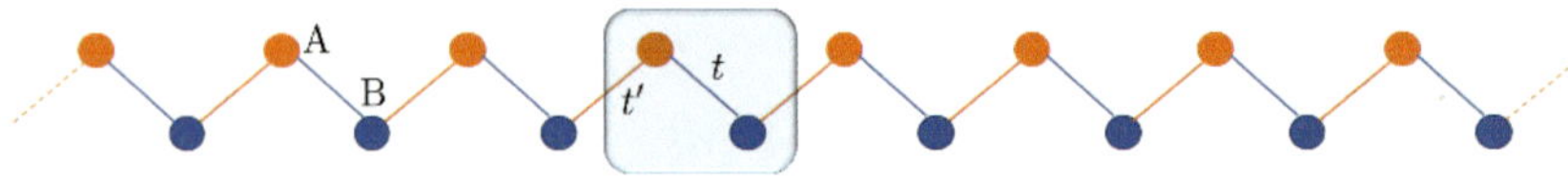

Figure 5.3 Cartoon showing the SSH model with staggered intra- and inter-cell hopping t and t'.

where $c_{\alpha,i}$ $\left(c^{\dagger}_{\alpha,i}\right)$ are the fermionic annihilation (creation) operator on $\alpha\,(= A, B)$ sublattice in the ith unit cell (see Figure 5.3). One can easily show that the above Hamiltonian is gapless (at half-filling and periodic boundary condition) only for $t = t^{'}$; otherwise, it is gapped. Hence, if there is a topological phase transition happening in the system, it only happens at $t = t^{'}$. Now, let us consider two limits $t^{'} = 0$ and $t = 0$. When $t^{'} = 0$, all the A and B atoms form dimer within the same unit cell and three are no edge state as shown in Figure 5.4 (a). While in the case of $t = 0$, all the A and B atoms in the bulk form dimer with the B and A atoms of the nearest unit cell as shown in Figure 5.4 (b), and, hence, the atoms on the edge are left out and

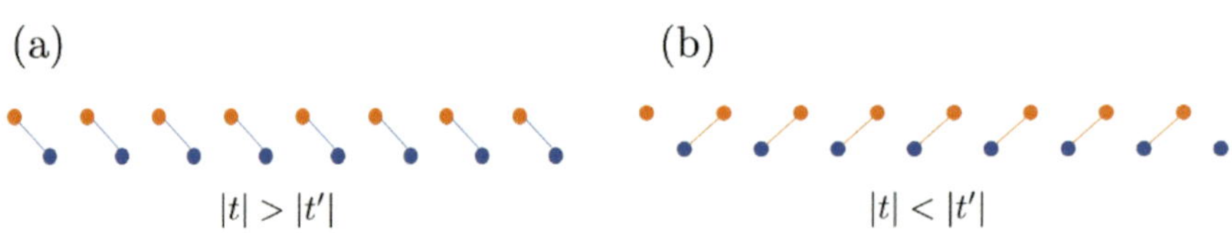

Figure 5.4 Cartoon showing (a) topologically trivial phase and (b) topologically non-trivial phase of finite SSH chain.

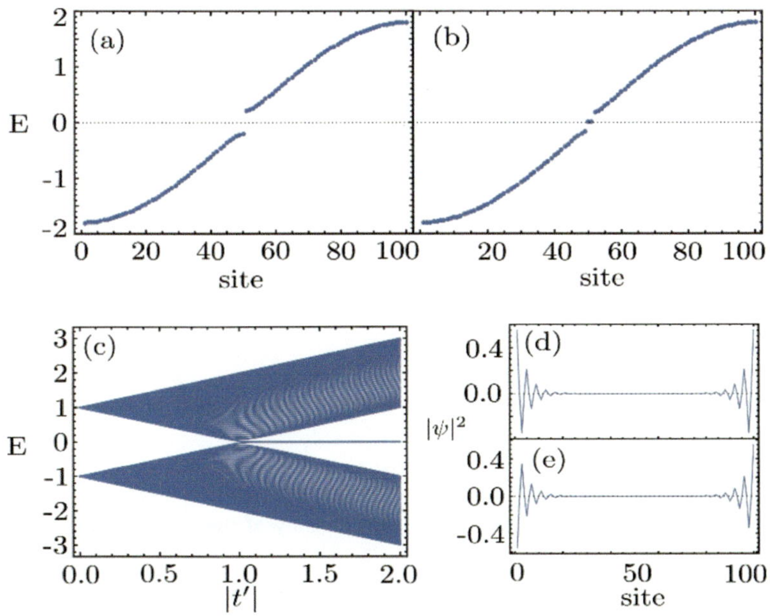

Figure 5.5 Spectrum of the Hamiltonian in Equation (5.8) for (a) $|t| > |t'|$ and (b) $|t| < |t'|$ and (c) as a function of $|t'|$ for $t = 1$. Spectrum in (b) shows two zero energy states which are localized on the edge as shown in (d) and (e).

form localize zero energy state. Figure 5.5 shows the formation of zero energy edge states as a function of t' in the Hamiltonian given by Equation 5.8.

5.2.3 Haldane Model

In 1988, Haldane [12] proposed a model which shows finite Hall conductance but does not require high magnetic field (or even Landau levels). It can occur in a standard Bloch band with a broken TRS but no net magnetic flux. The model was based on graphene (honeycomb lattice) with a real nearest neighbor (NN) hopping and complex next nearest neighbor (NNN) hopping as shown in Figure 5.6.

The Hamiltonian can be written as

$$\mathcal{H} = \sum_{<i,j>} t_1 c^{\dagger}_{A,i} c_{B,j} + \sum_{\ll i,k \gg} t_2 e^{i\phi_{i,k}} c^{\dagger}_{A,i} c_{A,k} + \sum_{\ll j,l \gg} t_2 e^{i\phi_{j,l}} c^{\dagger}_{B,j} c_{B,l} + \sum_{i,\alpha=A,\ B} (-M)^{\alpha} c^{\dagger}_{\alpha,i} c_{\alpha,i}. \tag{5.9}$$

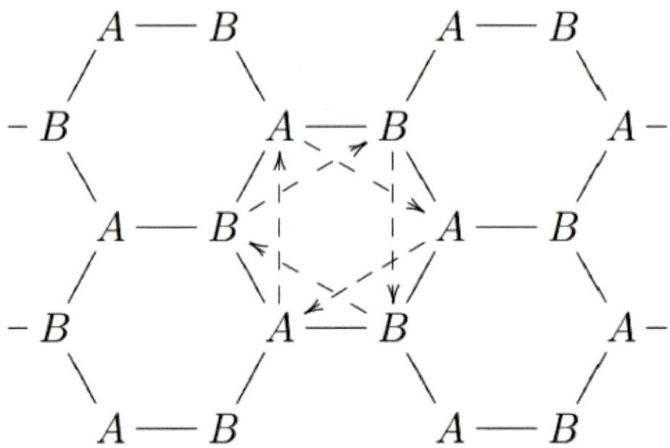

Figure 5.6 Honeycomb lattice used in Haldane model where A and B are the two sublattices with real NN hopping (shown by solid lines) and complex NNN hopping (shown by dashed lines with arrow). The arrow represents the directionality of the NNN hopping and hence breaks the TRS.

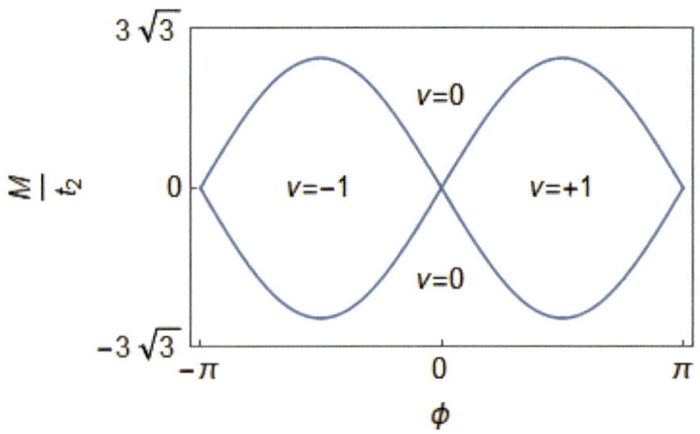

Figure 5.7 Phase diagram of the Haldane model (Equation (5.9)).

Using the formula in Equation (5.4), we can show that the Hamiltonian in Equation (5.9) has finite Chern number in some parameter regimes as shown in Figure 5.7. A straightforward generalization of the above Hamiltonian on the square lattice is given by

$$H_{CI} = \sum_{k,\alpha,\beta} h_{\alpha,\beta}(k)\, c^{\dagger}_{k,\alpha} c_{k,\beta}$$

$$h(k) = [\cos(k_x) + \cos(k_y) - \mu]\,\sigma_z + \sin(k_x)_{x} + \sin(k_y)\,\sigma_y. \qquad (5.10)$$

Again, using Equation (5.4), we can show that the Chern number of the above Hamiltonian is

$$\begin{aligned} \nu &= -1,\ 0 < \mu < 2 \\ &= +1, -2 < \mu < 0 \\ &= 0, \text{otherwise.} \end{aligned} \tag{5.11}$$

5.2.4 Quantum Spin Hall Effect

As we have seen, breaking TRS is necessary to see QHE; hence, for the TRS system, the Quantum Hall response σ_{xy} must be zero (as it is odd under TR). In 2005, Kane and Mele [21] came up with a model which is made up of two copies of TRS breaking Hamiltonian such that the total Hamiltonian preserves the TRS. They showed that each copy can have finite Chern number, but the sum of both the Chern number is always zero (as expected), but the difference between the two is not. They called it "*spin Chern number*." A model Hamiltonian can be written as

$$\mathcal{H}_{\mathrm{TRS}} = \begin{bmatrix} h(k) & 0 \\ 0 & h^*(-k) \end{bmatrix} \tag{5.12}$$

where $h(k)$ is the same 2×2 Hamiltonian in Equation (5.10) and $h^*(-k)$ is the time-reversed partner of $h(k)$. Since $\mathcal{H}_{\mathrm{TRS}}$ contains both copies, it is invariant under TRS, and, hence, the total Chern number is always 0. As we have already seen, $h(k)$ has finite Chern number for $-2 < \mu < 2$; hence, the difference of the Chern number is finite for $-2 < \mu < 2$. As a consequence of TRS, edge states now come in pairs and move in the opposite direction such that the total charge current is 0. The cartoon in Figure 5.8 shows the edge states in quantum spin Hall effect (QSHE) which is equivalent to the two counter-propagating edge modes in QHE.

5.3 Interacting Topological Insulator

So far, we have discussed topological phases in systems with free electrons, be it SSH model, IQHE, Haldane model, or TRS TI. In condensed matter physics, electron–electron interaction has played a crucial role and always fascinated the researchers around the globe by showing non-trivial effects. Also, interactions are intrinsically present in the system because of the fundamental nature of the electrons. Although, interactions can be neglected sometimes because of screening and/or if they are weak, but sometimes they

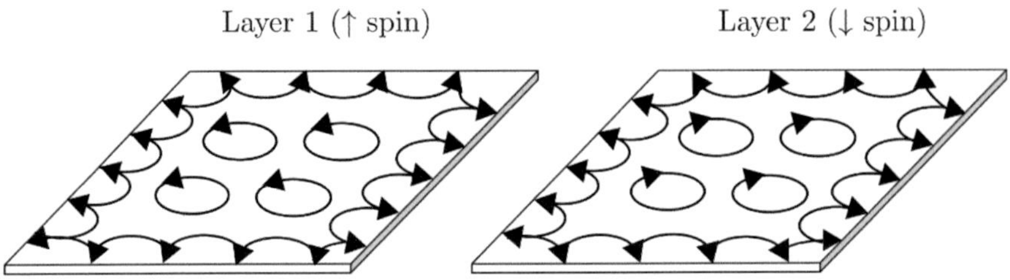

Figure 5.8 Cartoon showing the edge state in the quantum spin Hall system; two different states on each edge represents two different spin states of opposite chirality.

are unavoidable and need serious consideration. Some famous examples are superconductivity, Mott insulator, and half-metallic behavior, just to name a few. So, it is natural to discuss the effect of interaction on topological phases as well. The major questions one can raise are the following:

- Is a non-interacting topological phase stable against interaction?
- Can one drive topological transition using interaction?
- Can one get an interaction-induced topological phase which does not have a non-interacting counterpart?

The answer to all the above questions is YES! (Why would I ask these questions otherwise?) In the following sections, we will mainly focus on these questions with plenty of examples.

5.3.1 Topological Insulator with Interaction

In this section, we will focus on the first question – is a non-interacting topological phase stable against interaction? As already told, the answer is yes! The first thing we can remind ourselves is that the bulk of the non-interacting topological system is gapped and as long as the interaction is significantly smaller than the gap, the system is supposed to stay robust against the interaction. This is one of the reasons why many works are dedicated to the effect of interaction on the edge states in the form of helical Luttinger liquid [22]. A popular way of dealing with the interaction in the bulk is to determine the topological invariants in terms of single particle Green's function which can encode the effect of interaction in the form of self-energy [23–26]. It can be shown that the Chern number (in all dimensions) can be written in terms of the single particle Green's function and its derivative. As the $\mathcal{Z}_2$ invariants (in dimensions less than 4) can be deduced from the second Chern number using the technique known as dimension reduction [43], it can also be written in terms of the single particle Green's function. Another interesting approach is to define a new effective Hamiltonian which includes the zero-frequency free energy as

$$\mathcal{H}_{\mathcal{I}} = \mathcal{H}_0(k) + \sum (\omega = 0, \mathrm{k}), \tag{5.13}$$

where $\sum(\omega = 0, k)$ is the zero-frequency free energy [27]. Now topological invariants can be calculated for this new Hamiltonian. There are efficient numerical methods (e.g., dynamical mean field theory, etc.) which can provide free energy, and, hence, this method is also popular. Compared to the previously mentioned Green's function approach, it also reduces the complexity of integrating the product of Green's function and its derivative which is extremely hard to handle numerically.

5.3.1.1 SSH Hubbard model

As discussed in Section 5.2.2, the SSH model is one of the simplest models for TI in 1D, and, hence, it is the best model to start the discussion of the effect of interaction on topological properties. Several variants of SSH model have been studied in the presence of electron–electron interaction such as spin-less SSH extended Hubbard model [28], spin-full SSH Hubbard model [29–31], etc., and all of them turned out to be robust against weak or moderate interaction. For the sake of completeness, Hubbard term is given by

$$H_U = \sum_i n_{i\uparrow} n_{i\downarrow}, \tag{5.14}$$

while the extended Hubbard term is given by

$$H_U^{\text{ext.}} = \sum_{<i,j>} n_i n_j, \tag{5.15}$$

where $n_i = n_{i\uparrow} + n_{i\downarrow}$ and $<,>$ represents the NN term. Bosonic SSH Hubbard model with repulsive interaction [32] has also been studied and strong interaction seems to stabilize the topological phase.

5.3.1.2 Haldane Hubbard model

Spin-less Haldane model with extended Hubbard interaction has been studied numerically to verify the robustness of the topological phase in the presence of weak and moderate interaction, and for strong interaction, the system goes to a Mott phase with charge density wave (CDW) order [33]. Spin-full Haldane Hubbard model has also been shown to have a non-trivial topological phase for weak interaction strength [34]. For strong enough interaction, some exotic Chern insulator phase has also been proposed [35].

5.3.1.3 BHZ Hubbard model

The Bernevig–Hughes–Zhang (BHZ) model [36] is very close to the model discussed in the Hamiltonian in Equation (5.12). It also has a spin rotation symmetry breaking term (Δ) and the final Hamiltonian looks

$$\mathcal{H}_{\mathrm{BHZ}} = \begin{bmatrix} h(k) & \Delta \\ \Delta^{\dagger} & h^{*}(-k) \end{bmatrix} \tag{5.16}$$

where $h(k)$ is the same as in Equation (5.10). The effect of Hubbard interaction has been extensively studied on this model [37–41]. which leads to the conclusion that the system is robust against weak interaction and eventually breaks down with the increase in interaction strength (through first-order phase transition [42]) as the time reversal breaking magnetic order sets in.

5.3.1.4 Topological axion insulator

2D time-reversal symmetric systems are characterized by $\mathcal{Z}_2$ topological invariants which take the value either 0 or 1, and upon breaking the TRS, it becomes undefined. In case of 3D time-reversal symmetric systems, another topological invariant known as axion angle (equivalent to $\mathcal{Z}_2$ invariant) can be defined which takes the value either 0 or π (equivalent to 0 or 1 of $\mathcal{Z}_2$ topological invariant) [43]. On breaking the TRS, it is still defined and can take any value in between 0 and π. This is known as dynamical axion which results in novel magneto-electric effect in 3D TIs. The effect of interaction on 3D TIs have been studied to propose the presence of topological axion insulator and demonstrate the novel magneto-electric effect and it was proposed to be observed as axionic-polariton in experiments [44]. It has also been shown that this topological axion insulator has many exotic properties [45].

5.3.2 Interaction-induced Topological Insulator

In this section, we will discuss how interaction can drive the topological phase transition, i.e., how a topologically trivial phase turns into a topologically non-trivial phase due to interaction. One example would be Bosonic SSH model in 1D. Non-interacting bosons tend to condense to a superfluid phase which is topologically trivial. On the other hand, SSH model with hardcore bosons is topologically non-trivial. This can be understood easily as follows: in 1D with NN hopping, hardcore bosons and fermions are same, and, hence,

1D SSH model of hardcore bosons is topologically non-trivial. As hardcore bosons are bosons with strong on-site interaction, this is an example of an interaction-induced TI.

5.3.2.1 Topological Mott insulator

Another interesting interaction-driven phase was pointed out on the honeycomb model where the long-range Hubbard model (with the NNN terms) was used [46]. It was argued that the NNN term can cause spontaneous symmetry breaking and can lead to topologically non-trivial ground state. TRS may or may not be broken spontaneously, and, hence, the resulting phase can be equivalent to QH state or QSH state [46]. Unfortunately, it turns out that the long-range interaction can lead to other spontaneously symmetry broken phase such as CDW (long-range interaction is prone to CDW order) [47], etc., and this phase was never observed in the standard numeric [47–51]. Later on, a similar model was proposed for the system with quadratic bands touching each other. Since the density of states is very high at these points, a small interaction is good enough for the symmetry breaking but not good enough for CDW. It was later on confirmed numerically that this model has a stable anomalous QH phase and QSH phase.

5.3.2.2 Topological Kondo insulator

Topological Kondo insulator [52–54] is one of the most exciting interaction-induced topological phases because the candidate material proposed to host this topological phase, SmB_6, was well known among physicists for more than five decades. The possible reason why it was never reported to be a TI is because of the requirement of low temperature to see the effect of edge states. To understand topological Kondo effect, we first need to understand the ingredients of this phase, d- and f-orbitals. Both have opposite parity and the latter one has strong spin−orbit coupling. The f-orbitals are generally localized (hence, the bands are quiet dispersion less) and hence can be studied well with Heisenberg model while the d-orbitals form the conduction band. At low temperature, interaction leads to the hybridization of d- and f-orbitals (due to Kondo effect [55]) and results in topological phase in SmB_6. In general, this effect leads to a strong TI in 2D [56] while weak/strong TI in 3D [54].

5.3.2.3 Quantum spin hall density wave insulator

Quantum spin Hall density wave insulator is another example of a topological phase driven by spontaneous symmetry breaking. One starts with a quasi-1D model with NN hopping and strong spin−orbit coupling. Upon introducing electron–electron interaction, it can be shown that the system is unstable toward a new kind of density wave known as spin–orbit density wave which breaks lattice translation symmetry but preserves TRS, and, hence, the ground states are TR invariant [57]. It was shown that when the opposite chiral bands are significantly nested by a magic Fermi surface nesting vector $Q \sim (\pi, 0) / (0, \pi)$, the chirality between the momentum k and $k + Q$ gets reversed. This has a very intriguing result on the topological properties of the system, and it turns into a $\mathcal{Z}_2$ invariant TI with localized edge modes [58].

It is worth mentioning that all the interaction induced topological phases discussed in the present section have some non-interacting counterpart including topological Kondo insulating phase as well, and the effective Hamiltonian can be mapped to a non-interacting Hamiltonian without closing the gap and retaining the topological properties intact.

5.3.3 Interaction-induced Exotic Topological Phases

In this section, we will discuss emergent topological states due to the interaction which does not have any non-interacting counterpart. Most of the topological states discussed in the previous sections are symmetry protected topological (SPT) states, be it particle number conservation symmetry, TRS, or other. Once the symmetry is removed, the topological state is not defined, and hence the name SPT. However, in the present section, we will discuss examples that do not require any protecting symmetry; they have intrinsic topology built-in and are generally known as topologically ordered states. Key features of topologically ordered states are long-range entanglement, fractionalization of charge/statistics, among others. These key features are not possible in non-interacting systems as they are essentially a product state. Examples discussed below are FQHE, fractional Chern insulator (FCI), and topological Mott insulator. Although not discussed below, it is worth mentioning that a huge class of topological spin liquids also fall into the same category [59].

5.3.3.1 Fractional quantum hall effect

FQHE is the first example of topologically ordered states which is only possible due to interaction, and, hence, it secured the first place in our list of interaction-induced exotic topological phase. The key ingredients of this phase are a high magnetic field, partially filled Landau level, and, of course, interaction. There are many FQH states known till date, but let us start with the one which was observed first, $\nu = 1/3$. This means only 1 out of 3 degenerate states of the lowest LL is filled. Due to the huge degeneracy, the interaction between the particles is effectively very high and even a small interaction has been shown to have a drastic effect. It turns out that a range of different varieties of interaction potentials leads to similar kind to FQH state. Soon after the observation of FQHE, Laughlin came up with a variational wave function for the FQH state which describes and correctly predicted the properties of FQH states (like fractional charge) [60]. Laughlin's wavefunction was suitable for fillings $\nu = 1/q$ or $1 - 1/q$. Later on, Jain generalized the idea to a more general state ($\nu^* = n/(2np \pm 1)$) based on his theory of composite fermions [61, 62]. Composite fermions are quasi-particles which have even number of magnetic fluxes attached to each electron. In the mean field picture, this attached flux reduces the effective magnetic field seen by the quasi-particles. This effective magnetic field is exactly enough for the IQHE for composite fermions [62]. Hence, FQH for fermions is mapped to IQHE for composite fermions. This theory has been demonstrated (numerically) to be extremely successful for small systems.

5.3.3.2 Fractional Chern insulator

Since the lattice version of QHE was already present, whether there exists a lattice version of FQHE became an obvious question for researchers. To get a lattice version of FQHE, we need to recall the key ingredients of FQHE, huge degeneracy of the LL, fractional filling, and interaction. Fractional filling and interaction are easy to achieve (at least numerically), but getting a flat band with realistic parameters was not conceived until 2011 [63–65]. Using the flat band, FQHE on a lattice or FCI became popular. To realize FCI, one starts with a non-interacting band with finite Chern number with appropriate fractional filling ($= 1/3$, for FQH 1/3 equivalent state) and add interaction. Long-range interactions have been shown to be crucial to stabilize FCI state [66] which leads to an obvious competing state, CDW state. It turns out that the CDW state does compete with FCI state, but a careful choice of symmetry breaking term may lead to destabilizing CDW state and the system prefers FCI over CDW.

5.3.3.3 Bosonic quantum hall Effect

So far, we have discussed a lot about fermions; let us discuss a model for QHE of bosons. This model consists of two species of bosons (say A and B) in the presence of a strong magnetic field and interaction [67]. One can define a new boson from the old bosons by using mutual flux attachment (based on flux attachment Chern–Simons theory). The idea is simple; each species sees the magnetic field due to another species [67]. As the attached magnetic field (internal gauge field) cancels the external magnetic field (external gauge field), new bosons do not see any field and condense. This leads to the locking of the internal gauge field to the external gauge field. It was shown to demonstrate the IQHE of bosons. Later on, it was also shown that a similar consideration can lead to the FQHE of bosons as well [68, 69].

5.4 Conclusion

To conclude, we have walked through the different aspects of the broad area of TIs from non-interacting to interaction induced exotic topological phases. In the non-interacting case, we saw how the area of TI developed based on the topology of the band structure. Then we moved on to see the effect of interaction on topological phases and asked and answered few questions like whether topological phases are robust against interaction, can interaction lead to topological transition, and can interaction give rise to topological phases which are otherwise impossible to achieve in the non-interacting picture. The final results are, weak interaction does not usually destroy the topological phase; hence, topological phases are robust against weak interaction. Sometimes, interaction can drive the topological phase transition and lead to a phase which is adiabatically connected to a similar topological phase in the non-interacting picture. The most interesting part is that interaction also leads to phases which do not have any non-interacting counterpart and show extremely exotic properties such as fractionalization of statistics and emergence of non-Abelian quasi-particles among others.

In recent years, the area of topological semi-metal has also emerged and created a lot of buzz in the community; e.g., topological Weyl semi-metal which possesses gapless node in the bulk and topologically protected Fermi arc on the surface [70]. It has been shown that interaction is very important in the host materials of Weyl semi-metal [71, 72]; hence, it is an active area of research.

Acknowledgement

I would like to thank Tanmoy Das, H. R. Krishnamurthy, Subhro Bhattacharjee, Aabhaas V. Mallik, Netanel Lindner, and many others for fruitful discussions related to the broad area of topological phases.

References

[1] L. D. Landau, Ukr. J. Phys. 11, 19 (1937).
[2] S. Sachdev, Quantum Phase Transitions (Cambridge university press, 2011).
[3] P. W. Anderson, Phys. Rev. 109, 1492 (1958).
[4] V. L. Berezinski\i, J. Phys. C 6, 1181 (1973).
[5] D. Tong, (2016).
[6] C. Nayak, S. H. Simon, A. Stern, M. Freedman, and S. Das Sarma, Rev. Mod. Phys. 80, 1083 (2008).
[7] A. Y. Kitaev, Ann. Phys. (N. Y). 303, 2 (2003).
[8] K. v. Klitzing, G. Dorda, and M. Pepper, Phys. Rev. Lett. 45, 494 (1980).
[9] D. J. Thouless, M. Kohmoto, M. P. Nightingale, and M. den Nijs, Phys. Rev. Lett. 49, 405 (1982).
[10] B. Simon, Phys. Rev. Lett. 51, 2167 (1983).
[11] D. C. Tsui, H. L. Stormer, and A. C. Gossard, Phys. Rev. Lett. 48, 1559 (1982).
[12] F. D. M. Haldane, Phys. Rev. Lett. 61, 2015 (1988).
[13] C. L. Kane and E. J. Mele, Phys. Rev. Lett. 95, 3 (2005).
[14] L. Fu, C. L. Kane, and E. J. Mele, Phys. Rev. Lett. 98, 106803 (2007).
[15] X. L. Qi and S. C. Zhang, Rev. Mod. Phys. 83, (2011).
[16] M. Z. Hasan and C. L. Kane, Rev. Mod. Phys. 82, 3045 (2010).
[17] A. Bansil, H. Lin, and T. Das, Rev. Mod. Phys. 88, 21004 (2016).
[18] E. H. Hall, Am. J. Math. 2, 287 (1879).
[19] M. V. Berry, Proc. R. Soc. A Math. Phys. Eng. Sci. 392, 45 (1984).
[20] W. P. Su, J. R. Schrieffer, and A. J. Heeger, Phys. Rev. Lett. 42, 1698 (1979).
[21] C. L. Kane and E. J. Mele, Phys. Rev. Lett. 95, 226801 (2005).
[22] C. Wu, B. A. Bernevig, and S.-C. Zhang, Phys. Rev. Lett. 96, 106401 (2006).
[23] J. C. Budich and B. Trauzettel, Phys. Status Solidi - Rapid Res. Lett. 7, 109 (2013).
[24] V. Gurarie, Phys. Rev. B 83, 085426 (2011).
[25] Z. Wang and S.-C. Zhang, Phys. Rev. X 4, 011006 (2014).
[26] Z. Wang, X.-L. Qi, and S.-C. Zhang, Phys. Rev. Lett. 105, 256803 (2010).
[27] Z. Wang and B. Yan, J. Phys. Condens. Matter 25, 155601 (2013).
[28] J. Sirker, M. Maiti, N. P. Konstantinidis, and N. Sedlmayr, J. Stat. Mech. Theory Exp. 2014, P10032 (2014).

[29] L. Barbiero, L. Santos, and N. Goldman, Phys. Rev. B 97, 201115 (2018).
[30] S. R. Manmana, A. M. Essin, R. M. Noack, and V. Gurarie, Phys. Rev. B 86, 205119 (2012).
[31] D. Wang, S. Xu, Y. Wang, and C. Wu, Phys. Rev. B 91, 115118 (2015).
[32] F. Grusdt, M. Höning, and M. Fleischhauer, Phys. Rev. Lett. 110, 260405 (2013).
[33] C. N. Varney, K. Sun, M. Rigol, and V. Galitski, Phys. Rev. B 82, 115125 (2010).
[34] C. Hickey, P. Rath, and A. Paramekanti, Phys. Rev. B 91, 134414 (2015).
[35] J. Maciejko and A. Rüegg, Phys. Rev. B 88, 241101 (2013).
[36] B. A. Bernevig, T. L. Hughes, and S.-C. Zhang, Science (80-.). 314, 1757 (2006).
[37] J. C. Budich, B. Trauzettel, and G. Sangiovanni, Phys. Rev. B 87, 235104 (2013).
[38] S. Miyakoshi and Y. Ohta, Phys. Rev. B 87, 195133 (2013).
[39] Y. Tada, R. Peters, M. Oshikawa, A. Koga, N. Kawakami, and S. Fujimoto, Phys. Rev. B 85, 165138 (2012).
[40] T. Yoshida, R. Peters, S. Fujimoto, and N. Kawakami, Phys. Rev. B 87, 085134 (2013).
[41] T. Yoshida, S. Fujimoto, and N. Kawakami, Phys. Rev. B 85, 125113 (2012).
[42] A. Amaricci, J. C. Budich, M. Capone, B. Trauzettel, and G. Sangiovanni, Phys. Rev. Lett. 114, 185701 (2015).
[43] X. L. Qi, T. L. Hughes, and S. C. Zhang, Phys. Rev. B - Condens. Matter Mater. Phys. 78, 1 (2008).
[44] R. Li, J. Wang, X.-L. Qi, and S.-C. Zhang, Nat. Phys. 6, 284 (2010).
[45] G. K. Gupta, K. Dolui, A. Kumar, D. D. Sarma, and T. Das, (2018).
[46] S. Raghu, X.-L. Qi, C. Honerkamp, and S.-C. Zhang, Phys. Rev. Lett. 100, 156401 (2008).
[47] A. G. Grushin, E. V. Castro, A. Cortijo, F. de Juan, M. A. H. Vozmediano, and B. Valenzuela, Phys. Rev. B 87, 085136 (2013).
[48] S. Capponi and A. M. Läuchli, Phys. Rev. B 92, 085146 (2015).
[49] M. Daghofer and M. Hohenadler, Phys. Rev. B 89, 035103 (2014).
[50] A. Dauphin, M. Müller, and M. A. Martin-Delgado, Phys. Rev. A 86, 053618 (2012).
[51] N. Alidoust, C. Liu, S.-Y. Xu, I. Belopolski, T. Qi, M. Zeng, D. S. Sanchez, H. Zheng, G. Bian, M. Neupane, Y.-T. Liu, S. D. Wilson,

H. Lin, A. Bansil, G. Cao, and M. Z. Hasan, Phys. Rev. B 93, 245132 (2016).
[52] V. Alexandrov, M. Dzero, and P. Coleman, Phys. Rev. Lett. 111, 226403 (2013).
[53] M. Dzero, K. Sun, V. Galitski, and P. Coleman, Phys. Rev. Lett. 104, 106408 (2010).
[54] M. Dzero, J. Xia, V. Galitski, and P. Coleman, Annu. Rev. Condens. Matter Phys. 7, 249 (2016).
[55] J. Kondo, Prog. Theor. Phys. 32, 37 (1964).
[56] J. Werner and F. F. Assaad, Phys. Rev. B 88, 035113 (2013).
[57] T. Das, Phys. Rev. Lett. 109, 1 (2012).
[58] G. K. Gupta and T. Das, Phys. Rev. B 95, 161109 (2017).
[59] L. Balents, Nature 464, 199 (2010).
[60] R. B. Laughlin, Phys. Rev. Lett. 50, 1395 (1983).
[61] J. K. Jain, Phys. Rev. Lett. 63, 199 (1989).
[62] J. K. Jain, Composite Fermions (Cambridge University Press, 2007).
[63] T. Neupert, L. Santos, C. Chamon, and C. Mudry, Phys. Rev. Lett. 106, 236804 (2011).
[64] K. Sun, Z. Gu, H. Katsura, and S. Das Sarma, Phys. Rev. Lett. 106, 236803 (2011).
[65] E. Tang, J.-W. Mei, and X.-G. Wen, Phys. Rev. Lett. 106, 236802 (2011).
[66] F. D. M. Haldane, Phys. Rev. Lett. 51, 605 (1983).
[67] T. Senthil and M. Levin, Phys. Rev. Lett. 110, 46801 (2013).
[68] Y.-C. He, S. Bhattacharjee, R. Moessner, and F. Pollmann, Phys. Rev. Lett. 115, 116803 (2015).
[69] Y. Fuji, Y.-C. He, S. Bhattacharjee, and F. Pollmann, Phys. Rev. B 93, 195143 (2016).
[70] A. A. Burkov and L. Balents, Phys. Rev. Lett. 107, 127205 (2011).
[71] Y. Qi, P. G. Naumov, M. N. Ali, C. R. Rajamathi, W. Schnelle, O. Barkalov, M. Hanfland, S.-C. Wu, C. Shekhar, Y. Sun, V. Süß, M. Schmidt, U. Schwarz, E. Pippel, P. Werner, R. Hillebrand, T. Förster, E. Kampert, S. Parkin, R. J. Cava, C. Felser, B. Yan, and S. A. Medvedev, Nat. Commun. 7, 11038 (2016).
[72] Z. Wang, M. G. Vergniory, S. Kushwaha, M. Hirschberger, E. V. Chulkov, A. Ernst, N. P. Ong, R. J. Cava, and B. A. Bernevig, Phys. Rev. Lett. 117, 236401 (2016).

6

Atomic Dynamics and Structural Phase Transitions in Perovskite and Multiferroic Compounds using Inelastic Neutron Scattering and Ab-Initio Calculation

Sanjay K. Mishra[1], Mayanak K. Gupta[1], Ranjan Mittal[1,2], and Samrath L. Chaplot[1,2]

[1]Solid State Physics Division, Bhabha Atomic Research Centre, India
[2]Homi Bhabha National Institute, Anushaktinagar, India

Abstract

We have used inelastic neutron scattering and density functional theory (DFT) to study atomic dynamics and structural phase transitions in perovskites and multiferroic compounds. The experiments and *ab-initio* phonon calculations are used to understand the role of structural distortions and their correlation to phonon instabilities, leading to phase transitions in these materials. We have also showed the presence of spin phonon coupling in multiferroics and studied its role in phase stability at high pressure.

Keywords: Density Functional Theory, Phase Transitions, Atomic Dynamics, Perovskites and Multiferroic compounds

6.1 Introduction

The properties of a material are known to be governed by spatial arrangement (structure) and dynamics (phonons, diffusion, etc.) of its constituent atoms. Hence, a detailed knowledge of the structure and dynamics of materials is essential. To investigate the atomic-scale structure and dynamics of a material, an "ideal" probe should have a wavelength and energy scale similar to

DOI: 10.1201/9781003337102-6

the interatomic distance and excitations in the material, respectively. Indeed, both of these requirements are simultaneously met by thermal neutrons. The absence of charge on a neutron helps to avoid strong scattering by electrons and allows deep penetration into materials. Unlike X-ray, neutron scattering cross section does not depend on atomic number and also its magnetic moment enables to probe the magnetic structure and magnetic excitations (magnons) in materials. These unique properties of neutrons make it an important scientific and technological resource that provides essential information about the fundamental properties of materials of interest in industry and in our daily lives. More importantly, this information also gives us the pathways to improve on the important properties of new materials at design stage itself.

The perovskite structure compounds have been a subject of intense investigations by different experimental and theoretical techniques [1–3] due to its simple structure and diverse application. Most of the studies focused on the structure, lattice dynamics, dielectric, and piezoelectric response functions and their relation with thermodynamics properties. Virtually, all perovskites exhibit the high-symmetry cubic structure at high temperatures. Phonons have been known to play a key role in the understanding of structural phase transitions. The thermodynamical stability of a phase requires an atomic structure that minimizes the free energy and leads to stable phonons at given thermodynamic conditions. Sometimes, a change in the thermodynamic condition (pressure, temperature, etc.) may cause significant decrease in the phonon frequency such that it may approach zero. This phenomenon is known as phonon instability or condensation and may lead to a phase transition. The structural phase transitions in perovskite-type materials (ABO_3) usually originate from competing interactions between different phonon instabilities occurring in the cubic phase. These transitions are generally classified into two classes: ferrodistortive (FD) transition and antiferrodistortive (AFD) transition [4]. The FD and AFD phase transitions are driven by zone-center (wave-vector $q = 0$) zone-boundary phonon instability ($q \neq 0$), respectively. Known examples of these transitions are the cubic to tetragonal phase transition in $BaTiO_3$ and $PbTiO_3$ for the FD case, and in $SrTiO_3$ and $CaTiO_3$ for the AFD case [1–4].

The evolution of these phase transitions depends on the condensation sequence of certain phonon modes in the cubic Brillouin zone. Some of the important high symmetry points in the cubic Brillouin zones are $\Gamma(0\ 0\ 0)$, $X(\frac{1}{2}0\ 0)$, $M(\frac{1}{2}\frac{1}{2}0)$, and $R(\frac{1}{2}\frac{1}{2}\frac{1}{2})$ points. The zone-boundary $R_{25}(\frac{1}{2}\frac{1}{2}\frac{1}{2})$ phonon mode is three-fold degenerate and the $M_3(\frac{1}{2}\frac{1}{2}0)$ mode is non-degenerate. The triply degenerate R_{25} soft mode is made up of three components corresponding to the rotational degrees of freedom of the BO_6 octahedra around the three separate [001] axes. If one of the components condenses at the transition point, the resulting structure would be tetragonal I4/mcm. On the other hand, the coupled condensation of the three components would lead to a rhombohedral $R3c$ structure. However, when successive phase transitions are associated with both the M_3 and R_{25} soft modes, the sequence of the phase evolution is non-trivial.

Phonons can be probed via various experimental methods, including infrared absorption, and inelastic scattering of light, X-ray, and neutron. To understand the phase transition behavior of materials, some studies of the temperature dependence of the Raman and infrared spectra are reported in literature. However, these measurements are limited to the Brillouin zone center and do not give a comprehensive picture of the dynamics. In this context, the combination of inelastic neutron scattering (INS) and first-principle calculations forms an adequate framework to obtain accurately phonon frequencies. Both the experimental and computational techniques are helpful to understand the role of structural distortions and their correlation to phonon instabilities, leading to phase transitions in materials.

In the present study, we reviewed over results of atomic dynamics and structural phase transitions of several novel compounds, namely $NaNbO_3$ [5], $CaMnO_3$, $PrMnO_3$ [6] $GaFeO_3$ [7], and $YMnO_3$ [8], respectively. $NaNbO_3$ has been a rich model system for understanding of the mechanism of structural phase transitions. While $NaNbO_3$ is non-magnetic in nature, $RMnO_3$ (R = Ca and Pr), $GaFeO_3$, and $YMnO_3$ have been used to understand the spin–phonon coupling. This paper is organized as follows. Sections II and III provide details on the INS investigations and computational technicalities, respectively. The results obtained from our studies on various compounds are discussed in Section IV. This section deals with the temperature dependence of phonon spectra and also highlights the effect of magnetic ordering on the calculated phonon spectra. We have also calculated the total energy and free energy to understand the phase diagram of the compounds.

6.2 Experimental Details

Polycrystalline samples of $NaNbO_3$, $CaMnO_3$, $PrMnO_3$, $GaFeO_3$, and $YMnO_3$ were prepared by a solid state reaction method [5–8]. The phase purity of the samples was confirmed by X-ray powder diffraction recorded using Cu K radiation in the angular range of $20°-130°$ at room temperature.

The INS measurements on $NaNbO_3$, $GaFeO_3$, and $YMnO_3$ samples were carried out using the direct-geometry thermal neutron IN4C spectrometer at the Institut Laue Langevin (ILL), France. The spectrometer is based on the time-of-flight technique and is equipped with a large detector bank covering a wide scattering angle range of about 10°–110°. The polycrystalline samples were prepared by solid state reaction method. About 10 gms of polycrystalline samples have been used for the measurements. The measurements were performed at several temperatures in the range 150–1198 K. The low-temperature measurements were performed using a standard orange cryostat. For the high-temperature range, the sample was put into a quartz tube insert and mounted into a furnace. The other end of the quartz tube was kept open in the air. For these measurements, we have used an incident neutron wavelength of 2.4 Å (14.2 meV), with the neutron energy gain mode. The momentum transfer, Q, extended up to 7 $Å^{-1}$.

INS measurements on $CaMnO_3$ and $PrMnO_3$ were performed using the fine resolution chopper spectrometer (SEQUOIA) [9, 10] at the Spallation Neutron Source (SNS) at Oak Ridge National Laboratory. The data were collected using fixed incident neutron energy of 110 meV with the neutron loss mode, which allowed for the measurement of excitations up to energy transfers of ~100 meV. The low-temperatures scans (4–300 K) were carried out using the closed cycle helium refrigerator. For high-temperature measurements up to 1250 K, the sample was heated in air. The low-temperature ($T < 305$ K) measurements of $CaMnO_3$ were done with high energy resolution, providing resolution of ΔE = 2.0 meV at the elastic line. All other measurements were done in high intensity configuration, with energy resolution of ΔE = 5.5 meV at elastic line. We have used a furnace (called MiCAS) with a quartz tube insert that allows controlling the atmosphere of sample.

In the incoherent one-phonon approximation, the measured scattering function $S(Q,E)$, as observed in the neutron experiments, is related [11, 12] to the phonon density of states $g^{(n)(}E)$ as follows:

$$g^{(n)}(E) = A < \frac{e^{2W_k(Q)}}{Q^2} \frac{E}{n(E,T) + \frac{1}{2} \pm \frac{1}{2}} S(Q,E) > \quad (6.1)$$

$$g^n(E) = B \sum_k \{\frac{4\pi b_k^2}{m_k}\} g_k(E) \quad (6.2)$$

where the + and – signs correspond to energy loss and gain of the neutrons, respectively, and $n(E,\mathrm{T}) = [\exp(E/k_B T) - 1]^{-1}$. A and B are normalization constants and b_k, M_k, and $g_k(E)$ are, respectively, the neutron scattering length, mass, and partial density of states of the kth atom in the unit cell. The quantity within angular brackets represents the average over all Q values. $2W(Q)$ is the Debye–Waller factor. The phonon density of states is corrected from multiphonon and Debye–Waller contributions in a self-consistent way according to the formalism developed by Sjolander [13] in the incoherent approximation[14]. The "MUPHOCOR" program, available as a routine in the ILL Lamp package, allows doing the calculations up to the 5th multiphonon term and was developed by Reichardt [15].

6.3 Computational Details

The calculation of phonon spectra has been performed using the first-principle density functional theory (DFT) methods in both local density approximation (LDA) and generalized gradient approximation (GGA). The GGA was formulated by the Perdew–Burke–Ernzerhof (PBE) density functional [16] and LDA was based on the Ceperly–Alder parametrization by Perdew and Zunger [17]. Relaxed geometries and total energies were obtained using the projector-augmented wave formalism [18, 19] as implemented in the Vienna *ab-initio* simulation package (VASP) [20]. We have used the fully relaxed structures for phonon calculations in all these compounds. All results are well converged [21] with respect to *k*-mesh and energy cutoff for the plane-wave expansion. The total energy calculations have been done using an energy cutoff of 1100 eV. A grid of $8 \times 8 \times 8$ *k*-points was generated using the Monkhorst–Pack (MP) scheme [22]. The convergence criteria for self-consistent field and the ionic relaxation loops were taken as 10^{-8} and 10^{-4} eV Å^{-1}, respectively.

We have also included the magnetic interaction between magnetic ions in the case of $CaMnO_3$, $PrMnO_3$, $GaFeO_3$, and $YMnO_3$. The choice of magnetic structures is based on previously reported configurations. However, in the tetragonal phase of $PrMnO_3$, the magnetic structure is not known. So, we have calculated the enthalpy of the tetragonal phase of $PrMnO_3$ including different antiferromagnetic (AFM) configurations, namely A, C, and G types, and ferromagnetic (FM) configurations. The calculations show that FM structure is favored in comparison to the other magnetic structures. In the cubic phase, G-type AFM and FM structures are adapted for $CaMnO_3$ and $PrMnO_3$, respectively.

The valence electronic configurations of Ga, Fe, and O as used in calculations for pseudo-potential generation are s^2p^1, d^7s^1, and s^2p^4, respectively. Both non-spin-polarized and spin-polarized calculations were performed. The magnetic calculations have been carried out for the A-type AFM ordering in the $Pc2_1n$ phase. Moreover, since $GaFeO_3$ is known to be a Mott insulator, the on-site Hubbard correction is applied using the Dudarev approach [24] using $U_{eff} = 4$ eV [25–29].

For $YMnO_3$, the A-type AFM ordering in both the hexagonal phases was adopted in the magnetic calculations. The on-site Coulomb interaction was accounted for within the Dudarev approach [24] using $U_{eff} = U - J = 7.12$ eV [30, 31]. The labels "FM," "PNM," and "FNM" are abbreviations of fully magnetic, partially non-magnetic, and fully non-magnetic. An energy cutoff of 800 eV and a $6 \times 6 \times 2$ k-points mesh for the Brillouin zone integration are used and found to satisfy the required numerical convergence.

Phonons are extracted from subsequent calculations using the direct method as implemented in the Phonon software [32]. The super cell approach or direct method is used to calculate the phonon frequencies in the entire Brillouin zone implemented in PHONON5.12 software [32]. The super cell of dimension $2 \times 2 \times 2$ has been used to calculate the Hellman–Feynman forces in the orthorhombic and the tetragonal phases. For the calculation in the high-temperature cubic phase, which has a small unit cell containing five atoms, we have chosen a super cell of dimension $4 \times 4 \times 4$ to properly account for the long-range interaction. The forces are extracted from individual displacements of the symmetry inequivalent atoms in the super cell along three Cartesian directions ($\pm x$, $\pm y$, and $\pm z$). In the orthorhombic phases, 24 distinct configurations are generated by displacing the symmetrically inequivalent atoms. However, in tetragonal and cubic phases, 16 and 8 distinct displacements are found, respectively.

The phonon calculations are performed in various magnetic and non-magnetic configurations to see the effect of spin degrees of freedom on phonons. The experimental structure and the measured neutron inelastic scattering spectrum are found to be better described by the calculations when the magnetic degrees of freedom are included, reflecting that the lattice couples to the magnetic structure. These calculations enable us to quantify the strength of the coupling between spins and phonons. Hereafter, the labeling "FRM" and "FRNM" refer to fully relaxed magnetic and fully relaxed non-magnetic calculations. The non-magnetic calculations are done in fully relaxed and partially relaxed geometry. In the fully relaxed non-magnetic calculations, the atomic coordinates and lattice parameter have been relaxed. However, partially relaxed non-magnetic calculation (PNM) refers to only relaxing the atomic coordinates and keeping the lattice parameter fixed as used in magnetic calculations.

6.4 Results and Discussion

6.4.1 Phonon Dynamics and Phase Stability of Sodium Niobate, $NaNbO_3$

$NaNbO_3$ has been a rich model system for understanding of the mechanism of structural phase transitions. This system exhibits one of the most complex sequences of structural phase transitions (Figure 6.1) in the perovskite family [21, 23, 33]. We have employed systematic neutron diffraction measurements as a function of temperature (T = 12−1050 K) and pressure to investigate the temperature/pressure-induced structures and understand the sequence of structural phase transitions of $NaNbO_3$. Above 913 K, it has a paraelectric cubic phase (Pm−3m). On lowering the temperature, it undergoes transition to a series of AFD phases in this order: tetragonal (T2) P4/mbm, orthorhombic (T1) Cmcm, orthorhombic (S) Pbnm, orthorhombic (R) Pbnm, orthorhombic (P) Pbcm phases, and a rhombohedral *R*3*c* phase (see Figure 6.1). We unambiguously provided [23] experimental evidence for the coexistence of the ferroelectric (FE) *R*3*c* phase and the antiferroelectric (AFE) phase (Pbcm) over a wide range of temperatures. High pressure neutron diffraction measurements carried out up to 11 GPa at ambient temperature indicate transition from the Pbcm to the Pbnm phase [33]. These transitions are characterized by the appearance and disappearance of superlattice reflections in the powder diffraction patterns. The superlattice reflections originate from the condensation of zone-center and zone-boundary phonon modes. Thus, the stability of various crystallographic phases could be understood in terms of phonon instabilities.

The focus of our study is to correlate the specific phonon modes relevant to the observed structural distortions of AFE phase. We have also calculated the phonon spectra of $NaNbO_3$ in AFE orthorhombic (Pbcm), FE rhombohedral (*R*3*c*), and cubic (Pm-3m) phases. We have successfully identified the characteristic AFE Raman modes, which are responsible for stabilization of AFE phase. We established correlation between specific symmetry points in the cubic phase and the zone-center irreducible representations of the AFE orthorhombic phase. We also succeeded in assigning the Raman and infrared modes, which are essential for the understanding of the nature of structural phase transitions induced by the change of temperature, pressure, and composition. We have also reported results of INS measurements of phonon spectra in different crystallographic phases of sodium niobate from 300 to 1048 K. The results of INS experiments provide an opportunity to validate

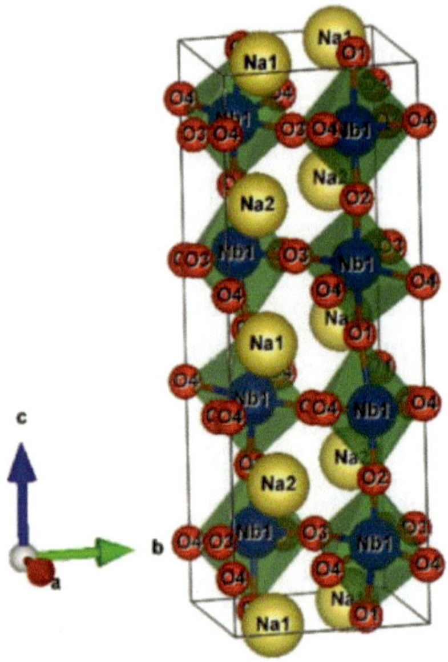

Cubic (Space group: *Pm-3m*)
Multiplicity: 1×1 ×1
↑ T=950 K
Tetragonal (Space group: *P4/mbm*)
Multiplicity: √2×√2 ×1
↑ T=865 K
Orthorhombic (Space group: *Cmcm*)
Multiplicity: 2×2 ×2
↑ T=810 K
Orthorhombic (Space group: *Pbnm*)
Multiplicity: √2 ×√2 ×12
↑ T=755 K
Orthorhombic (Space group: *Pbnm*)
Multiplicity: √2 ×√2 ×6
↑ T= 680 K
Orthorhombic (Space group: *Pbcm*)
Multiplicity: √2 ×√2 ×4
↓ T< 10 K
Rhombohedral (Space group: *R3c*)
Multiplicity: √2 ×√2 × 2√3

Figure 6.1 Structural phase transitions obtained in $NaNbO_3$ as a function of temperature after Mishra *et al.* [23]. The AFE orthorhombic phase (space group: Pbcm).

our calculations as well as important insights into the correlations between vibrational spectra and phase transitions. The details are given below.

Figure 6.2 (a) depicts the evolution of INS spectra for $NaNbO_3$ at T = 303, 783, 838, 898, and 1048 K. The spectra correspond to different crystallographic phases. Five features (broad peaks centered around 19, 37, 51, 70, and 105 meV) can be easily identified. At 303 K, additional well-resolved peaks below 37 meV are also observed. Their intensity decreases significantly with increasing temperature. Remarkably, the peak around 70 meV shifts significantly toward lower energies with increasing temperature, while the others do not change in a noticeable way. At 783 K, a prominent change is observed and the spectra become more diffusive than that at 303 K. The variation of the INS spectra is associated with the occurrence of phase transitions. Prominent changes in the phonon spectra are usually expected across strong first-order phase transitions, while weak first- or second-order phase transitions lead to minimal changes. Upon heating, sodium niobate undergoes a strong first-order phase transition at 680 K. On the other hand, all the other phase transitions above 783 K are of a weak first or second order

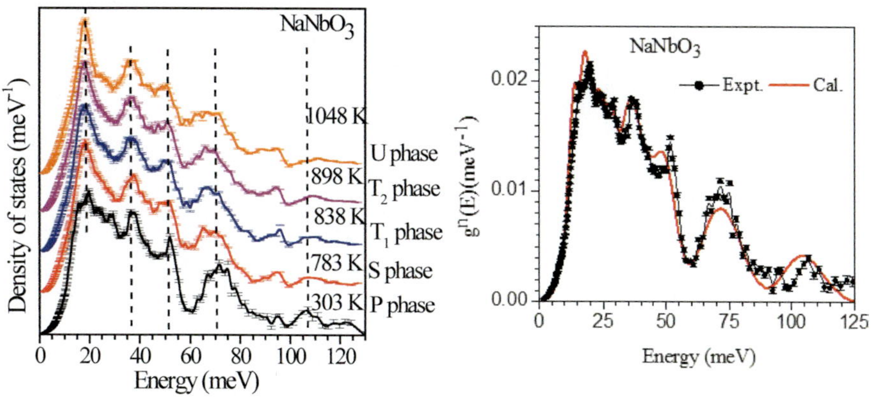

Figure 6.2 The temperature dependence of the phonon spectra of $NaNbO_3$ as observed by neutron inelastic scattering. (b) The experimental (dotted line at 303 K) and calculated (solid line at 0 K) phonon spectra for NaNbO3 in the AFE phase (Pbcm). The calculated spectra have been convoluted with a Gaussian of full width at half maximum (FWHM) of 15% of the energy transfer in order to describe the effect of energy resolution in the experiment [5].

in nature. Thus, a prominent change is only observed in the phonon spectra at 783 K, as compared to that at 303 K. Further heating above 783 K leads to phonon spectra resembling each other.

Figure 6.2 (b) compares the experimental and calculated neutron-weighted phonon density of states of NaNbO3 in the AFE phase. The calculations are found to be in fair agreement with the measurements. All the observed features are computationally well reproduced. Our *ab-initio* lattice dynamics calculations show that both the FE ($R3c$) and the AFE (Pbcm) phases are dynamically stable (all frequencies are real). All the calculations were performed at zero temperature. The calculated free energies of FE (−7.8102 eV/atom) and AFE (−7.8082 eV/atom) phases are comparable. The small energy ($\sim$ 2 meV/atom) difference between the two phases could explain the coexistence of both the phases observed in neutron powder diffraction experiments [23]. Machado *et al.* [34] have also computationally investigated the relative phase stability of sodium niobate and reported a similar behavior.

In order to detect these phonon instabilities using first-principle technique, we have calculated the phonon dispersions from the zone-center (Γ) to the zone-boundary points R and M (Figure 6.3). We found that the polar instability strength at the Γ point is stronger than the AFD instabilities at the R and M points, and it extends over a wider region of the Brillouin zone. As

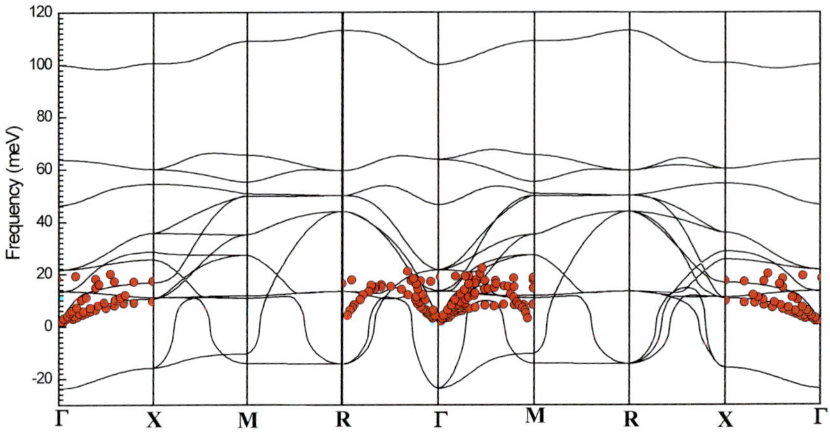

Figure 6.3 Computed phonon dispersion relations for cubic phase (Pm-3m) of $NaNbO_3$ compared to reported experimental INS single crystal data (red circles) [5, 35].

explain above, the *R*- and *M*-point soft phonons involve the antiphase and in-phase tilting of adjacent oxygen octahedra, respectively. Further, the strengths of the *M*- and *R*-point instabilities are quite similar. Interestingly, the branches along the Γ-*R* and Γ-*M* directions show dramatic changes when reaching the *R* and *M* points. When moving away from *M* to *R*, two unstable modes are detected. One of them is rather flat and the other one shows rapid stiffening and becomes stable. Moreover, one of the stable modes becomes unstable at *T* (½, ½, ¼) point.

Above 950 K, $NaNbO_3$ occurs in the cubic phase. On decreasing the temperature, it transforms to a tetragonal phase (P4/mbm). The first structural transformation is from cubic to tetragonal structure, where the unit cell is doubled in the plane perpendicular to the rotational axes of the M3 mode. By further lowering the temperature, condensation of the *R*-point phonon leads to the orthorhombic Cmcm (T2) phase. Unstable phonon branches along the *M*-*R* line contribute to the successive phase transitions. These phonons play an important role in stabilizing the different phases (*P*, *S*, and *R*) in $NaNbO_3$. In a previous study [23], we have proposed that the additional superlattice reflections are due to the condensation of the zone-boundary phonons at *T* (q = ½, ½, g). The orthorhombic structures of the *S*, *R*, and *P* phases result from the condensation of the phonon mode (q = ½, ½, g), with g = 1/12, 1/6, and 1/4. These orthorhombic phases originate from the modulation of the high symmetry cubic phase, associated with the phonon modes at q = (½, ½, g). Further, the freezing of the *R*-point phonon mode and a zone-center phonon stabilizes the low-temperature FE rhombohedral phase. Based on our calculation, we are able to assign the characteristic Raman modes in the AFE phase, which are due to the folding of the T (95 cm^{-1}) and $\triangle$(129 cm^{-1}) points of the cubic Brillouin zone, to the A1g symmetry.

6.4.2 Spin–Phonon Coupling and High Pressure Phase Transitions of $RMnO_3$

The perovskite transition metal oxides show a variety of interesting physical properties, such as dielectric, magnetic, optical, and transport properties. The study of perovskite manganite $RMnO_3$ (R = Ca, La, Pr, etc.) has been of great relevance [36–48]. Among these, CaMnO3 has received much interest in the past decade due to its structural [49], physical [50], magnetic [51], and thermo-electrical properties [52, 53].

$CaMnO_3$ crystallizes in the distorted orthorhombic structure (space group Pnma) and consists of single-valent Mn^{4+} ions and does not exhibit Jahn–Teller distortion. At ambient conditions, it is paramagnetic and at ~130 K, it undergoes a G-type AFM transition [54]. In a high-temperature X-ray diffraction study, Taguchi *et al.* [55] showed that oxygen-deficient $CaMnO_3$ undergoes orthorhombic to tetragonal phase (I4/mcm) transition at 1169 K and finally transforms to cubic phase (Pm-3m) at 1186 K. Doping of trivalent and tetravalent ions at Ca^{2+} site provides additional electrons into e_g orbital of Mn ions and causes a variety of phase transitions, e.g., from collinear G-type AFM insulator to canted G-type AFM metal, collinear C-type AFM insulator, etc. This suggests that we can tune the magnetic phase transition with a small amount of electron doping. For example, solid solution of $(1-x)$ $CaMnO_3$-$x$$PrMnO_3$ ($Ca_{1-x}Pr_xMnO_3$) exhibits a variety of ground states, namely, G-AFM insulator, CE-type charge ordered state, CE-type AFM spins ordering, and canted AFM state, which consists of a mixture of AFM and FM clusters, depending on temperature and composition. The other end member of solid solution, $PrMnO_3$, also crystallizes in the orthorhombic phase, but it exhibits [47] a strong Jahn–Teller distortion of MnO_6 octahedra associated with the ordering of e_g orbital and undergoes A-type AFM spin ordering below 95 K [47]. The difference between the magnetic structures of $CaMnO_3$ and $PrMnO_3$ is attributed to different occupancy of d electrons in Mn ions. Hence, it is intriguing to study the role of magnetic interaction on lattice vibrations.

The structural, physical, and magnetic properties of $RMnO_3$ (R = Ca and Pr) were investigated by a variety of experimental techniques [42, 56–61].

The physical properties of these materials are governed by a delicate interplay of charge, spin, orbital, and lattice degrees of freedom. To understand the relation among these interactions, first-principle calculations have been performed in cubic and orthorhombic phases. Using a DFT approach within the local spin density approximation, Bhattacharjee *et al.* [58] computed structural, dielectric, and dynamical properties of orthorhombic phase of $CaMnO_3$. They computed the zone-center phonon modes and proposed the assignment of experimental data. Zhang *et al.* [59] investigated geometry, ground-state electronic structure, and charge distributions of $CaMnO_3$. The effect of compressive and tensile strains on the magnetic phase transition in electron-doped $CaMnO_3$ was theoretically studied by Tsukahara *et al.* [60]. Further, the first-principle calculations of structural, electronic, and magnetic properties of the $PrMnO_3$ as well as the calculation of zone-center phonon modes in the cubic phase were reported by Bouadjemi [61].

Understanding of various functional properties of derived compounds $Pr_xCa_{1-x}MnO_3$ needs complete study of magnetic and dynamical properties of the parent compounds $CaMnO_3$ and $PrMnO_3$. We have reported results of INS measurements at different temperatures of $CaMnO_3$ and $PrMnO_3$. We have identified coherent magnetic neutron scattering in localized regions in reciprocal-space that arises from long-range correlated magnetic spin waves below magnetic transition temperature (T_N) and short-range stochastic spin−spin fluctuations above the magnetic transition temperature (T_N). We have also extracted the temperature dependence of phonon spectra for $CaMnO_3$ and $PrMnO_3$. The interpretation and analysis of the observed phonon spectra have been performed using *ab-initio* phonon calculations. We have extended the *ab-initio* calculations of the ground-state orthorhombic phase in the entire Brillouin zone. Differences in the nature of interactions between the two compounds lead to noticeable differences in the nature of the phonon spectra. The results of the temperature-dependent neutron experiments in the two compounds and its interpretation in terms of the magnetic and phonon excitations have been discussed.

Recently, Mota *et al.* [43] reported a high pressure study in which they observed changes in the X-ray diffraction patterns of PrMnO3 and disappearance of the Raman spectrum at 45 GPa. The authors [43] suspected that the changes might be related to the structural phase transition from orthorhombic to tetragonal structure. However, the disappearance of Raman peaks could also be due to insulator to metal transition. With a view to overcome this ambiguity, we have performed the first-principle calculations to investigate the effect of pressure on the structural distortions in the orthorhombic phase.

The calculated structure as a function of pressure in $PrMnO_3$ shows that the suppression of Jahn–Teller distortion and insulator to metal transition occurs simultaneously. We find that the tetragonal phase is energetically favored over the orthorhombic phase above 20 GPa but is dynamically unstable. Some additional results on unstable phonon modes in the cubic (Pm-3m) phase that lead to stabilization of the tetragonal phase at high temperature are presented.

$CaMnO_3$ is known to undergo paramagnetic to AFM transition at $T_N \sim 130$ K [54]. Other transitions are observed from orthorhombic to tetragonal phase at 1169 K and then to cubic phase at 1186 K. The neutron inelastic spectra of $CaMnO_3$ were measured from 7 to 1250 K, across the magnetic transition ($\sim$130 K) and the structural phase transitions at high temperatures. The data were collected over a wide range of momentum transfer (Q) from 0.5 to 7 Å^{-1}. The magnetic signal is expected to be more pronounced at low Q, and it decreases as Q increases, following the magnetic form factor. In contrast to this, in the case of phonons, the intensity increases with Q in the low-Q region showing a Q^2-like dependence. Hence, we have integrated the inelastic scattering function $S(Q,E)$ in two ranges, namely total Q range (0.5–7 Å^{-1}) and high Q (4–7 Å^{-1}); the latter one represents mostly the contribution from phonon.

The data collected at 7 K show an intense peak at about 20 meV in the low-Q data (Figure 6.4). This peak is largely contributed from magnetic excitations as it shows strong Q-dependence of intensity (stronger at low Q and weaker at high Q). Moreover, as the temperature increases, the contribution from spin excitations is expected to become weaker. Above the magnetic transition temperature (T_N = 130 K), we also see that the intensity of the peak at 20 meV is decreasing with increase in temperature up to 601 K. This indicates that the contributions from the paramagnetic scattering persist up to very high temperatures of 601 K. At temperatures of 951 K and above, the intensity does not change with increase in temperature, which reflects residual phonon contribution.

Figure 6.4 illustrates the experimental $S(Q,E)$ in the low-temperature range from 7 to 300 K. The data collected at 7 K clearly show the signature of spin-wave excitations, which are probably gaped at Q = 1.45 Å^{-1}, corresponding to magnetic Bragg peak (011) (the gap is not resolved with the current experimental setup, providing energy resolution of $\sim$2 meV at the elastic line). Further, we notice that at higher temperature (T = 110 K), the intensities of these excitations are strongly redistributed to lower energies, and, in particular, the peak position shifts from 19.5 meV at 7 K to 16.5 meV at 110 K. At $T > T_N$ (in paramagnetic state) at small Q values ($Q < 3.5$ Å^{-1}),

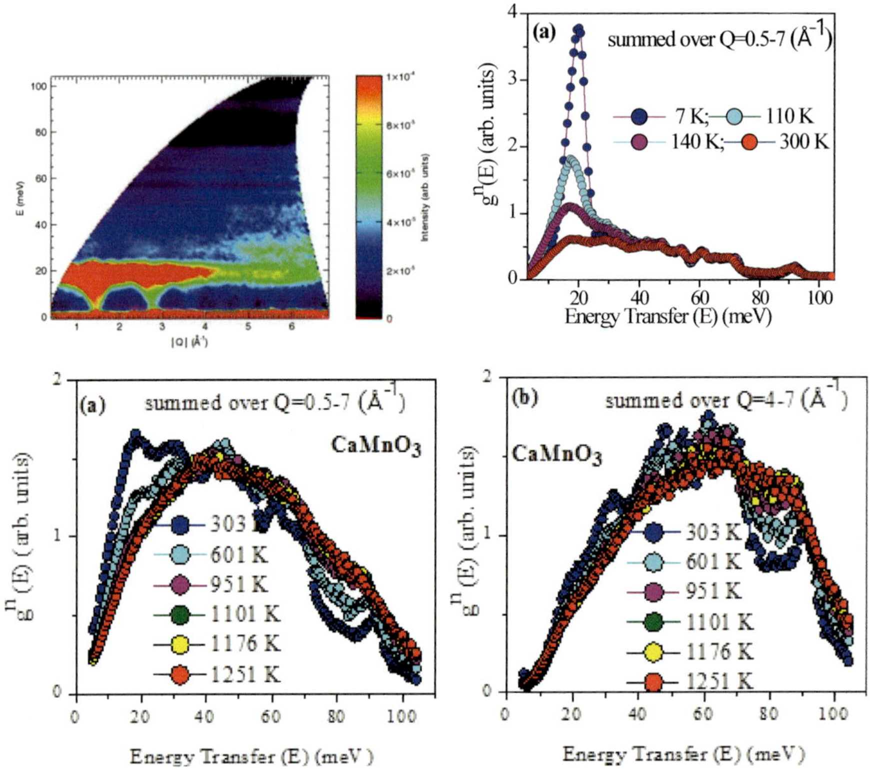

Figure 6.4 The (Q,E) contour plot of $S(Q,E)$ data for $CaMnO_3$ at $T = 7$ K measured at SEQUOIA with incident neutron energy of 110 meV is shown at the top. The neutron inelastic spectra of $CaMnO_3$ at low temperatures, the data were summed over (a) $Q = 0.5\text{–}7$ Å^{-1}. The temperature dependence (above 300 K) of the neutron inelastic spectra of CaMnO3, the data were summed over (c) $Q = 0.5\text{–}7$ Å^{-1} and (d) $Q = 4\text{–}7$ Å^{-1}, respectively. The peak at ~20 meV is due to spin-wave excitations and not due to phonons (1 meV = 8.0585 cm^{-1}) [6].

we observe spin-fluctuation excitations that extend from the elastic line up to ~20 meV. The INS spectrum at $T = 110$ K (just below T_N) shows that clear spin-wave excitations propagate with increasing intensity to higher energies starting at low energies and $Q = 1.47$ Å^{-1}. At $T = 140$ K ($T > T_N$), there is no long-range magnetic order in the sample, and the elastic magnetic peak at $Q = 1.47$ Å^{-1} disappears. The INS spectrum around this Q area shows much broader features. While at higher Q ($Q = 1.2\text{–}1.5$ Å^{-1}), they look like quasi-elastic neutron scattering (QENS) signal, and at smaller Q ($Q = 0.9\text{–}1.2$ Å^{-1}), they look like over-damped dispersed excitations. We notice that the excitations are localized around $Q = 0.8\text{–}1.6$ Å^{-1}; therefore, the scattering should be coherent. In the case of incoherent scattering, the signal will be almost Q-independent. At a higher temperature (300 K), the spectral features in this range look more like typical quasi-elastic signal, but they are still localized in Q-space; so this spectrum relates to coherent neutron scattering from strongly disordered spin–spin fluctuations. It appears that the nature of neutron scattering intensity $S(Q,E)$ changes above T_N from long-range correlated magnetic spin waves to short-range stochastic spin–spin fluctuations, and the intensity being still localized in Q-space, spreads over a broader range of Q. The intensity of these spin fluctuations corrected for the thermal population Bose factor leads to a peak in $g^{(n)}(E)$, which is plotted in Figure 6.4, that decreases with a temperature increase.

The excitation around 17–18 meV at Q = 6 Å^{-1} increases its intensity with temperature (at $T > T_N$). The intensity of the magnetic excitations would strongly decrease; so this excitation should be due to phonons. However, at lower Q ($\sim$4.5 Å^{-1}), the intensity at 17 meV dropped too abruptly, which may be due to strongly dispersed phonon modes. So, in this case, we cannot expect the intensity of this excitation at lower Q range to vary as Q^2.

The high-temperature neutron inelastic spectra show strong modification with increasing temperature. In particular, with an increasing temperature up to 601 K, the intensity of peaks below 40 meV decreases while the peak around 90 meV shows significant shift toward lower energies. The other peaks do not change in a noticeable way. We find that above 601 K, a prominent change is observed in the neutron inelastic spectra, where they are found to be more diffusive as compared to the spectra at low temperatures. These changes in the phonon spectra may be associated with phase transitions in $CaMnO_3$ that transforms from the orthorhombic to the tetragonal phase at 1169 K and finally to cubic phase at 1186 K [55].

6.4.2.1 High pressure phase stability of $CaMnO_3$ and $PrMnO_3$

Both $CaMnO_3$ and $PrMnO_3$ occur in the cubic phase at very high temperatures and undergo a sequence of phase transition with temperature and pressure. Our calculations identified the structural instabilities in the cubic phase and its relationship with phase transitions with temperature [6]. Pressure tunes the interplay between lattice and electronic degree of freedom to a much larger extent than any other parameter like temperature and magnetic field. In the present section, we discuss the effect of pressure on the structural distortions in the orthorhombic phase of $CaMnO_3$ and $PrMnO_3$. The main emphasis of this study is to investigate the pressure dependence of the likely Jahn–Teller distortion and existence of insulator to metal transition at high pressure. Mota *et al.* [43] reported a high pressure study of orthorhombic rare-earth manganites using a combination of synchrotron X-ray diffraction and Raman scattering techniques. The authors observed a change in the diffraction pattern of $PrMnO_3$ and disappearance of the Raman spectrum with pressure. The authors [43] suspected that the change in the diffraction pattern might be related to a structural phase transition from orthorhombic to tetragonal (I4/mcm) structure at 45 GPa. On the other hand, the disappearance of Raman peak could also be due to insulator to metal transition. The first-principle simulation with pressure helped to overcome this ambiguity.

We have calculated the pressure dependence of the equivalent pseudo-cubic lattice parameters for the orthorhombic phase of (a) CaMnO3 and (b) PrMnO3 in FRM structures and using GGA and compared it with experimental data reported in the literature [43, 62–64]. The computed (experimental) bulk modulus for CaMnO3 and PrMnO3 are found to be 185.4 (224 $\pm$ 25 [63, 64]) and 137.4 (139 $\pm$4 [43]) GPa, respectively. We found that compression is more anisotropic for $PrMnO_3$ as compared to $CaMnO_3$. We have computed the pressure dependence of the Jahn–Teller distortion, which is defined in terms of the spatial distribution of Mn-O bond lengths of the distorted MnO_6 octahedra. It is interesting to note that the computed Jahn–Teller distortion is found to be two orders lower for $CaMnO_3$ in comparison to that for $PrMnO_3$ and could be considered as zero. This result is satisfying as $CaMnO_3$ is known [44] to not possess any Jahn–Teller distortion.

The calculated variations of Mn-O distances with pressure are found to be highly anisotropic for $PrMnO_3$ but are isotropic for $CaMnO_3$. On increasing pressure on $PrMnO_3$, the long Mn-O2 distance decreases faster than the shorter Mn-O2 and Mn-O1 and becomes nearly equal at around 45 GPa. This nearly isotropic behavior of the Mn-O distance at high pressure is attributed to disappearance of Jahn–Teller distortion. Basically, the application of external pressure opens the Mn-O-Mn angles and, in turn, shortens the Mn-O bond lengths, leading to less distorted octahedra. The calculated magnetic moment in the equilibrium structure at Mn site is 2.6 μB, which is in agreement with the experimental values of free Mn3+ (3.0). We found that at about 45 GPa, the Mn magnetic moment decreases to 1.0 μB, which could be the signature of the insulator to metal transition.

Theoretically, we can predict the transition pressure in $PrMnO_3$ by comparing the enthalpy as a function of pressure for different phases. Figure 6.5 (a) depicts the computed pressure dependence of the difference in the enthalpy $\triangle H = H_{ortho} - H_{tetra}$ of the orthorhombic and tetragonal phases.

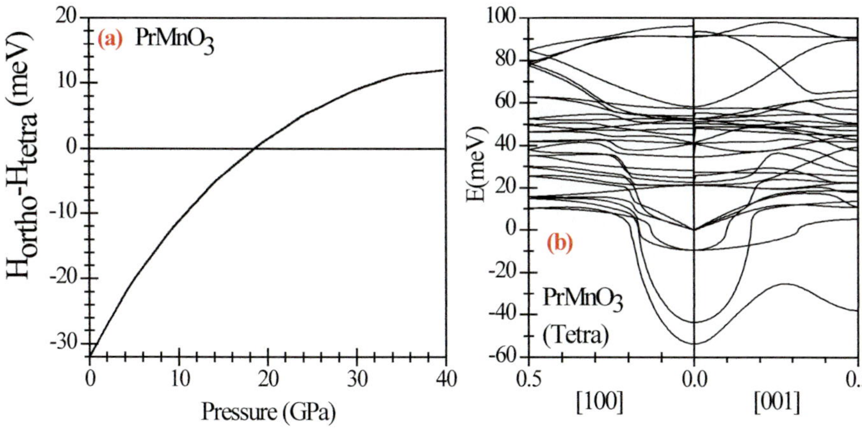

Figure 6.5 (a) Enthalpy difference (ΔH) between the orthorhombic (Pnma) and tetragonal (I4/mcm) phases of $PrMnO_3$ as calculated using *ab-initio* DFT calculation (in FRM structures and using GGA). The computed phonon dispersion relations for $PrMnO_3$ in tetragonal phase at P = 30 GPa are shown in (b) [6].

We find that the orthorhombic phase has the lowest enthalpy at low pressures below 19 GPa, as it is the well-known ground state of $PrMnO_3$ at T = 0 K. However, at pressure above 19 GPa, the tetragonal phase becomes favorable over the orthorhombic phase. The value of the pressure is in agreement with the pressure where shear strain becomes unstable (see Figure 6.7 of [13]). Mota *et al.* [43] reported that the strain analysis does not give any evidence for the suppression of the Jahn–Teller distortion at this pressure. If we recall pressure dependence of the Jahn–Teller distortion, we find that the computed Jahn–Teller distortion decreases rapidly up to this pressure and then it decreases slowly. Based on combined diffraction and Raman data, Mota *et al.* [43] suspected that $PrMnO_3$ may undergo orthorhombic to tetragonal structural phase transition at around 45 GPa. As mentioned above, the calculated pressure dependence of the enthalpy suggests that the tetragonal phase is energetically favorable over the orthorhombic phase above P = 19 GPa. In order to investigate the dynamical stability of tetragonal phase, we computed the phonon dispersion relation (Figure 6.5 (b)) in the entire Brillouin zone at 30 GPa. The presence of unstable phonon modes in the tetragonal phase clearly suggests that although the tetragonal phase is energetically favored over orthorhombic phase, it is dynamically unstable. Therefore, the calculations do not support the possibility of orthorhombic to tetragonal phase transition in $PrMnO_3$ around 45 GPa.

As discussed above, the suppression of the Jahn–Teller distortion occurs at 45 GPa. Further, we find that the magnetic moment at Mn site is also quenched at the same pressure. The suppression of the Jahn–Teller distortion is expected to decrease the volume at high pressures. Thus, change in the volume of the unit cell in the experimental data at around 45 GPa may be associated with disappearance of Jahn–Teller distortion followed by insulator to metal transition. Our band structure calculation also suggests that $PrMO_3$ becomes poor metal above 45 GPa. Similar observation is also reported in the case of $LaMnO_3$ [65], which shows the suppression of the Jahn–Teller distortion and insulator to metal transition by application of pressure. In $LaMnO_3$, the Jahn–Teller distortion and orbital ordering are known to be completely suppressed well below the insulator to metal transition. In contrast to $LaMnO_3$, in the present case, we notice that the suppression of Jahn–Teller distortion and insulator to metal transition occur simultaneously.

6.4.3 Spin–Phonon Coupling and Phase Transitions in Multiferroic Compounds $GaFeO_3$ and $YMnO_3$

Materials showing more than two ferroic properties (magnetism, electricity, and elasticity) simultaneously come under the umbrella of multiferroics whose characteristics include the emergence of simultaneous electric and magnetic orderings, offering, therefore, opportunities for multifunctional device applications. This justifies the intense research going on in this class of materials, and the keen interest they are subject to, at both the fundamental and practical sides [66–80]. Magnetism in transition metals containing materials is induced by the active spin components in the d-shell levels. On the other hand, ferroelectricity occurs generally in the absence of d-electrons. Hence, it is intriguing to observe multiferroicity since this phenomenon involves a simultaneous emergence of both the properties. Over the last few decades, various multiferroic materials have been discovered, which exhibit magnetoelectric (ME) coupling. However, most of the ME multiferroics possess magnetic and FE transition temperatures far from the ambient one. For example, in the case of $BiMnO_3$ [79], the Curie temperature, T_C, is about 100 K, while the Neel temperature, T_N, is close to 750 K. This results in a weak ME coupling at ambient conditions. Practically, the weak coupling materials are not potentially useful. However, there are few mechanisms allowing to tune these properties simultaneously, such as magnetic ferroelectricity induced by frustrated magnetism, lone pair effect, charge-ordering, and local non-centro-symmetry. For instance, the charge ordering driven magnetic ferroelectricity is observed in a large number of rare earth oxides [76, 77]. Understanding the mechanism of multiferroicity is of considerable importance for the design of new multiferroics at controllable conditions (temperature and pressure). Hence, the electric and magnetic properties attributed to the dynamics of ions and electrons need to be studied and explored. The results obtained on multiferroic compounds $GaFeO_3$ and $YMnO_3$ are given below.

6.4.3.1 $GaFeO_3$

$GaFeO_3$ belongs to the class of multiferroic compounds and shows the ME coupling at a low temperature. It does not contain lead or bismuth species, making it ecologically and biologically attractive. At room temperature, the structure [80] is chiral orthorhombic ($Pc2_1n$), while its parental oxides Fe_2O_3 and Ga_2O_3 occur [81] in the rhombohedral and monoclinic phases, respectively. The orthorhombic structure of $GaFeO_3$ has eight formula units per unit cell, with two different symmetry inequivalent sites of iron and gallium atoms: Fe1, Fe2, and Ga1 and Ga2, respectively. The tetrahedral sites are occupied by Ga1, while Ga2, Fe1, and Fe2 occupy octahedral sites. The electric polarization is found [80] to be along the *b*-axis at ambient conditions. Ideally, the magnetic structure of $GaFeO_3$ is expected to reflect an AFM ordering since the magnetic moments of Fe1 and Fe2 cations are antiparallel. However, due to the observed disorder on the Fe and Ga sites [80], a ferri-magnetic transition is observed below 225 K, instead [80, 82]. The magnetization axis was found to be along *c*-axis.

6.4.3.1.1 Temperature dependence of phonon spectra

The phonon spectra of $GaFeO_3$ were measured up to 1198 K, across the magnetic transition (~225 K). The magnetic signal is expected to be more pronounced at low Q, and to vanish at high Q, following the magnetic form factor. Therefore, two Q-domains were considered; i.e., high Q (4–7 $Å^{-1}$) and low Q (1–4 $Å^{-1}$).

The temperature dependence of the Bose factor corrected $S(Q,E)$ plots of $GaFeO_3$ are shown in Figure 6.6. At low temperatures (up to 315 K), the low-Q data show a larger elastic line as compared to the high-Q spectra. Presently, given the lack of detailed magnetic measurements, we speculate that this quasi-elastic scattering may originate from spin fluctuations that disappear at high temperatures. In the high-temperature range, only phonons contribute significantly to the spectra, and, therefore, the width of the elastic line is similar in both the Q ranges.

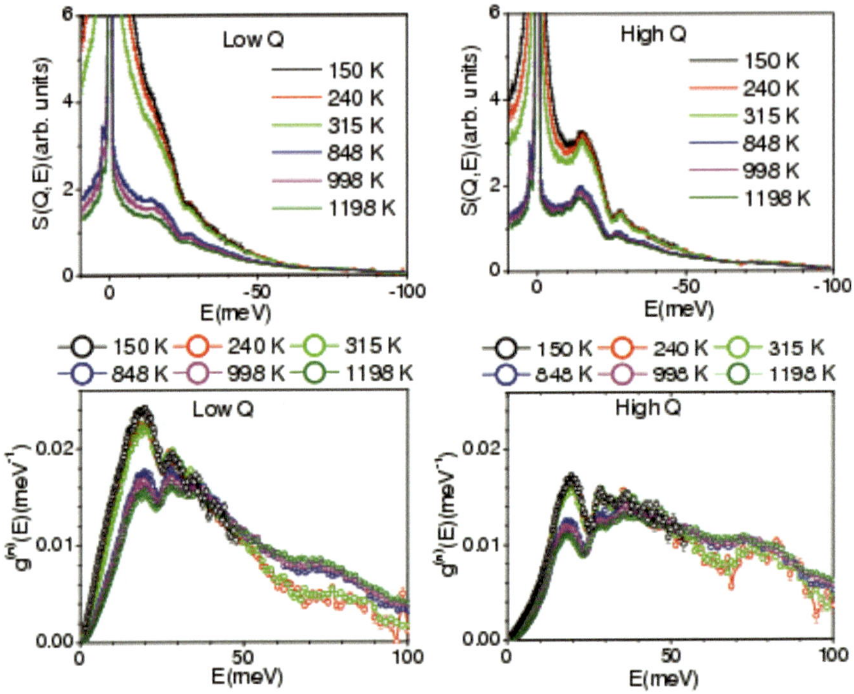

Figure 6.6 Temperature-dependent inelastic neutron spectra of $GaFeO_3$. Top panel: the low-Q and high-Q Bose factor corrected $S(Q,E)$, where both the energy loss (0−10 meV) and the energy gain (−100 to 0 meV) sides are shown. Bottom panel: the low-Q and high-Q, unity-normalized, phonon density of states, $g^{(n)}(E)$, inferred from the neutron energy gain mode $S(Q,E)$ data, within the incoherent approximation [7].

The phonon spectra inferred from the $S(Q,E)$ data, within the incoherent approximation, are also shown in Figure 6.6. The phonon spectra consist of several peaks located around 20, 30, 55, and 80 meV. We find that both the high-Q as well as the low-Q data show large variation in the intensity as a function of temperature. At low energy (below 40 meV), the low-Q data are more intense in comparison to the high-Q data. Further, for the low-Q part, at 150 K below the magnetic transition temperature (~225 K), there is a large intensity of the low energy inelastic spectra (~20 meV) as compared to the data collected at higher temperatures. This is expected to be due to a strong magnetic signal. At 848 K, it is found that in both the low-Q as well as the high-Q data, there is a considerable decrease of the intensity of the low energy peaks around 20 meV. Although $GaFeO_3$ undergoes a paramagnetic to ferri-magnetic transition [80] on cooling around 225 K, the paramagnetic scattering persists in the low energy range around 20 meV, at 240 and 315 K. The intensity in the higher energy range, above 55 meV, of the high-Q data does not show significant temperature dependence, confirming a pure phonon contribution in this spectral regime. Above 848 K, there is a loss of intensity, due to paramagnetic scattering, and only phonons contribute in this range. $GaFeO_3$ does not show any structural phase transition at high temperature. However, polyhedral (GaO_4, GaO_6, and FeO_6) distortions are found to increase upon heating up to 1198 K [83]. This might be an additional reason for the broadening of the phonon spectra above 60 meV at high temperatures, besides the increased thermal amplitudes.

6.4.3.1.2 Magnetic ordering and the calculated phonon spectra in $GaFeO_3$

The microscopic origin of polarization in multiferroic materials is attributed to the hybridization of the electronic orbitals producing a polar charge distribution and ionic displacements from the related centro-symmetric positions. Hence, it is important to study the lattice dynamics in order to understand the ME properties of multiferroics. Detailed electronic structure calculations of $GaFeO_3$ have been reported in the literature [84, 85]. However, phonon studies over the whole Brillouin zone were missing. Calculations of (electronic) structure and dynamics would help to gain newer and deeper insights into the various physical properties and possible phase transitions of this kind of materials.

The calculated Fe magnetic moments in the equilibrium structure in the $Pc2_1n$ phase at Fe1 and Fe2 sites are 4.1 and 4.1 μ_B, respectively, which are in good agreement with the reported experimental values [80] of 3.9 and 4.5 μ_B. Neglecting the spin degrees of freedom in the calculations leads to a collapse of the b-lattice parameter, with a value decreasing from 9.29 to 8.77 Å. However, by considering Fe magnetism, the calculated value of b-lattice parameter is brought to agreement with the observation.

Figure 6.7 compares the experimental and calculated phonon spectra. The FNM calculation results in a shift of all the modes to higher energies. This is due to the fact that the b-axis is underestimated in FNM calculations, leading to an overestimation of the phonon energies. Interestingly, the model calculations FM and PNM provide a very good agreement with the experimental spectra. We notice, however, some differences in the low energy part of the phonon spectra. The difference comes, in fact, from the value of the Fe magnetic moment in the two numerical models. The main effect of the Fe spin degrees of freedom is to soften the calculated phonon energies around

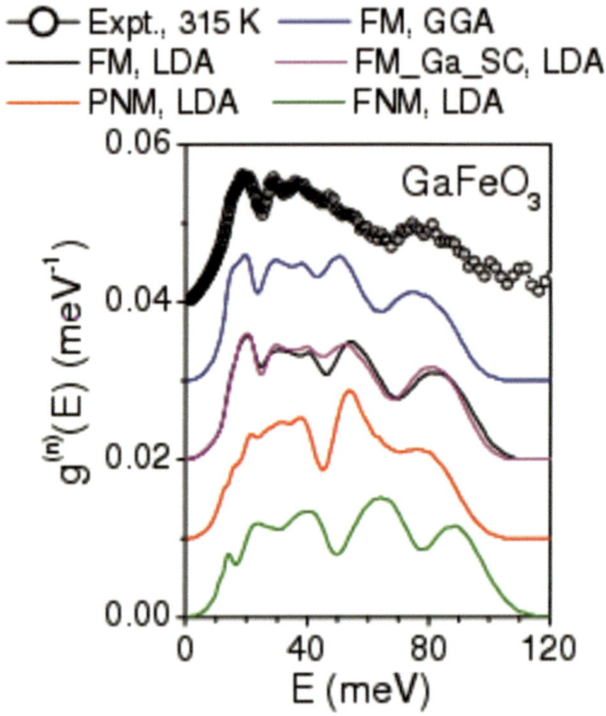

Figure 6.7 The calculated and experimental neutron inelastic scattering spectra of $GaFeO_3$. The experimental data consist of the "High-*Q*" data collected at 315 K. The calculated spectra have been convoluted with a Gaussian of FWHM of 15% of the energy transfer in order to describe the effect of energy resolution in the experiment. All the spectra are normalized to unity in the entire spectral range. FM, FNM, and PNM refer to fully relaxed magnetic, fully relaxed non-magnetic, and partially relaxed non-magnetic calculations, respectively. "FM_Ga_SC" refers to the fully relaxed magnetic calculations including the semicore electrons of Ga atoms. For better visibility, the experimental and calculated phonon spectra are shifted vertically with respect to each other [7].

30 meV, bringing them closer to the experimental values. This demonstrates the role of magnetic interactions in $GaFeO_3$, in a similar way to the recent phonon studies in other systems [86, 87].

The FM-based calculated phonon spectra (Figure 6.3) lead to peaks centered around 20, 30, 55, and 80 meV. The experimental spectra show peaks at 20 and 30 meV and clear humps at 55 and 80 meV. $GaFeO_3$ is known to have a Ga-Fe disorder as determined from diffraction measurements [80]. However, our phonon calculations were done in the ordered phase. The structural disorder could lead to a large variation of the Ga/Fe-O bonds and would then result in a broadening of the peaks, as experimentally observed.

The difference in the phonon spectra (Figure 6.7) from the various calculations can be understood from the estimated atomistic contributions in terms of the partial density of states from LDA calculations. The difference is primarily due to the nature of the chemical bonding, in the magnetic and non-magnetic configurations, as well as the related volume effect. We find that the vibrations of Fe and Ga atoms extend up to 45 meV, while the dynamics of the oxygen atoms spread over the entire spectral range, up to 100 meV. The vibrational spectra due to the two Ga symmetry inequivalent atomic sites remain nearly invariant in all the three calculation types, while the Fe vibrations show a considerable change. The intensity of vibrational density of states of the Fe2 atoms is enhanced around 20 meV. The vibrations of Fe1 as calculated around 30 meV in the non-magnetic calculations are found to soften magnetically and exhibit a peak around 20 meV. The FNM calculations predict the oxygen vibrations to extend up to about 100 meV. The overestimation in the range of vibrations is primarily due to the non-inclusion of the Fe magnetic moment, which results in a contraction of the unit cell. The FM and PNM model calculations show that the vibrations of all the oxygen atoms soften in the energy range 60–100 meV. A further interesting finding consists of the vibrations of the O5 atoms, as extracted from the "FM" calculation type. The O5 atoms are connected only to the Fe1 and Fe2 atoms. The O5 vibrations around 60 meV in non-magnetic calculation are related to the Fe magnetism. This dynamic is found to shift to lower energies at about 30 meV in the FM calculations.

Given the known effect of the density functional approximation (LDA or GGA) on the volume description (LDA tends to underestimate the volume value and GGA shows the opposite trend), we compare the FM calculated phonon spectra from LDA and GGA approaches. The unit cell volume from LDA and GGA calculations is estimated to be 405.3 and 434.0 $Å^3$, respectively. The experimental value is 413.9 $Å^3$ [80]. The low energy part of the phonon spectra, which is sensitive to Fe magnetism, is nearly the same in both LDA and GGA. Above 50 meV, some variations are, however, noticed. The GGA calculated phonons above 50 meV are found to be slightly at lower energies as compared to LDA calculated phonons. Both the exchange-correlation methods lead to an overall good matching with the observations.

Under the orthorhombic ($Pc2_1n$) symmetry, $GaFeO_3$ possesses 120 zone-center modes corresponding to the irreducible representations: $\Gamma = 30A_1 + 30A_2 + 30B_1 + 30B_2$. The LDA and GGA approximations lead basically to the very similar phonon energies. Several modes are found to significantly differ when comparing the magnetic and non-magnetic cases. This confirms a spin–phonon coupling behavior. The change in energies of the modes below 25 meV is mainly due to the magnetic interactions, while the high energy phonons are influenced by the volume effect.

Further, fully relaxed magnetic calculations (FM_Ga_SC) including the semicore electrons of the Ga atoms ($d^{10}s^2p^1$) are performed. The detailed comparison of the "FM" and "FM_Ga_SC" calculated phonon spectra (Figure 6.7) shows that the low energy phonons below 40 meV are not affected by the inclusion of the semicore d-shell electrons of the Ga atoms. The only noticeable difference is detected in the high energy modes that soften by about 1 meV. Both the calculations are in good agreement with the experimental data. The calculated partial densities of states indicated that the changes in the high-energy range are associated with the change in the partial contribution of O atoms.

6.4.3.1.3 High pressure phase stability of $GaFeO_3$

The high pressure measurements [88], up to 70 GPa (increasing and decreasing cycles), revealed a very rich phase diagram of $GaFeO_3$. Arielly and coworkers [91] reported the emergence of a new orthorhombic phase (space group Pbnm) above 25 GPa upon increasing pressure. The transition was found to fully establish at 45 GPa. In this phase, all the Ga atoms have eight coordinations. However, in the $Pc2_1n$ phase, two different Ga sites are distinguishable: one with a six-fold symmetry and the other possessing a four-fold coordination. Increasing further the pressure to about 53 GPa results in another first-order transition with a significant drop of the volume. However, the system remains in the same orthorhombic space group (Pbnm). At this pressure value (53 GPa), the magnetic interactions weaken due to the broadening of the iron d-bands. Mossbauer measurement reveals that the Neel temperature is close to 5 K at about 77 GPa. Further decreasing the pressure to the ambient value, the hexagonal *R*3*c* phase was found to be the stable one, which is different from the originally starting orthorhombic $Pc2_1n$ phase, at ambient conditions.

In literature [88], only the lattice parameters of $GaFeO_3$ are available in the Pbnm and *R*3*c* phases. The related atomic coordinates are missing. We have, therefore, started from the atomic coordinates of $LuFeO_3$ and $LiNbO_3$, as provided in [89] and [90], respectively. Mossbauer spectroscopy reveals the existence of magnetic ordering in $GaFeO_3$ [88] even at high pressures. The crystal structure of $GaFeO_3$ in Pbnm and *R*3*c* phases has been calculated by relaxing the atomic coordinates as well as lattice parameters. The total energy has been calculated in both the phases in various AFM configurations represented by the A, C, and G ordering types. Computationally, we found that the Pbnm phase is the most stable when adopting the G-type AFM ordering, while the *R*3*c* phase stabilizes with the A-type antiferromagnetism. The calculated structural details under the Pbnm and *R*3*c* phases at 25 GPa and ambient pressure, respectively, have been reported by us [7], which are in good agreement with the available experimental lattice parameters.

The high pressure equation of state, total energy (Φ), and enthalpy ($H = \Phi + PV$) of various phases of $GaFeO_3$ were estimated for the fully relaxed magnetic (FM) configuration. The GGA calculated enthalpy showed that the high pressure Pbnm phase is more stable than the $Pc2_1n$ phase at ambient pressure. Figure 6.8 (a) presents the enthalpy difference from LDA calculations, for the $Pc2_1n$ and *R*3*c* phases with respect to the Pbnm phase. Above 23 GPa, the Pbnm phase is found to be stable when comparing to $Pc2_1n$. The application of pressure leads to a change in the correlation between the electronic motions and affects the magnetic interaction. A quenching of the Fe magnetic moment in the Pbnm phase is found at 47 GPa, which triggers a sudden drop of the volume and increases the total energy (Figure 6.8 (b)). This is in agreement with the high pressure data [88] that shows a similar behavior around 53 GPa. The values of the magnetic moments on the Fe atoms remain at about 4.1 μ_B from ambient pressure to below 47 GPa and then decrease to 1.0 μ_B at this transition.

Further, the total energy (Φ) and enthalpy ($H = \Phi + PV$) of various phases of $GaFeO_3$ are estimated from LDA for the fully relaxed magnetic (FM_Ga_SC) configuration including the semicore electrons of the Ga atoms. Here again, we found that computationally the Pbnm and *R*3*c* phases are most stable when adopting the G-type AFM ordering and A-type antiferromagnetism, respectively. The FM_Ga_SC calculated enthalpy difference in the various phases shows that the Pbnm phase is stable above 30 GPa (Figure 6.8 (c)) when comparing to $Pc2_1n$. The pressure increase leads to a quenching of the Fe magnetic moment in the Pbnm phase at 36 GPa, resulting in an increase of the total energy (Figure 6.8 (d)) and a sudden drop of the volume. We also find that the Fe magnetic moment is quenched in the *R*3*c* phase, when the pressure is raised to 45 GPa. The FM_Ga_SC calculated relative change of the unit cell volume as a function of pressure (Figure 6.9) in various phases is found to be in a qualitative agreement with the experimental data [88].

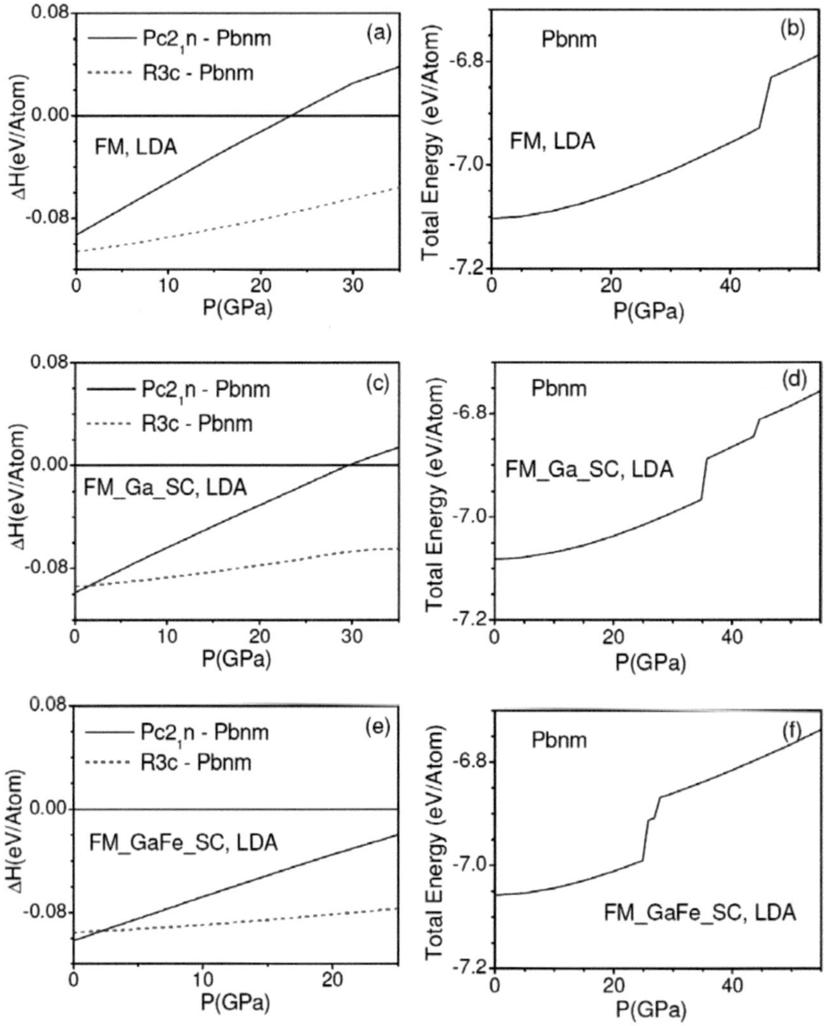

Figure 6.8 (a, c, e) The calculated enthalpy ($H = \Phi + PV$) difference in the $Pc2_1n$ and $R3c$ phases with respect to the Pbnm phase of $GaFeO_3$ as a function of pressure within the local density approximation (LDA). (b, d, f) The calculated total energy (Φ) in the Pbnm phase of $GaFeO_3$ as a function of pressure within the LDA. The explanation of the labelings FM, FM_Ga_SC, and FM_GaFe_SC are described in text [7].

The environment of the Fe in $GaFeO_3$ is strongly asymmetric; therefore, polarization of the low-lying semicore states could influence the total energy. In this context, we have also calculated total energy (Φ) and enthalpy

(H = Φ + PV) within the LDA framework for the fully relaxed magnetic configuration (FM_GaFe_SC) including the semicore electrons of both the Ga and Fe atoms. As in the above two types of calculations, the G-type and A-type AFM ordering is found to be stable for Pbnm and $R3c$ phases, respectively. The comparison of the enthalpy of the $Pc2_1n$ and Pbnm phases shows (Figure 6.8 (e)) that the former phase is stable up to 26 GPa. Further increase in pressure leads to stability of $GaFeO_3$ in the Pbnm phase. This is due to quenching of the Fe magnetic moment, which leads to an increase of the total energy and a sudden drop of the volume. In the $R3c$ phase, the Fe magnetic moment is also found to quench at 28 GPa.

The equation of state is found to be qualitatively different as obtained from calculations performed in FM, FM_Ga_SC, and FM_GaFe_SC configurations. We find that the comparison of enthalpy in the $Pc2_1n$ and Pbnm phases, in the FM and FM_Ga_SC calculations, reveals a stability of $GaFeO_3$ in the Pbnm phase above 36 and 26 GPa, respectively, while no such phase transition is found in the FM_GaFe_SC calculations. As shown in Figure 6.9 , quenching of the magnetic moment at high pressure in the Pbnm structure induces a number of phase transitions in the different calculations. However, in the $R3c$ phase, no quenching of Fe magnetic moment was found in FM calculations. We notice that the Fe moment was found to quench (Figure 6.9) in both FM_Ga_SC and FM_GaFe_SC calculations, leading to a phase transition.

The FM calculated enthalpy value under the $Pc2_1n$ phase is -7.196 eV/atom, while in the $R3c$ phase, this is estimated to be -7.209 eV/atom, indicating that the $R3c$ phase is more stable as compared to $Pc2_1n$. The calculated energy difference between the two phases is rather small ($\sim$13 meV/atom). On the other hand, the FM_Ga_SC and FM_GaFe_SC LDA calculated enthalpy values indicate that, at ambient pressure, the $Pc2_1n$ phase is energetically favorable by $\sim$4 and $\sim$6 meV/atom, respectively, in comparison to the $R3c$ phase. However, the $Pc2_1n$ phase is found to be stable only below 2 GPa. As discussed and highlighted above, the inclusion of the semicore electrons in the atomistic pseudo-potentials has only a minor influence on the obtained dynamical properties, while these seem to induce some qualitative changes in the equation of state.

The calculated phase diagram is qualitatively in a good agreement with the observation. It should be mentioned that it is difficult to identify experimentally the high pressure equilibrium phases due to the large hysteresis. Figure 6.9 shows the comparison between the LDA-calculated and experimental relative change of the unit cell volume in various phases of $GaFeO_3$

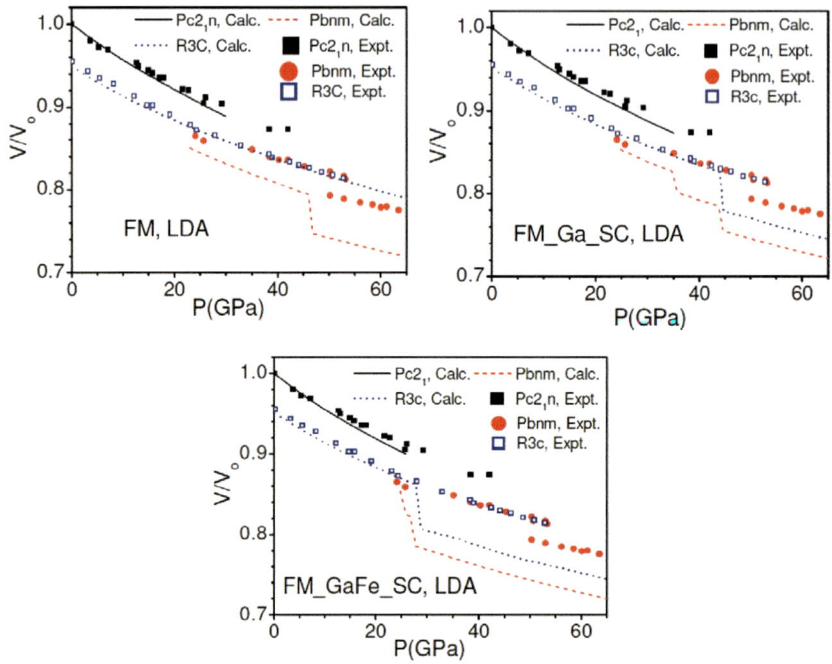

Figure 6.9 The LDA-calculated equation of state of various phases of $GaFeO_3$ and a comparison with available experimental data [91]. *V* refers to the volume per formula unit at pressure *P*. V_o refers to the volume per formula unit of $Pc2_1n$ phase at ambient pressure. The explanation of the labeling FM, FM_Ga_SC, and FM_GaFe_SC is given in the main text [7].

as a function of pressure. A very good agreement is noticed between our calculations and the measurements [88] in the $Pc2_1n$ and *R*3*c* structures; however, the volume in the Pbnm phase is underestimated. Table 6.3 gathers the LDA and GGA calculated elastic constants. The estimated bulk modulus values from LDA and GGA calculations, in the $Pc2_1n$ phase, are 207 and 178 GPa, respectively. The LDA determined value is found to be in a better agreement with the experimental bulk modulus value (226 GPa) [88]. As expected, the GGA underestimates the elastic constants by about 15% with respect to LDA, given that GGA tends to overestimate the calculated unit cell volume. This results in lowering the calculated bulk modulus values.

6.4.3.2 Yttrium manganese oxide, $YMnO_3$

Yttrium manganese oxide ($YMnO_3$) keeps attracting a keen interest as it is known to exhibit ferroelectricity and antiferromagnetism simultaneously

[92]. At ambient conditions, the compound has a hexagonal $P6_3cm$ structure. Above 1258 ± 14 K, an FE to paraelectric phase transition occurs, and the system crystallizes [93] under the hexagonal $P6_3/mmc$ space group (Figure 6.10). The exact nature of this FE transition is still under debate [73, 94–102].

Although ferroelectricity in $YMnO_3$ is believed to be due to the tripling of the unit cell, there is, however, an ambiguity about a second transition observed at ~920 K in this material. It was suggested [103] that this transition can be considered as a hidden order in which a residual symmetry, displayed by [103] the trimerization order parameter, is spontaneously broken. In this context, a generic $P6_3cm \leftrightarrow P3c1 \leftrightarrow P\text{-}3c1$ phase diagram was proposed, corresponding to the observation in another isostructural compound ($InMnO_3$).

Most of the dynamical probes of manganites are based on Raman and infrared techniques [104, 105]. These are limited to the zone-center modes. Available INS measurements on single crystals are limited to low energies, below 20 meV [106, 107]. However, a better understanding of thermodynamical behavior of a material requires a complete description of phonon spectra in the entire Brillouin zone (zone center and zone boundary). INS can be used to probe phonons in the entire Brillouin zone. In this context, we have presented the temperature-dependent INS measurements of phonon spectra in $YMnO_3$, over the temperature range 50−1303 K, i.e., in the low-temperature ($P6_3cm$) as well as in the high-temperature ($P6_3/mmc$) phase of $YMnO_3$. Our measurements are accompanied by first-principle density-functional-based magnetic lattice dynamic calculations for the sake of analysis and interpretation of the neutron data. Phonon dispersion relations in the entire Brillouin zone have been calculated in both the high- and low-temperature hexagonal phases of $YMnO_3$. Unstable phonon modes in the high-temperature phase are related to the stabilization of the low-temperature phase.

6.4.3.2.1 Temperature dependence of phonon spectra of $YMnO_3$

The INS measurements were performed across the magnetic and structural transition temperatures. In addition to the temperature dependence of the INS spectra, we present the data in terms of the Q-dependence as well in order to map out both magnetic and lattice (phonon) excitations. The magnetic signal is expected to increase as Q decreases, following the magnetic form factor [108]. However, in the case of phonons, the form factor exhibits a Q^2-like dependence (multiplied by the Debye−Waller factor $\exp(-2W(Q))$; hence, the intensity of the phonon signal increases with Q in the range of

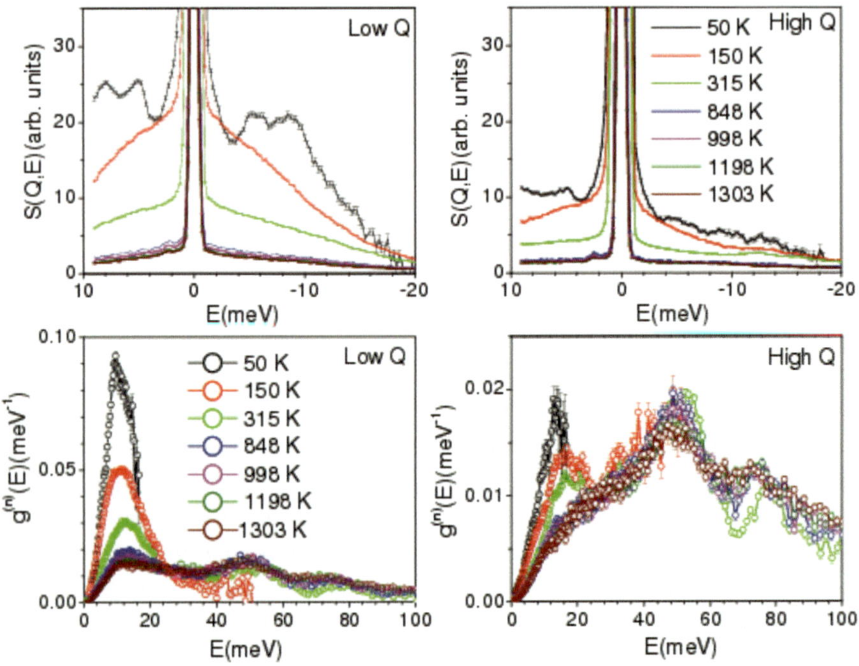

Figure 6.10 Temperature-dependent neutron inelastic spectra of $YMnO_3$. Top panel: the low-Q and high-Q Bose factor corrected $S(Q,E)$, where both the energy loss (0–10 meV) and the energy gain (−20 to 0 meV) sides are shown. Bottom panel: the low-Q and high-Q, unity-normalized, neutron inelastic spectra, $g^{(n)}(E)$, inferred from the neutron energy gain mode $S(Q,E)$ data, within the incoherent approximation [8].

our measurements up to 7 $Å^{-1}$. So, also, the phonon signal increases with temperature due to increased population of the phonons. On the other hand, the magnetic signal decreases with the increase of temperature. Hereafter, the Q-dependence is limited to compare INS spectra in the low-Q (1–4 $Å^{-1}$) and high-Q (4–7 $Å^{-1}$) regions.

The measured temperature dependence of the dynamical scattering function $S(Q,\omega)$ within the low-Q and high-Q regions is shown in Figure 6.10. At 50 K, the observed peak around 10 meV is of magnetic nature, given its Q-dependence (stronger at low Q and absent at high Q). Above 70 K, the compound undergoes [92, 109] a paramagnetic transition. Our data at 150 and 315 K show an appreciable broadening of the elastic peak that may result from the paramagnetic scattering. The increase in the temperature would lead to a decay of the paramagnetic fluctuations, which reduces the width of the elastic peak. Above 848 K, the width of the elastic line stays close to the instrumental resolution, indicating that the paramagnetic scattering vanishes at high temperatures. The neutron inelastic spectra are also depicted in Figur 6.10, for the low-Q and high-Q regions. For the low-Q data, the low-energy excitations below 20 meV show a pronounced change as a function of the temperature. However, such a trend is not observed in the high-Q data, confirming a more phonon-like character in this spectral range. Both the magnetic and phonon excitations are present in the low-Q part, whereas the high-Q data contain mainly phonon contributions. A prominent peak is observed in the low-Q data of the neutron inelastic spectra at 50 K and decaying strongly with temperature. The INS [106, 107] carried out on a single crystal of $YMnO_3$ indicates that the dispersion of spin waves lies within the energy range 5−15 meV. This is in agreement with our measurements, where we also find a large change in the intensity in the low-Q data, across the Neel temperature ($T_N \sim 70$ K). The stretching modes around 80 meV, at 315 K, are found to soften as the temperature increases to 848 K. This might primarily be due to the increase of the Mn-O bond lengths leading to such a softening. The crystal structure and the phonon spectrum of the compound do not show large changes even above the Neel temperature of 70 K, when the AFM ordering is destroyed, which suggests that some kind of local magnetic order persists above 70 K. Our earlier work on $GaFeO_3$ shows [7] that magnetic signal is present well above the magnetic ordering temperature of 225 K. We, therefore, believe that in $YMnO_3$ also, above 70 K in the paramagnetic phase local magnetic order as well as paramagnetic fluctuations may be present, which give rise to the magnetic signal in the low-Q region at 150 and 315 K.

The structural phase transition ($P6_3cm$ to $P6_3/mmc$) in $YMnO_3$, occurring at 1258 ± 14 K, involves a structural distortion and an FE to paraelectric transition. It is well known that the phase transition from the FE to paraelectric phase is driven by a softening of a zone-boundary phonon mode at the k-point. We did not observe any significant change across the transition (up to 1303 K). It should be noted that the neutron inelastic spectra measurements were performed on a polycrystalline sample and are averaged over the whole Brillouin zone. Therefore, it would be crucial to detect small changes in specific dispersion branches, which may only be measured efficiently using a single crystal.

6.4.3.2.2 Effect of the magnetic ordering on the calculated phonon spectra

The purpose of the three model calculations (FM, FNM, and PNM) is to study the effect of the cell volume and magnetic interactions on the phonon spectra.

The calculated magnetic moment per Mn atom is 3.99 and 3.95 μ_B, from LDA and GGA, respectively. However neutron diffraction measurements [110, 111] reported that, below T_N, the Mn moment has a value of ~3 μ_B. $YMnO_3$ is a non-collinear magnetically frustrated two-dimensional system [92]. The Mn-Mn distance in the *a-b* plane and along *c*-axis is 3.5 and 6.05 Å, respectively. Therefore, the magnetic interaction in the *a-b* plane is much stronger than that along *c*-axis, which indicates low dimensionality of the magnetic structure in $YMnO_3$. The nearest neighbor AFM exchange interaction and hexagonal magnetic lattice in the *a-b* plane results in a magnetically frustrated [112] spin configuration. The spin frustration parameter (θ_{CW}/T_N, where θ_{CW} and T_N are Curie–Weiss (CW) and Neel temperature, respectively) [95] for $YMnO_3$ is close to 10. The large value of the spin frustration parameter for $YMnO_3$ [113] can be partly due to the low dimensionality of the Mn networks in addition to the geometrical frustration effects. The deviation from the calculated value might be due to magnetic fluctuations associated with frustration and/or low dimensionality.

The calculated phonon spectra are compared with the measured ones (at 315 K) in Figure 6.11. The FM-type calculation using GGA reproduces the low energy features of the measured phonon spectra. The peak in the experimental spectra at about 52 meV is estimated around 45 meV. Further, the Mn-O stretching modes are underestimated around 67 meV, while, experimentally, these are observed around 80 meV. The GGA determined structural parameters clearly show that both the "a" and "c" lattice parameters are overestimated by ~2%. This results in the overestimation of the various Mn-O bond lengths, which leads to the underestimation of the energies of the Mn-O stretching modes. Results of the FM-LDA calculations are found to be close to the experimental data. This is in good agreement with the correct estimation of the structural parameters as well as bond lengths using this model calculation. Consequently, in the following, we adopt the LDA density functional.

The calculated phonon spectra (Figure 6.11, using the three model calculations, PNM, FM, and FNM, are closely similar up to ~25 meV. A significant change is observed at higher energies. To gain deeper insights into the phonon spectra, we extracted the atomistic partial contributions to the calculated phonon density of states. The Y and Mn atoms are found to contribute mainly below 50 meV, while the O atoms are dynamically active within the range 40–90 meV. In all the three model calculations, the contributions from Y and Mn atoms remain nearly unperturbed in the entire energy range. On the other hand, the differences in the phonon spectra are

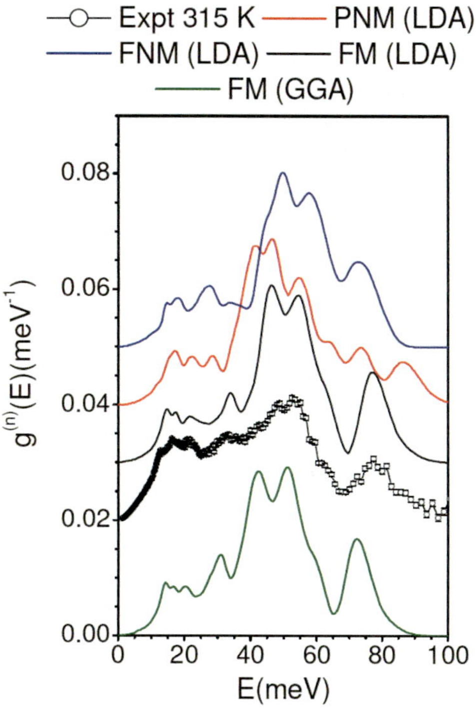

Figure 6.11 The calculated and experimental neutron inelastic spectra of $YMnO_3$. The experimental data were collected at 315 K and averaged over the high-*Q* region. The calculated phonon spectra have been convoluted with a Gaussian of FWHM of 10% of the energy transfer in order to describe the effect of energy resolution in the experiment. For better visibility, the experimental and calculated phonon spectra are shifted vertically with respect to each other. Multiphonon contribution as calculated using the Sjølander formalism has been subtracted for comparison with the calculations [8].

found to originate from the dynamics of the oxygen atoms. Such differences from the various calculations are primarily due to the nature of the chemical bonding, in the magnetic and non-magnetic configurations, as well as a volume effect in this case as well. By comparing the FNM and FM phonon calculations, we highlighted an effect of magnetism and cell volume on the phonon spectra.

By comparing the FM and PNM calculated phonon spectra, we can identify the specific modes sensitive to magnetic interactions. In both FM and PNM calculations, the calculated lattice parameters are similar. However, there are slight changes in the related atomic positions. We find that the calculated Mn-O1 and Mn-O2 bond lengths show a large difference in both the calculations. The FM calculated bond lengths are closer to the experimentally refined ones. The calculated values of Mn-O2 are 1.879 and 1.806 Å in the FM and the PNM calculations, respectively. The significant shortening of the Mn-O2 bond length as compared to the refined value of 1.891 Å [93] results in a hardening of the modes (Figure 6.11) around 80 meV in the PNM calculations. The overestimation of Mn-O1 bond lengths in the PNM calculations leads to a softening of the modes involving O atoms, in comparison with the FM calculations. Further, we find that the PNM-calculated low-energy modes around 40 meV for the planar O3 and O4 atoms soften significantly in comparison (Figure 6.11) to the "FM" calculations. The difference in the calculated phonon spectra in both FM and PNM is due to the fact that in the PNM calculations, the Mn^{3+} magnetic moment is zero, while the FM calculated magnetic moment of Mn^{3+} is 3.99 μ_B. Therefore, the super exchange interactions between Mn cations, mediated by O atoms, influence the partial phonon spectra involving oxygen. It is worth to notice that the Mn phonon modes seem to be insensitive to the magnetic moment exhibited on the Mn site.

The FNM calculated structure shows that the "c" lattice parameter has a value of 12.01 Å, while the experimentally refined value is 11.40 Å [93]. The "a" lattice parameter is found to be underestimated (5.838 Å) in comparison to the experimental value of 6.1415 Å. All the four FNM calculated Mn-O bond lengths are more isotropic. This leads to an underestimation of the energies of the Mn-O stretching modes (Figure 6.12) around 80 meV in the FNM calculations in comparison to the FM calculations. Figure 6.12 shows the change in the energy of the estimated zone-center modes in the PNM and FNM configuration, with respect to the FM calculations. There is a noticeable deviation when comparing energies from magnetic and non-magnetic calculations. The maximum shift in phonon energies is observed for high energy Mn-O stretching modes. This supports a spin–phonon coupling behavior, that is, the coupling of the magnetic structure with phonons. The change in the energies of the modes around 30 meV is mainly due to magnetic interactions, while the high energy stretching modes are most influenced by the estimated Mn-O bond lengths.

6.4.3.2.3 Phonon spectra in the room temperature ($P6_3cm$) and high-temperature ($P6_3/mmc$) hexagonal Phases

The calculated phonon spectra of $YMnO_3$ were subject of previous works [30, 102, 114]. The estimated phonon dispersions have been reported [114] in

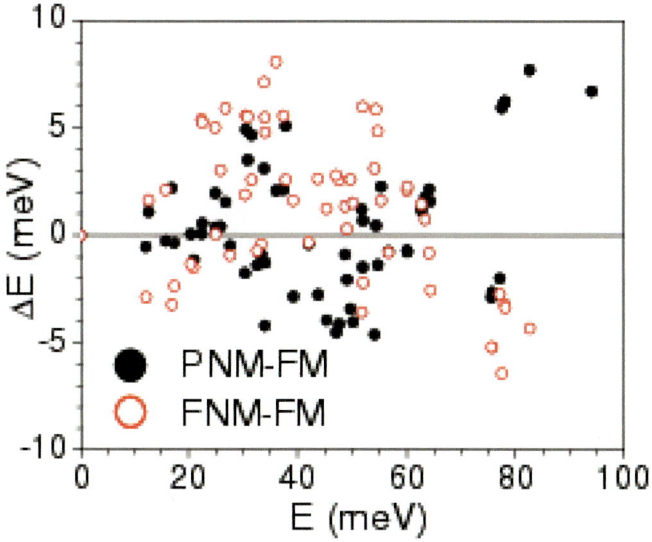

Figure 6.12 The calculated shift of the zone-center optic phonon modes in "PNM" and "FNM" configurations with respect to the "FM" model calculation [8].

the entire Brillouin zone, at room temperature in the $P6_3cm$ hexagonal phase. However, only calculations of phonon modes at the zone center and *K*-point were reported [102] in the high-temperature hexagonal phase ($P6_3/mmc$). The group theoretical analysis along with the first-principle phonon calculations at zone center is used to understand [102] the mechanism of multiferroicity in $YMnO_3$. Alina *et al.* [115] calculated the zone-center phonon modes using LDA, and they reassigned the Raman modes. We went beyond obtaining only the zone-center modes by extracting also phonon dispersion relations (Figure 6.13), in the entire Brillouin zone and along various high-symmetry directions in both the hexagonal phases ($P6_3cm$ and $P6_3/mmc$). We discuss also the relationship between the phonon modes at the FE ($P6_3cm$) to the paraelectric ($P6_3/mmc$) phase transition.

The calculated phonon dispersion relations in both the room-temperature (space group $P6_3cm$) and the high-temperature (space group $P6_3/mmc$) phases are shown in Figure 6.13. The phonon modes, in the low-temperature phase, are found to be stable in the entire Brillouin zone. However, phonon instability is clearly noticed in the high-temperature phase, at the symmetry point *K* (1/3 1/3 0). The unstable mode is highly anharmonic in nature, and it becomes stable at higher temperatures due to anharmonicity.

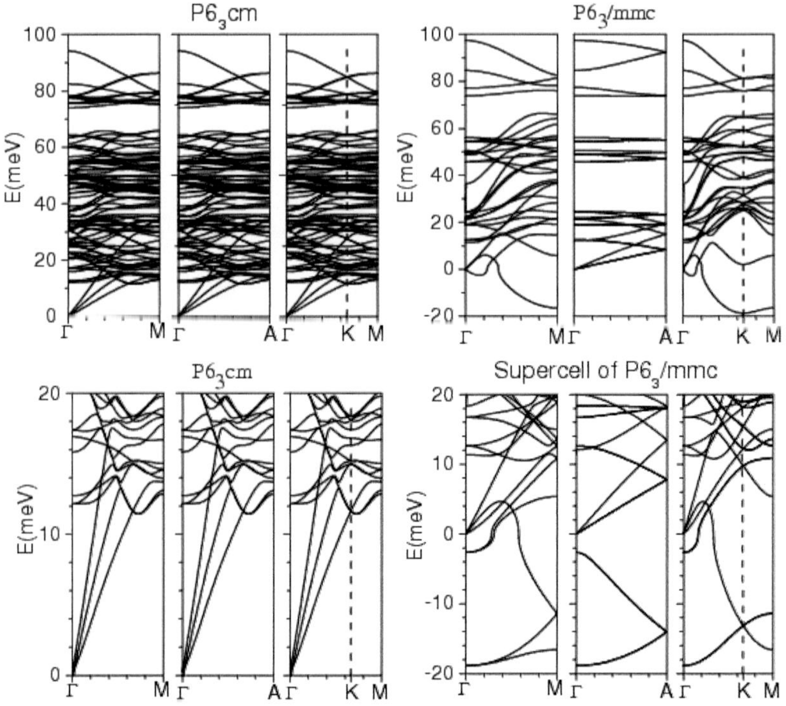

Figure 6.13 The calculated phonon dispersion relations along the high-symmetry directions of the ambient-temperature (space group $P6_3cm$) and the high-temperature (space group $P6_3/mmc$) hexagonal phases of $YMnO_3$. The zoom of the calculated phonon dispersion relations in the ambient-temperature ($P6_3cm$) and high-temperature phase ($P6_3/mmc$) with a super cell of $\sqrt{3}\times\sqrt{3}\times1$ are also shown. The high-symmetry points are: K (1/3 1/3 0), A (0 0 1/2), and Γ (0 0 0). The size of the hexagonal unit cell is nearly the same in the $P6_3cm$ and super cell of $P6_3/mmc$ [8].

It has been proposed [94, 106] that the condensation of the unstable phonon mode at *k*-point drives the transition to the low-temperature structure of $YMnO_3$. This mode is not polar in nature. However, ferroelectricity in the improper FE $YMnO_3$ arises from the coupling of the unstable *k*-point mode with a stable mode at the Γ-point. The latter mode is polar in nature and, therefore, contributes to the ferroelectricity in the room-temperature phase. The eigenvectors of these modes have been extracted from our *ab-initio* calculations. The displacement pattern of the unstable mode at the *k*-point is shown in Figure 6.14. At the *k*-point, the mode consists of an unequal displacement of two Y atoms in opposite direction, along with an out-of-phase rotation of MnO_5 bipyramid units, around the *c*-axis. The unequal

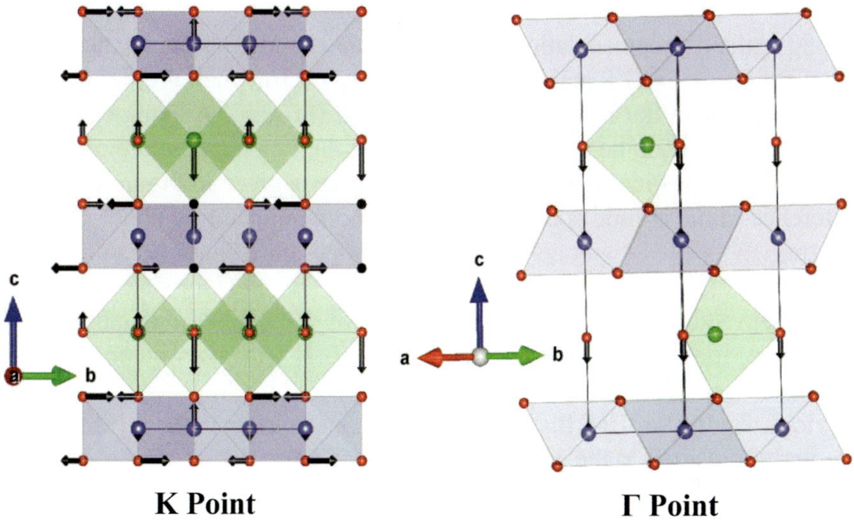

Figure 6.14 The displacement patterns of the lowest phone mode at K and Γ points in the high-temperature phase (space group $P6_3/mmc$) of $YMnO_3$. The lengths of arrows are related to the displacements of the atoms. Key: Y, blue spheres; Mn, green spheres; and O, red spheres [8].

amplitude of motions of the O atoms induces a distortion of the MnO_5 units. The displacement pattern of the stable mode at the Γ-point consists of a motion of O atoms belonging to the plane formed by the Mn atoms of the MnO_5 units.

We have also calculated the phonon dispersion relation of the high-temperature phase ($P6_3/mmc$) of $YMnO_3$ (Figure 6.15) with a super cell of $\sqrt{3}\times\sqrt{3}\times1$ which is equivalent to room-temperature hexagonal phase ($P6_3cm$). The structure as used in the phonon calculations is given in Table 6.1. The comparison of the two structures shows that in the room-temperature phase ($P6_3cm$), the atomic positions are slightly distorted in comparison to the structure obtained from the $\sqrt{3}\times\sqrt{3}\times1$ super cell of the high-temperature phase ($P6_3/mmc$). The group theoretical classification at zone-center of low-temperature phase ($P6_3cm$) is $10A_1+5A_2+5B_1+10B_2+30E_1+30E_2$, while the classification at k-point in the high-temperature phase ($P6_3/mmc$) is given by $2K_1+2K_2+3K_3+3K_4+12K_5+8K_6$. The freezing of the unstable k-point mode in the high-temperature phase will lead to transition to the room temperature phase. Group theoretical analysis shows that the unstable modes at the k-point (1/3 1/3 0) and K^*-point (−1/3 −1/3 0) (of imaginary energy = 18.7 i meV) belong to the K_3 representation in P6$_3$/mmc and condenses into stable modes of A1 (of energy = 21.0 meV) and B2 (of energy = 16.9 meV) representations at the Γ-point in the room-temperature phase (P6$_3$cm). A symmetry-based correlation between the zone-center modes in the low-temperature phase and the k-point modes in the high-temperature phase is given in Figure 6.15. The difference in the atomic coordinates of the room temperature phase (P6$_3$cm) and the $\sqrt{3}\times\sqrt{3}\times 1$ super cell of the high-temperature phase (P6$_3$/mmc) is a measure of the distortion required to stabilize the ambient temperature phase. The eigenvector of the unstable K_3 mode of the high-temperature phase (P6$_3$/mmc) for the super cell is given in Table 6.1. The eigenvector of the K_3 mode is in good agreement with the distortion vector, which corroborates the previous results [95, 104] that freezing of the unstable K_3 mode is responsible for phase transition to the ambient temperature phase.

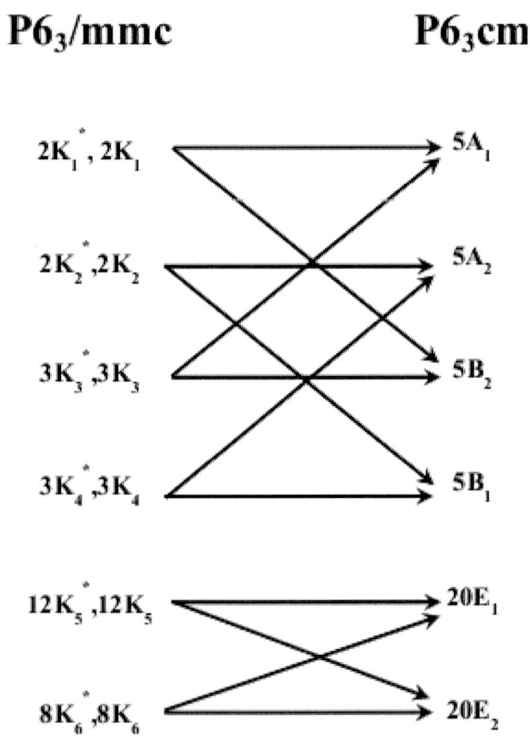

Figure 6.15 A symmetry-based correlation between the zone-center modes in the low-temperature phase (P6$_3$cm) and the k-point (1/3,1/3,0) and K^*-point (−1/3, −1/3, 0) modes in the high-temperature phase (P6$_3$/mmc) [8].

Table 6.1 The calculated structures in the ambient temperature and super cell ($\sqrt{3}\times\sqrt{3}\times1$) of high-temperature phase of $YMnO_3$. The super cell ($\sqrt{3}\times\sqrt{3}\times1$) of high-temperature phase is equivalent to the room-temperature hexagonal phase ($P6_3$ cm). The distortion vector is obtained from the difference in atomic coordinates of the ambient temperature ($P6_3$ cm) and super cell of high-temperature phases. The amplitude of the eigenvector of O1 atom is scaled to match with the distortion vector.

		Ambient-temperature phase ($P6_3$cm)	Super cell of high-temperature phase ($P6_3$/mmc)	Distortion vector in fractional coordinates	Eigenvector of the unstable K_3 mode in the high-temperature phase ($P6_3$/mmc)
	a (Å)	6.095	6.165		
	b (Å)	6.095	6.165		
	c (Å)	11.416	11.223		
O1	*x*	0.306	0.333	−0.027	−0.027
	y	0.000	0.000	0.000	0.000
	z	0.164	0.1665	−0.003	0.000
O2	*x*	0.640	0.667	−0.027	−0.027
	y	0.000	0.000	0.000	0.000
	z	0.336	0.334	0.002	0.000
O3	*x*	0.000	0.000	0.000	0.000
	y	0.000	0.000	0.000	0.000
	z	0.476	0.500	−0.024	−0.039
O4	*x*	0.333	0.333	0.000	0.000
	y	0.667	0.667	0.000	0.000
	z	0.021	0.000	0.021	0.020
Mn	*x*	0.333	0.333	0.000	0.000
	y	0.000	0.000	0.000	0.000
	z	0.000	0.000	0.000	0.000
Y1	*x*	0.000	0.000	0.000	0.000
	y	0.000	0.000	0.000	0.000
	z	0.275	0.250	0.025	0.019
Y2	*x*	0.667	0.667	0.000	0.000
	y	0.333	0.333	0.000	0.000
	z	0.731	0.750	−0.019	−0.018

6.5 Conclusion

We have used the state-of-the-art DFT methods to compute the interaction between atoms in various crystalline materials. In order to validate the theoretical calculations, we have carried out measurements of phonon density of states from polycrystalline sample using neutron inelastic scattering technique. The DFT methods have been used to compute and analyze the various features of the vibrational excitations in the solids.

In $NaNbO_3$, we have performed INS experiments to determine the phonon density of states as a function of temperature across the various phase transitions. Using the *ab-initio* DFT calculations, we have identified the various soft phonon modes at specific points in the Brillouin zone that are associated with various phase transitions as a function of temperature.

For AFM $RMnO_3$ (R = Ca, Pr), the INS data (5−1250 K) have been obtained, which indicate the presence of spin−phonon coupling in these compounds. Theoretical calculation for $PrMnO_3$ suggests that suppression of Jahn−Teller distortion and insulator to metal transition occur simultaneously. We have performed measurements of neutron inelastic scattering spectra of the multiferroic materials $GaFeO_3$ and $YMnO_3$ over a wide temperature range (50−1300 K) covering all the relevant characteristic phase transition temperatures. The *ab-initio* phonon calculations highlighted unambiguously a spin−phonon coupling in $GaFeO_3$ and $YMnO_3$. The enthalpy calculations in various phases of $GaFeO_3$ showed that the quenching of the Fe magnetic moment leads to the observed high pressure structural phase transition at 47 GPa. The calculated phonon dispersion relation in $YMnO_3$ indicates phonon instability in the high-temperature ($P6_3/mmc$) hexagonal phase at the symmetry point *K* (1/3 1/3 0). The unstable phonon mode may lead to stabilization of the low-temperature ($P6_3cm$) phase.

References

[1] M. E. Lines and A. M. Glass, Principles and Applications of Ferroelectrics and Related Materials (Oxford University Press, 2001, Oxford, 2001).

[2] Y. Xu, in Ferroelectric Materials and their Applications, edited by Y. Xu (Elsevier, Amsterdam, 1991), pp. ix.

[3] 1J. L. G. F. L.G. Tejuca, Properties and Applications of Perovskite-Type Oxides Dekker, New York, 1992), Chemical Industries.

[4] A. D. Bruce and R. A. Cowley, Advances in Physics 29, 219 (1980).
[5] S. K. Mishra, M. K. Gupta, R. Mittal, M. Zbiri, S. Rols, H. Schober, and S. L. Chaplot, Physical Review B 89, 184303 (2014).
[6] S. K. Mishra, M. K. Gupta, R. Mittal, A. I. Kolesnikov, and S. L. Chaplot, Physical Review B 93, 214306 (2016).
[7] M. K. Gupta, R. Mittal, M. Zbiri, R. Singh, S. Rols, H. Schober, and S. L. Chaplot, Physical Review B 90, 134304 (2014).
[8] M. K. Gupta, R. Mittal, M. Zbiri, N. Sharma, S. Rols, H. Schober, and S. L. Chaplot, Journal of Materials Chemistry C 3, 11717 (2015).
[9] G. E. Granroth, A. I. Kolesnikov, T. E. Sherline, J. P. Clancy, K. A. Ross, J. P. C. Ruff, B. D. Gaulin, and S. E. Nagler, Journal of Physics: Conference Series 251, 012058 (2010).
[10] G. E. Granroth, D. H. Vandergriff, and S. E. Nagler, Physica B: Condensed Matter 385–386, Part 2, 1104 (2006).
[11] D. L. Price and K. Skold, in Methods in Experimental Physics, edited by K. Sköld, and D. L. Price (Academic Press, 1986), pp. 1.
[12] J. M. Carpenter and D. L. Price, Physical Review Letters 54, 441 (1985).
[13] A. Sjölander, Arkiv för Fysik 14, 315 (1958).
[14] V.S. Oskotskii, Sov. Phys. - Solid State 9,420 (1967).
[15] P. W. Reichardt, 13.03.01p06L (1984). Primärbericht, 13.03.01, p06L ((1984)).
[16] J. P. Perdew, K. Burke, and M. Ernzerhof, Physical Review Letters 77, 3865 (1996).
[17] J. P. Perdew and A. Zunger, Physical Review B 23, 5048 (1981).
[18] G. Kresse and D. Joubert, Physical Review B 59, 1758 (1999).
[19] A. D. Corso, Phys. Rev. B 82, 075116 (2010).
[20] G. Kresse and J. Furthmüller, Computational Materials Science 6, 15 (1996).
[21] S. K. Mishra, R. Mittal, V. Y. Pomjakushin, and S. L. Chaplot, Physical Review B 83, 134105 (2011).
[22] H. J. Monkhorst and J. D. Pack, Physical Review B 13, 5188 (1976).
[23] S. K. Mishra, N. Choudhury, S. L. Chaplot, P. S. R. Krishna, and R. Mittal, Physical Review B 76, 024110 (2007).
[24] S. L. Dudarev, G. A. Botton, S. Y. Savrasov, C. J. Humphreys, and A. P. Sutton, Physical Review B 57, 1505 (1998).
[25] I. A. Vladimir, F. Aryasetiawan, and A. I. Lichtenstein, Journal of Physics: Condensed Matter 9, 767 (1997).
[26] M. J. Han, T. Ozaki, and J. Yu, Physical Review B 73, 045110 (2006).

[27] M. P. J. Punkkinen, K. Kokko, W. Hergert, and I. J. Väyrynen, Journal of Physics: Condensed Matter 11, 2341 (1999).
[28] G. Rollmann, A. Rohrbach, P. Entel, and J. Hafner, Physical Review B 69, 165107 (2004).
[29] A. Bandyopadhyay, J. Velev, W. H. Butler, S. K. Sarker, and O. Bengone, Physical Review B 69, 174429 (2004).
[30] J. E. Medvedeva, V. I. Anisimov, M. A. Korotin, O. N. Mryasov, and A. J. Freeman, Journal of Physics: Condensed Matter 12, 4947 (2000).
[31] O. Gunnarsson, O. K. Andersen, O. Jepsen, and J. Zaanen, Physical Review B 39, 1708 (1989).
[32] K. Parlinksi, PHONON, 2003.
[33] S. K. Mishra, M. K. Gupta, R. Mittal, S. L. Chaplot, and T. Hansen, Applied Physics Letters 101, 242907 (2012).
[34] R Machado, M. Sepliarsky, and and M G Stachiotti, Phys. Rev B 84, 134107 ((2011).).
[35] I. Tomeno, Y. Tsunoda, K. Oka, M. Matsuura, and M. Nishi, Physical Review B 80, 104101 (2009).
[36] L. Chaix et al., Physical Review Letters 112, 137201 (2014).
[37] N. Ganguli and P. J. Kelly, Physical Review Letters 113, 127201 (2014).
[38] Y. Jia, R. V. Chopdekar, E. Arenholz, A. T. Young, M. A. Marcus, A. Mehta, and Y. Takamura, Physical Review B 92, 094407 (2015).
[39] M. Hoffmann, V. S. Borisov, S. Ostanin, I. Mertig, W. Hergert, and A. Ernst, Physical Review B 92, 094427 (2015).
[40] Y. W. Windsor et al., Physical Review Letters 113, 167202 (2014).
[41] P. Barone, D. Di Sante, and S. Picozzi, Physical Review B 89, 144104 (2014).
[42] J. Hong, A. Stroppa, J. Íñiguez, S. Picozzi, and D. Vanderbilt, Physical Review B 85, 054417 (2012).
[43] D. A. Mota, A. Almeida, V. H. Rodrigues, M. M. R. Costa, P. Tavares, P. Bouvier, M. Guennou, J. Kreisel, and J. A. Moreira, Physical Review B 90, 054104 (2014).
[44] S. Bhattacharjee, E. Bousquet, and P. Ghosez, Physical Review Letters 102, 117602 (2009).
[45] U. Aschauer, R. Pfenninger, S. M. Selbach, T. Grande, and N. A. Spaldin, Physical Review B 88, 054111 (2013).
[46] N. Hamdad and B. Bouhafs, Physica B: Condensed Matter 405, 4595 (2010).

[47] J. Hemberger, M. Brando, R. Wehn, V. Y. Ivanov, A. A. Mukhin, A. M. Balbashov, and A. Loidl, Physical Review B 69, 064418 (2004).
[48] L. Bocher, M. H. Aguirre, R. Robert, D. Logvinovich, S. Bakardjieva, J. Hejtmanek, and A. Weidenkaff, Acta Materialia 57, 5667 (2009).
[49] K. R. Poeppelmeier, M. E. Leonowicz, J. C. Scanlon, J. M. Longo, and W. B. Yelon, Journal of Solid State Chemistry 45, 71 (1982).
[50] M. R. N. D. Sousa, C. Silveira, I. Matos, A. B. Lopes, and M. E. M. , Jorge, Mater. Chem. Phys. 109, 311 (2008).
[51] X. J. Fan, H. Koinuma, and a. T. Hasegawa, Physica B 329-333, 723 (2003).
[52] N. Kumar, H. Kishan, A. Rao, and and V. P. S. Awana, J. Alloys Compd. 502, 283 (2010).
[53] Y. Wang, Y. Sui, P. Ren, L. Wang, X. Wang, W. Su, and and H. Fan, Inorg. Chem. 49, 3216 (2010).
[54] M. Nicastro and C. H. Patterson, Physical Review B 65, 205111 (2002).
[55] H. Taguchi, M. Nagao, T. Sato, and M. Shimada, Journal of Solid State Chemistry 78, 312 (1989).
[56] M. Molinari, D. A. Tompsett, S. C. Parker, F. Azough, and R. Freer, Journal of Materials Chemistry A 2, 14109 (2014).
[57] R. Kajimoto, H. Mochizuki, H. Yoshizawa, S. Okamoto, and S. Ishihara, Physical Review B 69, 054433 (2004).
[58] B. Satadeep, B. Eric, and G. Philippe, Journal of Physics: Condensed Matter 20, 255229 (2008).
[59] F. P. Zhang, Q. M. Lu, X. Zhang, and J. X. Zhang, Journal of Alloys and Compounds 509, 542 (2011).
[60] H. Tsukahara, S. Ishibashi, and K. Terakura, Physical Review B 81, 214108 (2010).
[61] B. Bouadjemi, S. Bentata, A. Abbad, W. Benstaali, and B. Bouhafs, Solid State Communications 168, 6 (2013).
[62] W. Paszkowicz, S. Woodley, P. Piszora, B. Bojanowski, J. Piętosa, Y. Cerenius, S. Carlson, and C. Martin, Appl. Phys. A 112, 839 (2013).
[63] J. Piętosa, W. Paszkowicz, R. Minikayev, J. Nowak, C. Lathe, and C. Martin, Powder Diffraction 26, 262 (2011).
[64] D. Sánchez, J. A. Alonso, and and M. J. Martínez-Lope, J. Chem. Soc., Dalton Trans., 4422 (2002).
[65] W. Cochran, Adv. Phys. 9, 387 (1960).
[66] P. Silvia and E. Claude, Journal of Physics: Condensed Matter 21, 303201 (2009).

[67] G. Catalan and J. F. Scott, Advanced Materials 21, 2463 (2009).
[68] C.-W. Nan, M. I. Bichurin, S. Dong, D. Viehland, and G. Srinivasan, Journal of Applied Physics 103, 031101 (2008).
[69] V. V. Pavlov, A. R. Akbashev, A. M. Kalashnikova, V. A. Rusakov, A. R. Kaul, M. Bayer, and R. V. Pisarev, Journal of Applied Physics 111, 056105 (2012).
[70] W. Eerenstein, N. D. Mathur, and J. F. Scott, Nature 442, 759 (2006).
[71] J. H. Lee et al., Nature 466, 954 (2010).
[72] T. Kimura, T. Goto, H. Shintani, K. Ishizaka, T. Arima, and Y. Tokura, Nature 426, 55 (2003).
[73] B. B. Van Aken, T. T. M. Palstra, A. Filippetti, and N. A. Spaldin, Nat Mater 3, 164 (2004).
[74] R. Ramesh and N. A. Spaldin, Nat Mater 6, 21 (2007).
[75] N. Hur, S. Park, P. A. Sharma, J. S. Ahn, S. Guha, and S. W. Cheong, Nature 429, 392 (2004).
[76] N. Ikeda et al., Nature 436, 1136 (2005).
[77] D. V. Efremov, J. van den Brink, and D. I. Khomskii, Nat Mater 3, 853 (2004).
[78] M. Zbiri, H. Schober, N. Choudhury, R. Mittal, S. L. Chaplot, S. J. Patwe, S. N. Achary, and A. K. Tyagi, Applied Physics Letters 100, 142901 (2012).
[79] T. Kimura, S. Kawamoto, I. Yamada, M. Azuma, M. Takano, and Y. Tokura, Physical Review B 67, 180401 (2003).
[80] T. Arima et al., Physical Review B 70, 064426 (2004).
[81] K. Kelm and W. Mader, Zeitschrift für anorganische und allgemeine Chemie 631, 2383 (2005).
[82] R. B. Frankel, N. A. Blum, S. Foner, A. J. Freeman, and M. Schieber, Physical Review Letters 15, 958 (1965).
[83] M. Somdutta, G. Ashish, and G. Rajeev, Journal of Physics: Condensed Matter 23, 445403 (2011).
[84] S. Daniel, Journal of Physics: Condensed Matter 24, 185502 (2012).
[85] M. J. Han, T. Ozaki, and and J. Yu, Phys. Rev B. 75, 060404 (2007).
[86] Z. Mohamed et al., Journal of Physics: Condensed Matter 22, 315701 (2010).
[87] R. Mittal, M. K. Gupta, S. L. Chaplot, M. Zbiri, S. Rols, H. Schober, Y. Su, T. Brueckel, and T. Wolf, Physical Review B 87, 184502 (2013).
[88] W. M. X. R. Arielly, E. Greenberg, G. Kh. Rozenberg, M. P. Pasternak, G. Garbarino, S. Clark, and R. Jeanloz, , Phys. Rev. B 84, 094109 (2011).

[89] H. Boysen and F. Altorfer, Acta Crystallographica Section B 50, 405 (1994).
[90] R. Mittal, S. L. Chaplot, and N. Choudhury, Progress in Materials Science 51, 211 (2006).
[91] R. Arielly, W. M. Xu, E. Greenberg, G. K. Rozenberg, M. P. Pasternak, G. Garbarino, S. Clark, and R. Jeanloz, Physical Review B 84, 094109 (2011).
[92] E. F. Bertaut and M. Mercier, Physics Letters 5, 27 (1963).
[93] A. S. Gibbs, K. S. Knight, and P. Lightfoot, Physical Review B 83, 094111 (2011).
[94] A. S. K. I. G. Ismailzade, Sov. Physc. Solid State 7 (1965).
[95] T. Katsufuji, S. Mori, M. Masaki, Y. Moritomo, N. Yamamoto, and H. Takagi, Physical Review B 64, 104419 (2001).
[96] T. Katsufuji et al., Physical Review B 66, 134434 (2002).
[97] G. R. Nenert, Y.; Stokes, H. T.; Palstra, T. T. M. , arxiv:condmat/0504546, 5 (2005).
[98] G. Nénert, M. Pollet, S. Marinel, G. R. Blake, A. Meetsma, and T. T. M. Palstra, Journal of Physics: Condensed Matter 19, 466212 (2007).
[99] J. Kim, K. C. Cho, Y. M. Koo, K. P. Hong, and N. Shin, Applied Physics Letters 95, 132901 (2009).
[100] J. Kim, Y. M. Koo, K.-S. Sohn, and N. Shin, Applied Physics Letters 97, 092902 (2010).
[101] T. Lonkai, D. G. Tomuta, U. Amann, J. Ihringer, R. W. A. Hendrikx, D. M. Többens, and J. A. Mydosh, Physical Review B 69, 134108 (2004).
[102] C. J. Fennie and K. M. Rabe, Physical Review B 72, 100103 (2005).
[103] A. Cano, Physical Review B 89, 214107 (2014).
[104] M. N. Iliev, H. G. Lee, V. N. Popov, M. V. Abrashev, A. Hamed, R. L. Meng, and C. W. Chu, Physical Review B 56, 2488 (1997).
[105] C. Toulouse et al., Physical Review B 89, 094415 (2014).
[106] S. Petit, F. Moussa, M. Hennion, S. Pailhès, L. Pinsard-Gaudart, and A. Ivanov, Physical Review Letters 99, 266604 (2007).
[107] S. Pailhès, X. Fabrèges, L. P. Régnault, L. Pinsard-Godart, I. Mirebeau, F. Moussa, M. Hennion, and S. Petit, Physical Review B 79, 134409 (2009).
[108] G. E. Bacon, Neutron diffraction (Oxford University Press, Oxford, 1975).
[109] T. Chatterji, S. Ghosh, A. Singh, L. P. Regnault, and M. Rheinstädter, Physical Review B 76, 144406 (2007).

[110] A. Muñoz, J. A. Alonso, M. J. Martínez-Lope, M. T. Casáis, J. L. Martínez, and M. T. Fernández-Díaz, Physical Review B 62, 9498 (2000).

[111] P. J. Brown and T. Chatterji, Journal of Physics: Condensed Matter 18, 10085 (2006).

[112] X. Fabrèges, S. Petit, I. Mirebeau, S. Pailhès, L. Pinsard, A. Forget, M. T. Fernandez-Diaz, and F. Porcher, Physical Review Letters 103, 067204 (2009).

[113] J. Park, J. G. Park, G. S. Jeon, H.-Y. Choi, C. Lee, W. Jo, R. Bewley, K. A. McEwen, and T. G. Perring, Physical Review B 68, 104426 (2003).

[114] K. Z. Rushchanskii and M. Ležaić, Ferroelectrics 426, 90 (2012).

[115] A. Prikockytė, D. Bilc, P. Hermet, C. Dubourdieu, and P. Ghosez, Physical Review B 84, 214301 (2011).

7

Strongly Correlated Doped Graphene Systems – Synthesis, Electronic Properties, and Applications

Anju K. Nair[1], Paulose Thomas[2], Kala M. S.[3], and Nandakumar Kalarikkal[4]

[1]Mar Thoma College, Tiruvalla, India
[2]Mar Thoma College for women Perumbavoor, India
[3]St. Teresa's College, Ernakulam, India
[4]International and Inter University Centre for Nanoscience and Nanotechnology, Mahatma Gandhi University, India

Abstract

Graphene, the two-dimensional wonder material of the century with its unique structural, chemical, and electronic properties, has revolutionized the everyday life of mankind because of its potential applications. The range and scope of these applications can be greatly enhanced with graphene hybrids, which combine graphene/doped-graphene with metal nanostructured materials and hence are an emerging field of research. Heteroatom-doped graphene is an innovative strategy that is further amazing. Facts have proved that it is an excellent platform for realizing the investigation of innovative correlated systems with excellent electronic properties, hence investigating profound shifts in optical characteristics and exciton physics. Additionally, it provides a framework for further investigation into strongly correlated systems. An innovative and novel matter-correlated phase that emerges after doping offers an attractive experimental framework designed for the architecture of new exciton condensed matter, spatial confinement physics, and two-dimensional correlation. The unique features of both graphene and nanostructures provide

DOI: 10.1201/9781003337102-7

hybrid materials their unique synergistic properties, greatly enhancing their application potential.

Keywords: graphene, heteroatom, doping, strong correlation

7.1 Introduction

One of the world's most active areas of research in the last decade has been in the domains of nanoscience and nanotechnology. Owing to the distinctive characteristics of nanomaterials as compared to that of bulk materials, advancements and breakthroughs in nanotechnologies have generated a wide range of possibilities in a variety of fields, including optics, power generation, product engineering, nanoelectronics, digital technology, bioengineering, and atmospheric engineering [1–4]. The most thoroughly researched colloid materials in the fields of nanotechnology and nanoscience are noble metal nanoparticles (NPs). The investigation of distinctive features that are morphological characteristics has accelerated as a result of ongoing advancements in the development of NPs with regulated morphologies. Nonspherical NPs are excellent prospects for improved biochemical, catalysis, and localized field applications due to their exceptional properties. The creation of innovative smart materials for numerous developing applications involves a wide range of nanomaterials with optimal dimensions, morphology, and composition [2, 3]. A fast growing area of research is the utilization of NPs as building blocks and their assembly into functional structures, which is particularly relevant for realistic development. As these designs are put together, new properties develop organically. One such feature is the creation of collective behavior brought about by interactions between various electrical, magnetic, and chemical processes.

7.2 Graphene

Graphene and its derived materials have recently received a lot of interest as a new category of two-dimensional carbon-based nanostructures [5]. As a result of its distinctive optical, electrical, biochemical, thermo, and structural qualities, it has been referred to as the "next generation material" [6]. A layer of carbon atoms with sp2 bonds that are one atom thick and closely packed into a 2D honeycomb structure is called graphene (Figure 7.1) and was first isolated from graphite through mechanical exfoliation in 2004 by

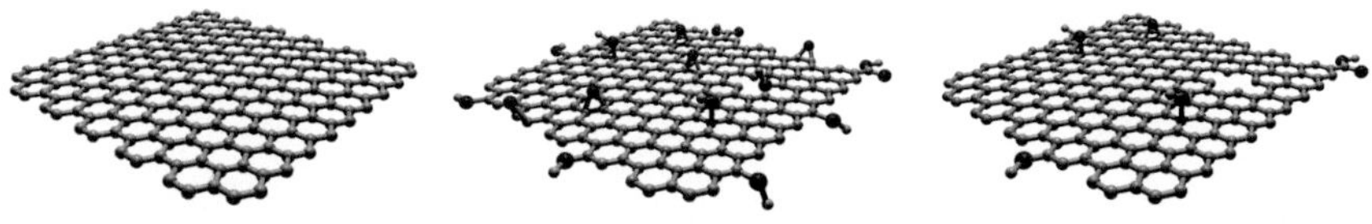

Figure 7.1 Single layer of graphene, graphene oxide, and reduced graphene oxide. Copyright reserved to the Royal Society of Chemistry (Creative Commons Attribution 3.0 Unported License) [7].

Novoselov and Geim, which was awarded the 2010 Noble Prize in physics [7, 8]. The charming properties of pristine graphene at the moment are largely understood and properly identified via widespread research in the past years. Graphene exhibits many specific physical properties including very high mobility of charge carriers (200,000 cm^2 V^{-1} s^{-1}), better thermal conductivity ($\approx$ 5000 Wm^{-1} K^{-1}), ambipolar electric field effect, strange quantum hall effect (QHE), high transparency (optical transmittance is 97.7%), very high surface area (2630 m^2/g), and massless relativistic charge carriers, all of which make graphene excellent for plenty of possible applications such as single molecule gas detectors, super-capacitors, lithium-ion batteries (LIBs), catalysis, sensors, and transparent conductors [9, 10].

7.2.1 The structure and electronic properties of graphene

The electrical characteristics of graphene reflect its structural adaptability. When regarded as a triangular structure, the pattern has a basis with two atoms at each unit cell (Figure 7.2) [11]. A trigonal planar structure is created by the sp2 hybridization of 1s orbital and 2p orbitals, which results in the formation of bonds between carbon atoms that are 1.42 Å apart [11–14]. All carbon allotropes' strength in the lattice arrangement is due to the bond. As a result of the Pauli principle, these bands have a filled shell and, in this way, frame a profound valence band. The unaltered p orbital can form a covalent bond with nearby carbon atoms and may be perpendicular toward the planar arrangement, prompting the development of a π bond. Because of this, every p orbital has one extra electron and the π bond is 1/2 filled [15]. Because of their high tight-binding properties, half-filled bands in transition elements are crucial to the science of strongly correlated structures because they produce better collective impacts, magnetic property, and insulating nature owing to correlation gaps [16, 17].

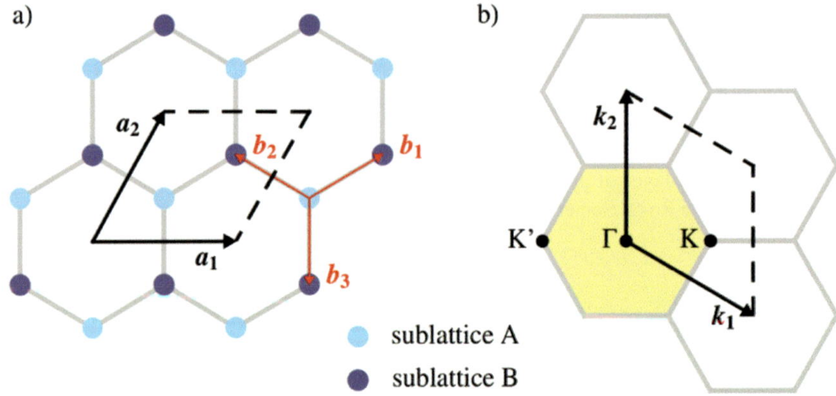

Figure 7.2 Graphene honeycomb lattice. (a) The unit cell's two carbon atoms' equivalent sublattices A and B, which are determined by the primitive lattice vector a1 and a2. (b) A reciprocal lattice with the high symmetry points, K and K', the Brillouin zones (yellow region), and the reciprocal lattice vectors k1 and k2. Copyright is reserved under creative common license attribution [11].

Band-structure studies of graphene suggested being a semi-metal with unusual linearly dispersing electronic excitations called Dirac electrons [14]. The massless and chiral nature of graphene's low-energy excitations is one of its most intriguing properties. The chemical nature closely overlaps the Dirac points in the neutral graphene. This particular dispersion replicates the mechanics of quantum physics for massless elementary particles and is simply significant at low energies. When exposed to magnetic fields, Dirac fermions behave differently from regular electrons, leading to the emergence of novel physical features such as the strange integer quantum Hall effect (IQHE) measured tentatively [18–20]. Graphene may be tweaked chemically or functionally in a variety of ways due to its amazing structural and electrical flexibility. By creating graphene-based frameworks with appealing and superconducting features that are unique in their 2D properties, the controlling of graphene characteristics can be expanded in new ways [21].

7.3 Heteroatom-doped Graphene

Graphene can be viewed as either a metal with diminishing Fermi surface or a semiconductor with zero bandgap. Because graphene's applications in fields like nanoelectronics, electrochemistry, and energy storages are severely constrained by its lack of an intrinsic bandgap, it is appealing to create one in graphene in order to further the aforementioned applications. Much work has recently been done to enhance performance, thereby exploring the electric properties of graphene. A variety of new or improved physio-chemical, electromagnetic, photonic, electrochemical, radiative, and physical features can be added to graphene through heteroatom doping. As a result, the range of materials based on graphene and their potential for a wide range of applications has been greatly expanded [22–24]. Incorporating heteroatoms into the carbon lattice must allow electrons modulating to modify their opto-electronic and chemical characteristics, which is advantageous for many applications [25–27]. Both *in situ* doping even during fabrication of carbonaceous nanomaterials and post-treating already-formed carbon nanomaterials with heteroatom-containing precursors can be used to create heteroatom-doped carbon nanomaterials. Post-doping of graphene typically only modifies the surface features without altering the bulk characteristics. Contrarily, *in situ* doping can uniformly include heteroatoms throughout the entire structure [28–30]. By forming covalent bonds with graphene during substitutional doping, the doping molecules can disturb the sp2 carbon network and result in the creation of sp3 defect areas [31].

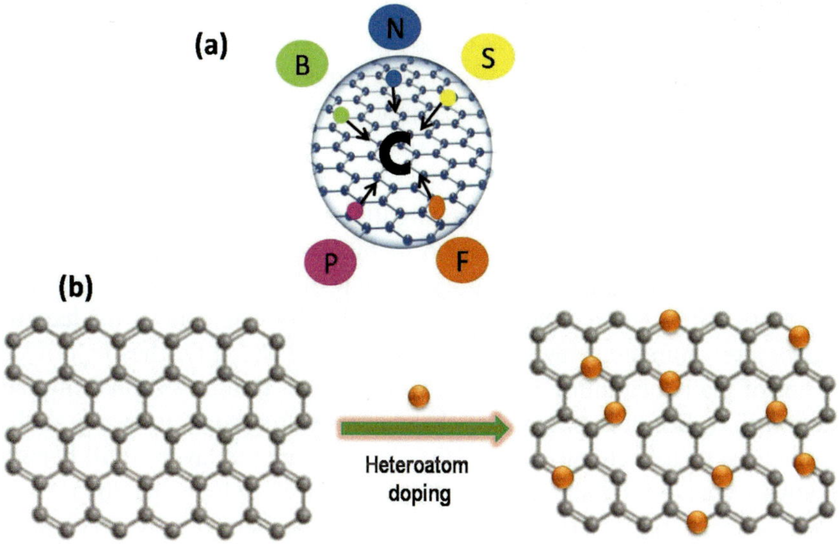

Figure 7.3 (a) Heteroatom doping of graphitic carbon structure (e.g., B, N, S, P, and F). (b) Post-treatment doping of the graphitic carbon structure with heteroatoms.

The doping system using heteroatoms, together with N, F, S, B, and/or P, is simply depicted in Figure 7.3 (a), whilst Figure 7.3 (b) suggests feasible graphitic structure modifications triggered by doping. Since heteroatoms differ from carbon atoms in dimension and electronegativity, adding them to nanostructured materials should result in electronic modification, which would alter the free dispersion and electronic properties. This, along with the doping-induced defects, can thus modify the chemical properties of the carbon nanomaterials, opening the door to a wide range of new and more effective purposes, including super-capacitors, battery systems, electrocatalysis for the oxygen reduction reaction (ORR), oxygen evolution reaction (OER), and so on [30, 31]. The Dirac point, which is located in the electronic structures of the substance and exhibits electronic behavior akin to that of high-energy electrons, is one of the distinctive electronic features of graphene that make it well-recognized. Exotic states like greater chiral geometric conductance, which can currently be investigated using controlled doping techniques, are included in the theoretical expectations of doped graphene [32].

7.3.1 Boron-doped graphene

Boron ($2s^2 2p^1$) is an element with only one less valence electron than carbon, which places it very near to carbon ($2s^2 2p^2$) and is greatly agreeable for graphene doping [23, 24]. When compared to out-of-plane bonding, it is discovered that the in-plane substitution doping (i.e., in-plane BC3) is a stable bonding. Due to in-plane doping, B atom in the lattice becomes sp2-hybridized, allowing for the preservation of graphene's planar structure. However, the lattice parameters are slightly changed so that the B–C bond is longer than the C–C bond. Instead of the in-plane B-doping, significant deformation of such graphene configuration results from the B-doping at the unoccupied sites with a form of various BC4 units where all of the hanging carbons are saturated. However, because of its instability, the majority of the out-of-plane beginning positions would return to their original in-plane position [32]. Boron doping cause the charges to rearrange unevenly within the material causes charge polarization in the graphitic layers (will become negatively charged) by adding one fewer outer electron to the nearby C atom, which is beneficial for electrocatalysis, and p-type behavior by decelerating the Fermi level nearer the Dirac points [32, 33]. The increased conductance was seen as a result of graphitic B-doping, which adds more holes to the graphene's valence band and increases the density in BG [34].

7.3.2 Nitrogen-doped graphene

In the periodic table, nitrogen is an element that is close to carbon. N-doped graphene differs significantly from boron-doped graphene due to the electron-rich nature of N. The polarization in the carbon lattice structure caused by the nitrogen atom's higher electronegativity than that of carbon has an impact on the electrical, optical, and appealing properties of graphene sheets [35–37]. A nitrogen atom typically has one of the three most common atomic bonding configurations within the carbon lattice when it is introduced into graphene. While pyrrolic N (sp3 hybridization) alludes to N particles that contribute two p electrons to the framework, pyridinic N (sp2 hybridization) contributes one p electron to the framework and is bonded with two C atoms at the borders of graphene. Quaternary N that is also sp2 hybridized replaces C atoms in the hexagonal-like structures [38, 39]. Sun et al. demonstrated that graphitic N might improve conductivity, which supports electron transport during the charge–discharge process, whereas pyridinic and pyrrolic N in NG are essential for boosting pseudocapacitance by utilizing redox reactions. By suppressing the surrounding density of states (DOS), n-doping can create a bandgap very close to the Dirac points, giving graphene n-type semiconducting characteristics. Similarly, it was shown that by altering the N-doping level, the n-type behavior of NG may be carefully controlled [40].

7.3.3 Sulfur-doped graphene

Because the C–S bond length in sulfur-doped graphene (1.78 Å) is 25% longer than the C–C bond, sulfur doping results in a stable structure where each S particle is 1.1 Å over the graphene plane [41]. S could be adsorbed on the graphene surface (S1), substitute for C atoms at edges (S2 and S3), form S/S oxide (S4 and S5), or connect two graphene sheets by framing an S cluster ring, according to Xia et al.'s analysis of sulfur-doped graphene SG using the density functional theory (DFT) method [42]. The most stable structure is produced by the adsorption of S on graphene, as shown by the arrangement liveliness; S seeks to replace C atoms along the zigzag edge, and the Stone–Wales deformities can promote S-doping due to the altered local charge distribution in the carbon lattice. Unlike the effects of N- and B-doping, the C–S bond has very little polarization due to the relative electronegativities of S (2.58) and C (2.55) [42, 43]. Most S-doped samples showed higher resistivity than their undoped counterparts, which Pumera's group attributed to the S and O functions' ability to trap free carriers [44]. It is interesting to note that Wang et al. showed that thiophene S-doping (S2 and S3) can improve graphene conductivity due to its more grounded electron donor capability and more viable reduction generated by S-doping [45].

7.3.4 Co-doped graphene

Co-doping with various foreign atom types may result in novel traits or have synergistic effects. Due to their similar sizes, B and N atoms have an adverse effect on graphene doping. Sen et al. have demonstrated through DFT calculations that co-doping graphene with N and B atoms can increase the effectiveness of N-doping because of the stability provided by B–N bonding [46, 47]. The ability of B and N atoms to lower the valence band and raise the conduction band, which causes electron delocalization and activates graphene, has been established. It was found that the bond structure specifically depended on the type of nitrogen dopant used in the boron-nitrogen doped graphene. Using pyrimidine and thiophene as pyrolytic precursors, Guan et al. produced SNG at 700 °C, where S-doping primarily took the form of sulfide groups (C-S-C) and, to a lesser extent, oxidized S clusters (C-SOx-C) with pyrrolic/graphitic N structures [49]. A hydrothermal synthesis of SNG using GO, mercaptoacetic acid, and DL-penicillamine as doping sources was discovered by Luo and colleagues [49].

7.4 Synthesis Methods of Doped Graphene

For the manufacture of graphene materials, a wide range of methods have been used, from which various doping systems could be chosen. The current methods for heteroatom doping can be divided into post-treatment techniques and *in situ* technologies. Graphene mix and heteroatom doping are achieved *in situ* using techniques like chemical vapor deposition (CVD), ball milling, and bottom-up methodology. Wet chemical procedures, thermal annealing of graphene oxides (GO) using heteroatom precursors, plasma, and arc-discharge processes are other post-treatment techniques.

7.4.1 Chemical vapor deposition

To create large, dependable, defect-free, single-layered, or few-layered graphene films, numerous CVD techniques have been used. The CVD method provides a useful method for doping heteroatoms during the assembly of graphene sheets, notably for directly incorporating heteroatoms into the

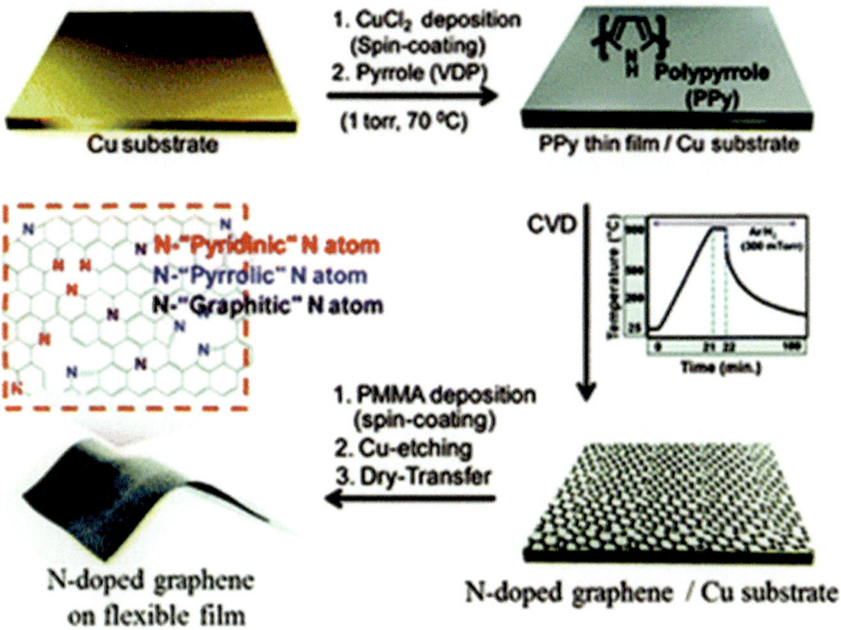

Figure 7.4 Polypyrrole-derived N-doped graphene. Copyright reserved to American Chemical Society [54].

graphitic carbon lattices [50]. Doping can take place by adding precursors that are solid, liquid, or vaporous and contain foreign atoms to the growth furnace along with the carbon sources. Sometimes the precursors of the alien atom(s) and carbon are similar. It is also possible to co-dope a variety of species, with the goal of creating cooperative synergy between the co-dopants [51]. Ajayan et al. used methane and ammonia borane as the carbon and B and N sources, respectively, to explain dual B- and N-doped graphene [52, 53]. In the study by Kwon et al., Cu^{2+}-assisted polymerization of pyrrole monomers on a Cu substrate was followed by CVD development (Figure 7.4) [54]. The developed N-doped few-layered graphene contains 3.14 at% N with overwhelming pyrrolic N likely acquired from the polypyrrole precursor.

7.4.2 Ball milling

By delaminating graphite and severing the C–C bonds, ball milling is a successful method for producing large amounts of graphene nanosheets at a reasonable cost [55]. Jeon et al. successfully fixed uncleaved N_2 to the fractured edges of graphene nanoplates by the ball-milling process of graphite under N_2 gas maintained for more than 48 hours [56]. In a different study, the individual zigzag and armchair edges create aromatic 5-membered pyrazole and 6-membered pyridazine rings. Additionally, by ball-milling graphite in the presence of sulfur with uniformly distributed sulfur elements at a level of 4.94 at%, edge-sulfurized graphene nanoplatelets were organized [57].

7.4.3 Bottom-up synthesis approach

The amazing heteroatom-doped graphene has been arranged using a bottom-up method called the Wurtz-type reductive coupling (WRC) reaction. Effectively manufactured B-doped few-layer graphene had a doping level of 2.56 at%, which could be adjusted by changing the quantity of B precursors [58]. Additionally, lithium nitride (Li_3N) and CCl_4 were combined to create gram-scale N-doped graphene by heating them to 120 °C for 12 hours [59]. For the WRC reaction, a variety of heteroatom precursors can quickly control the doping level [59].

7.4.4 Thermal annealing

When the proper doping precursors are present, thermal annealing of GO or reduced GO (rGO) at high temperatures successfully accomplishes both the recoupling of the sp2 carbon network and heteroatom doping. Due to the high oxygen component of rGO, B particles are doped differently from CVD method, becoming BC_3 and BC_2O. Compared to B–O bonding, the arrangement of B–C bonding is best at a higher temperature [60]. More frequently, annealing GO with high-purity ammonia gas results in N-doping. Li et al. found that N-doping starts at 300 °C and reaches the highest degree of doping of $\sim$5 at% at 500 °C [61]. It is suggested that the best range for the general stability of all N species is 500–600 °C (amino, pyridinic, pyrrolic, and graphitic N) [62]. Dai et al. thermally annealed GO in the presence of boric acid and NH_3 at 1000 °C to produce B, N co-doped graphene [63]. Despite the annealing temperature, the predetermined precursors also have a vital impact on the doping proficiency and arrangement. Other nitrogen precursors besides NH_3 have also been used, such as melamine, polypyrrole, cyanamide, and dicyandiamide. By utilizing porous silica, Yang et al. restricted the usage of GO sheets for N- or S-doping [64–66]. It was shown that S-doping is weaker than N-doping and that S-doping at defect locations creates structures like thiophene.

7.4.5 Wet chemical methods

GO may be easily distributed in water and other solvents because of its amphiphilic nature, and the oxygen functional groups on its surface provide helpful chemical handles to react with heteroatom precursors. For the simultaneous decrease of GO and N-doping in solution, hydrazine has been tested [67]. Another often used N precursor is ammonia solution (NH_4OH), which is highly reactive at typically low temperatures [68]. Sun et al. prepared N-graphene via urea as the precursor [39]. They suggested that for the underlying creation of amide and amine intermediates, NH_3 persistently reacts with the oxygen functional groups of GO. These intermediates then quickly rearrange due to the lack of hydration and decarbonylation to build more stable pyridine- and pyrrole-like structures. Heteroatoms have also been added to 3D graphene gels, which have an extremely large surface area and a macroporous structure, using wet chemical techniques [69–71].

7.5 Electronic Properties of Doped Graphene

The best way to change graphene's electrical characteristics is through chemical doping. The advancement of graphene in the field of nano-based electronics is limited since it possesses a zero bandgap. In order to broaden the spectrum of its applications, it is essential to finish the research on opening the graphene bandgap. The dissipation system at lattice defects and ionized impurities are substantially affected by the electrical properties of graphene, especially the carrier transport [72]. Doping using heteroatoms, which is analogous to silicon-based technology, is a viable method for adjusting and controlling the electrical characteristics of graphene. As a result, by opening a bandgap, doping with boron or nitrogen molecules enables graphene to convert into a p- or n-type semiconductor separately [73–75].

In their study, Zang et al. considered the electronic characteristics of N-doped graphene. After being placed on a silicon oxide substrate and formed into microribbons, the N-doped graphene film was then produced in bottom-gate field-effect transistors (FETs). A divergence in electron-hole transport caused by enhanced doping in graphitic N-governed graphene suggests that the dispersion of ionized fused N atoms plays a crucial role in determining the carrier transport actions [76]. Nearly half of the electron will come from a bonded graphite N atom, increasing the electron concentration and moving the Fermi level into the conduction band (Figure 7.5(a) inset). The gate oxide layer has a thickness of 300 nm. The electrical characteristics of graphene based on graphite N and pyrrole N are displayed in Figure 7.5.

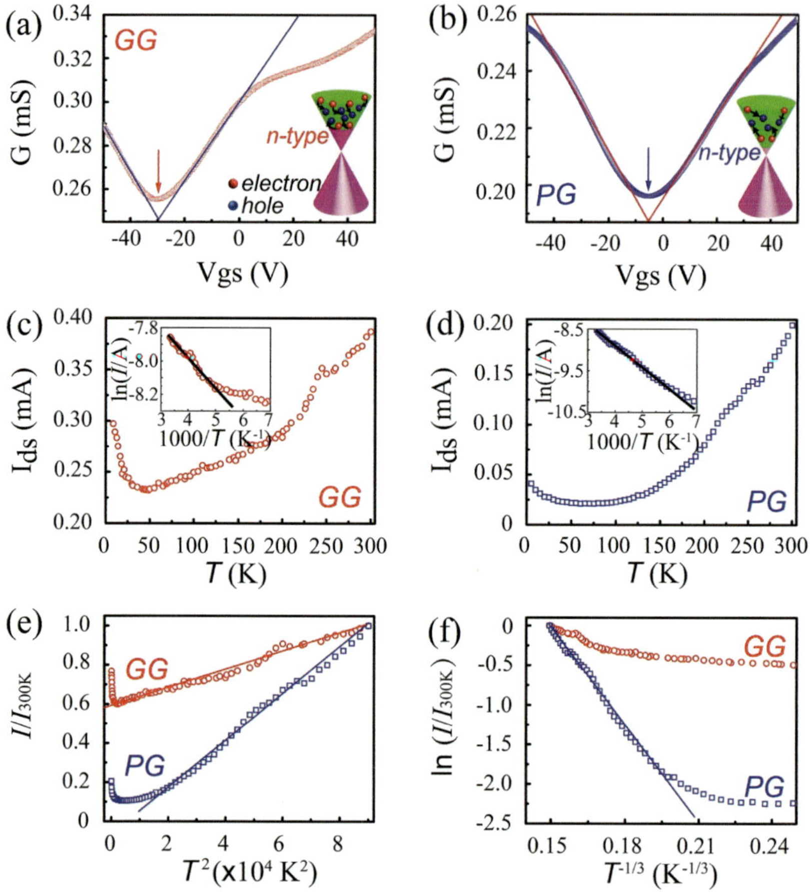

Figure 7.5 (a) Conductance (G) as a function of gate voltage (V_{gs}) for GG with G_{min} at -30 V, V_{ds} = 1 V. Inset is the energy band of the GG, and the red and blue spheres represent the electron and hole, respectively. (b) Conductance (G) as a function of gate voltage (V_{gs}) for PG with G_{min} at -5.2 V, V_{ds} = 1 V. Inset is the energy band of the PG. Drain–source current (I_{ds}) with V_{ds} = 1 V and floating gate (V_{gs} = 0 V) of (c) GG and (d) PG, respectively. The inset shows the corresponding Arrhenius plot and yields activation energy. (e) Normalized I_{ds} by that at 300 K (I_{ds}/I_{300K}) as a function of T^2. (f) The I_{ds}/I_{300K} as a function of $T^{-1/3}$ with logarithmic coordinates. Copyright reserved to Creative Commons Attribution 4.0 International License [76].

In a different study, Vyalikh et al. created a useful technique for converting N into graphite form by placing Au under N-graphene and then annealing it for a prolonged period of time. According to studies on the effects of graphene nitrogen, there is a bandgap of around ~0.2 eV and an energy in fermilevel (EF) shift of about 300 meV in graphene. Additionally, it can directly estimate the mobile charge that is transported from graphite N to the graphene conduction band and discover that the mobile charge is almost half of each N atom's electron [77].

DFT was used by Mahendran et al. to examine the electronic characteristics of nitrogen- and boron-doped graphene. The electrical structure of graphene was significantly altered by the addition of boron and nitrogen atoms. The doped structure yields the highest occupied molecular orbital (HOMO) and the lowest unoccupied molecular orbital (LUMO). As the number of boron atoms rises, so does the energy gap of HOMO-LUMO. To analyze the adsorption of H_2, a box with 12 Å along the *c*-axis was also taken into consideration in addition to the electrical characteristics. According to the quantity of doped atoms, the adsorption binding energy on boron- and nitrogen-doped graphene rises or falls [78].

In order to comprehend how N- and S-dopants affect the electronic structure of graphene, Lee et al. employed DFT to calculate the band structures of several N- and S-doped graphenes. The energy band rapidly descends below the Fermi level as the amount of S atoms in the N/S co-doped graphene grows, culminating in n-type behavior. The "p-type" behavior will therefore change to the "n-type" behavior as the S concentration rises as the graphene structure is altered through the N- and S-doping techniques. Because of their stronger electronegativity relative to C, Mulliken population analysis reveals that the doped N atoms next to the single vacancy are negatively charged, but the doped S atoms close to the electronegative charge similar to C are positively charged [78].

The effects of the concentration and placement of boron atoms as impurities in graphene nanoribbons are investigated using DFT and Landauer methods. In order to achieve this, we created graphene nanoribbons doped with boron atoms in three different concentrations, and we computed the DOS, current, and thermal power. When boron atoms are present in graphene as impurities, the DOS curves of pure and boron-doped graphene are compared to demonstrate how this causes an energy gap close to the Fermi energy [79].

Because it is thought that the hexagonal symmetry of graphene dramatically promotes the balance between magnetic and superconductivity, graphene is a good option for the investigation of Van Hove's superconductivity. The Fermi level, which has the greatest filled electron energy level at absolute zero, is quite far from the Van Hove point in the original single-layer graphene. However, by intercalating chemically doped materials, for instance, the Van Hove point can be attained. In this case, a certain kind of atom is put in between the graphene monolayer and the substrate, which causes the inserted atom to transfer charge carriers to the graphene. These doping methods have now created the groundwork for a methodical investigation of graphene close to the Van Hove point. The anticipated many-body interactions in this situation will alter the electronic energy band structure in addition to bringing about more significant alterations in the system with the emergence of the pertinent phase.

7.6 Applications of Doped Graphene Nanostructures

The goal of developing composite materials is to maximize their potential for practical applications. The potential uses of the aforementioned graphene-noble metal nanohybrids are described in length in this section.

7.6.1 Surface-enhanced Raman scattering applications

Surface-enhanced weak Raman signals of analytes are significantly increased by the physical phenomena known as Raman scattering (SERS) when they are in close proximity to the surface of plasmonic NPs [80, 81]. Generally speaking, chemical and electromagnetic enhancements are responsible for the SERS mechanism [82, 83]. It is generally known that metal nanoparticles, because of their potent plasmonic properties, can boost electromagnetic fields [84]. Some structural features of 2D materials, such as the π–π structure, which traps more molecules, and the high surface area, which concentrates more molecules, play significant roles in enhancing the Raman signals of detection molecules [85, 86].

Heteroatom-doped graphene additionally encourages the SERS activity in addition to pristine graphene or GO. The charge transfer between the bound analytes and graphene may be accelerated by the highly polarized heteroatom–C bonds caused by doping, which will subsequently improve the SERS signal. A DFT simulation research showed that substitutional B-doping can enhance the pyridine's SERS signal by 3–4 orders of magnitude [87]. This is attributed to the charge transfer that took place between the positively charged B particle of graphene and the negatively charged N molecule of pyridine. Additionally, the detection of biomolecules at the maximum is made easier by the hybridization of metal nanomaterials and 2D materials (graphene or doped graphene), making hybrid nanostructures attractive SERS substrates.

In the design and development of hybrid nanostructured SERS substrates, significant advancements have been accomplished. In the beginning, Ag nanospheres were thermally evaporated on top of CVD-grown graphene to create hybrid metal/graphene systems, which have been employed as SERS substrates [88]. Other shaped metal NPs have also been added to the graphene surfaces of hybrid SERS substrates because it has been discovered that the morphology of metal NPs significantly influences SERS signals [89]. The detection of pesticides was made possible by the greatly improved SERS signal of crystal violet (CV) on Au nanorods coated with graphene or GO [90]. When a GO-Ag nanocage hybrid substrate is used, the Raman signals of nitro explosives can be improved by four orders of magnitude [91, 92]. Additionally, vertical graphene sheets with metal NPs added have been used as a highly effective 3D SERS substrate with great repeatability [93, 94].

7.6.2 Energy conversion: oxygen reduction reaction in fuel cells

Research into new environmentally friendly energy sources and their practical uses has received growing attention over the past few decades as a result of the rapidly rising demand for energy and the depletion of fossil fuel resources. In reality, a fuel cell is an electrochemical device that uses chemical processes at the interface of the electrolyte and electrode to transform the chemical energy of a fuel into electrical energy [95–97]. The oxygen reduction at the cathode and the fuel oxidation at the anode are the two primary processes in a fuel cell. Platinum (Pt) has the highest electrocatalytic activity for the electro-oxidation of small organic fuels at the anode and the oxygen reduction at the cathode among the metal catalysts for anode and cathode reactions. However, a number of clear drawbacks severely restrict the use of Pt as a catalyst in fuel cells [98]. Specifically, the nanostructured materials meet the requirements of a catalyst candidate excellently with a very large surface area, improved surface active sites, and low metal loading. Heteroatom-doped graphene (particularly B- and N-doped graphene) has been observed to show very high electrocatalytic activity for ORR [98–100]. Because of the charge polarization of the heteroatom–C bond, B-, N-, and P-dopants encourage the adsorption of oxygen and O–O bond breakage [101–104]. In rare circumstances, a rise in both charge polarization and spin density may simultaneously lead to ORR [105]. Graphene co-doped with heteroatoms (such N-B and N-P) displays the synergistic effects of various co-dopants at the ORR parameters [106]. The longevity of the electrode and catalytic activity can both be improved by heteroatom-doped graphene decorated with a homogeneous and dense pattern of precious metal catalysts. S-doped graphene/Pt nanowire composite was employed by Wang et al. as both the cathode for ORR and the anode for methanol oxidation processes (MOR) [107]. In comparison to the commercial Pt/C catalyst, it exhibits a higher current density and a lesser decrease potential for ORR. N-dopants encourage Pt-Ru immobilization, improving MOR stability. Pt and Pt-Ru modified N-doped graphene compounds have also been found to have extraordinarily good MOR electrocatalytic performance [107].

7.6.3 Energy storage: lithium-ion batteries

Devices having a high energy density are called LIBs. However, they have a weak cycling stability and a relatively low power density. Due to its low affinity for Li atoms and high energy barrier for Li to pass through the graphene sheet, pure graphene is not appropriate for storing lithium [108, 109]. To optimally tune Li storage and diffusion for graphene-based electrodes, heteroatom doping may be helpful [110]. An appealing way to increase the storage capacity of electron-donating Li [111] is to dope graphene with B atoms, which is an electron-deficient framework. Due to the electrostatic attraction between Li and N, graphene that has been doped with electron-rich graphitic N exhibits more effective delithiation [109]. As an LIB anode, Wang et al. created a 3D N, S co-doped graphene network [112]. Such LIB exhibits an incredible rate performance and a large reversible capacity, which is almost three times the theoretical capacity of graphite and significantly higher than pristine graphene, thanks to the synergistic effects of the 3D structure and co-dopants. To improve the LIB performance, metal oxides and nanostructured metals (such as Co_3O_4, TiO_2, Ag, and Au) can be anchored by heteroatom doping [113–118]. For instance, compared to pure graphene paper or SnO_2 NPs, a sandwich paper made of N-doped graphene and SnO2 exhibits a greater capacity [113].

7.6.4 Electrochemical sensing

The use of doped graphene materials can enhance the performance of electrochemical sensors because the electrochemically active sites created by heteroatom doping can enable charge transfer, analyte adsorption and activation, and the anchoring of functional groups or molecules [119]. Heteroatom-doped graphene can directly reduce hydrogen peroxide (H_2O_2), a crucial signaling molecule in cells and a consequence of several biological activities. An enzyme-free H_2O_2 sensor based on an N-doped graphene electrode was disclosed by Shao et al. [120]. Since the O–O bond of H_2O_2 is weakened by nitrogen-induced charge delocalization and the reduction reaction benefits from N-dopants' propensity to donate electrons, the sensor displayed good sensitivity and a broad linear detection range. An innovative Ag NPs based on N-doped graphene hybrid material was created by Tian et al. and employed as an electrochemical sensor for the detection of H_2O_2 [121].

7.7 Conclusion

Electronics, energy storage and conversion, catalysis, sensing, and biological research applications have all seen significant changes as a result of graphene. Notably, graphene has been found to exhibit a wide range of intriguing qualities, such as a high specific surface area, exceptional mechanical strength and flexibility, unmatched thermal and electrical conductivity, and superior electronic properties. Uncovering the full potential of graphene and its derivatives still requires enormous work. Either a metal with vanishing Fermi surfaces or a semiconductor with no bandgap can be used to describe graphene. It is intriguing to induce a bandgap in graphene to enhance the aforementioned applications because the lack of an intrinsic bandgap severely restricts the use of pristine graphene in fields like nanoelectronics, energy storage, and electrocatalysis. Heteroatom doping can give graphene a variety of new or enhanced optical, structural, physicochemical, and electromagnetic properties. In order to use graphene and its derivatives for a variety of purposes, such as sensing, nonlinear optics, solar cells, nanoelectronics, energy applications, and so forth, it is crucial and important to tune the properties of these materials through functionalization with organic and inorganic materials.

References

[1] M. Loos, in *Carbon Nanotube Reinforced Composites*, 2015, **3**, 1–36.
[2] Z. L. Wang, *Charact. Nanophase Mater*, 2001, **1**, 1–12.
[3] G. M. Whitesides, *Small*, 2005, **1**, 172–179.
[4] M. Meyyappan, *IEEE Nanotechnol. Mag.*, 2009, **3**, 1–36.
[5] K. Geim, *Science*, 2009, **324**, 1530–1534.
[6] S. Stankovich, D. A. Dikin, G. H. B. Dommett, K. M. Kohlhaas, E. J. Zimney, E. A. Stach, R. D. Piner, S. T. Nguyen and R. S. Ruoff, *Nature*, 2006, **442**, 282–286.

[7] G. Reina, J. M. González-Domínguez, A. Criado, E. Vázquez, A. Bianco and M. Prato, *Chem. Soc. Rev.*, 2017, **46**, 4400.
[8] A. K. Geim and K. S. Novoselov, *Nat. Mater.*, 2007, **6**, 183–191.
[9] K. S. K. S. Kim, Y. Zhao, H. Jang, S. Y. Lee, J. M. Kim, J. H. H. Ahn, P. Kim, J. Y. Y. Choi, and B. H. Hong, *Nature*, 2009, **457**, 706–710.
[10] K. S. Novoselov, E. McCann, S. V. Morozov, V. I. Falko, M. I. Katsnelson, U. Zeitler, D. Jiang, F. Schedin and K. Geim, *Nat. Phys.*, 2006, **180**, 177–180.
[11] F. Wendler, A. Knorr, and E. Malic, *Nanophotonics*, 2015, **4**, 224–249.
[12] F. Molitor, J. Güttinger, C. Stampfer, S. Droscher, A. Jacobsen, T. Ihn and K. Ensslin, *J. Phys. Condens. matter.*, 2011, **23**, 243201-243207.
[13] M. F. Craciun, S. Russo, M. Yamamoto and S. Tarucha, *Nano Today*, 2011, **6**, 42–60.
[14] P. Avouris, *Nano Lett.*, 2010, **10**, 4285–4294.
[15] L. Pauling, *J. Chem. Educ.*, 1992, **69**, 519-529.
[16] X. F. Li, K. Y. Lian, Q. Qiu and Y. Luo, *Nanoscale*, 2015, **7**, 4156–4162.
[17] S. Onoda and M. Imada, *Phys. Rev. B*, 2003, **67**, 161102-1611109.
[18] Y. B. Zhang, Y. W. Tan, H. L. Stormer and P. Kim, *Nature*, 2005, **438**, 201–204.
[19] X. Wu, Y. Hu, M. Ruan, N. K. Madiomanana, J. Hankinson, M. Sprinkle, C. Berger and W. A. De Heer, *Appl. Phys. Lett.*, 2009, **95,** 1121-1125.
[20] Z. Jiang, Y. Zhang, Y. W. Tan, H. L. Stormer and P. Kim, *Solid State Commun.*, 2007, **143**, 14–19.
[21] C. Soldano, A. Mahmood and E. Dujardin, *Carbon N. Y.*, 2010, **48**, 2127–2150.
[22] X. Wang, G. Sun, P. Routh, D. H. Kim, W. Huang and P. Chen, *Chem. Soc. Rev.*, 2014, **43**, 7067–7098.
[23] J. L. McChesney et al., Phys. Rev. Lett, 2010 104, 136803.
[24] S. Link et al., Phys. Rev. B, 2019, 100, 121407
[25] J. Duan, S. Chen, M. Jaroniec and S. Z. Qiao, *ACS Catal.*, 2015, **5**, 5207–5234.
[26] L. K. Putri, W. J. Ong, W. S. Chang and S. P. Chai, *Applied Surface Science*, 2015, **358**, 2–14.
[27] A. Ariharan, B. Viswanathan and V. Nandhakumar, *Graphene*, 2016, **5**, 39–50.
[28] J. Zhang, J. Li, Z. Wang, X. Wang, W. Feng, W. Zheng, W. Cao and P. Hu, *Chem. Mater.*, 2014, **26**, 2460–2466.

[29] L. S. Panchakarla, K. S. Subrahmanyam, S. K. Saha, A. Govindaraj, H. R. Krishnamurthy, U. V. Waghmare and C. N. R. Rao, *Adv. Mater.*, 2009, **21**, 4726–4730.

[30] L. Dai, Y. Xue, L. Qu, H. J. Choi and J. B. Baek, *Chem. Rev.*, 2015, **115**, 4823–4892.

[31] B. Guo, L. Fang, B. Zhang and J. R. Gong, *Insciences J.*, 2011, **10**, 80–89.

[32] L. Xian L, D. M. Kennes, N. Dejean, M. Altarelli and A. Rubio, *Nano Lett.* 2019, **19**, 84934-4940.

[33] Y. B. Tang, L. C. Yin, Y. Yang, X. H. Bo, Y. L. Cao, H. E. Wang, W. J. Zhang, I. Bello, S. T. Lee, H. M. Cheng and C. S. Lee, *ACS Nano*, 2012, **6**, 1970–1978.

[34] L. Zhao, M. Levendorf, S. Goncher, T. Schiros, L. Pálová, A. Zabet-Khosousi, K. T. Rim, C. Gutiérrez, D. Nordlund, C. Jaye, M. Hybertsen, D. Reichman, G. W. Flynn, J. Park and A. N. Pasupathy, *Nano Lett.*, 2013, **13**, 4659–4665.

[35] D. Usachov, O. Vilkov, A. Grüneis, D. Haberer, A. Fedorov, V. K. Adamchuk, A. B. Preobrajenski, P. Dudin, A. Barinov, M. Oehzelt, C. Laubschat and D. V. Vyalikh, *Nano Lett.*, 2011, **11**, 5401–5407.

[36] C. Zhang, L. Fu, N. Liu, M. Liu, Y. Wang and Z. Liu, *Adv. Mater.*, 2011, **23**, 1020–1024.

[37] H. Wang, T. Maiyalagan and X. Wang, *ACS Catal.*, 2012, **2**, 781–794.

[38] Z. Jin, J. Yao, C. Kittrell and J. M. Tour, *ACS Nano*, 2011, **5**, 4112–4117.

[39] L. Sun, L. Wang, C. Tian, T. Tan, Y. Xie, K. Shi, M. Li and H. Fu, *RSC Adv.*, 2012, **2**, 4498.

[40] X. Li, H. Wang, J. T. Robinson, H. Sanchez, G. Diankov and H. Dai, *J. Am. Chem. Soc.*, 2009, **131**, 15939–15944.

[41] P. A. Denis, R. Faccio and A. W. Mombru, *ChemPhysChem*, 2009, **10**, 715–722.

[42] L. Zhang, J. Niu, M. Li and Z. Xia, *J. Phys. Chem. C*, 2014, **118**, 3545–3553.

[43] I. Y. Jeon, S. Zhang, L. Zhang, H. J. Choi, J. M. Seo, Z. Xia, L. Dai and J. B. Baek, *Adv. Mater.*, 2013, **25**, 6138–6145.

[44] H. L. Poh, P. Šimek, Z. Sofer and M. Pumera, *ACS Nano*, 2013, **7**, 5262–5272.

[45] Z. Wang, P. Li, Y. Chen, J. He, W. Zhang, O. G. Schmidt and Y. Li, *Nanoscale*, 2014, **6**, 7281-7289.

[46] D. Sen, R. Thapa and K. K. Chattopadhyay, *ChemPhysChem*, 2014, **15**, 2542–2549.

[47] X. H. Li and M. Antonietti, *Angew. Chemie - Int. Ed.*, 2013, **52**, 4572–4576.

[48] Z. H. Sheng, L. Shao, J. J. Chen, W. J. Bao, F. Bin Wang and X. H. Xia, *ACS Nano*, 2011, **5**, 4350–4358.

[49] J. Xu, G. Dong, C. Jin, M. Huang and L. Guan, *ChemSusChem*, 2013, **6**, 493–499.

[50] X. Wang, G. Sun, P. Routh, D. H. Kim, W. Huang and P. Chen, *Chem. Soc. Rev.*, 2014, **43**, 7067–7098.

[51] C. K. Chang, S. Kataria, C. C. Kuo, A. Ganguly, B. Y. Wang, J. Y. Hwang, K. J. Huang, W. H. Yang, S. B. Wang, C. H. Chuang, M. Chen, C. I. Huang, W. F. Pong, K. J. Song, S. J. Chang, J. H. Guo, Y. Tai, M. Tsujimoto, S. Isoda, C. W. Chen, L. C. Chen and K. H. Chen, *ACS Nano*, 2013, **7**, 1333–1341.

[52] L. Ci, L. Song, C. Jin, D. Jariwala, D. Wu, Y. Li, A. Srivastava, Z. F. Wang, K. Storr, L. Balicas, F. Liu and P. M. Ajayan, *Nat. Mater.*, 2010, **9**, 430–435.

[53] Y. Xue, D. Yu, L. Dai, R. Wang, D. Li, A. Roy, F. Lu, H. Chen, Y. Liu and J. Qu, *Phys. Chem. Chem. Phys.*, 2013, **15**, 12220-12227.

[54] O. S. Kwon, S. J. Park, J. Y. Hong, A. R. Han, J. S. Lee, J. S. Lee, J. H. Oh and J. Jang, *ACS Nano*, 2012, **6**, 1486–1493.

[55] V. Leon, M. Quintana, M. A. Herrero, J. L. G. Fierro, A. Hoz, M. Prato and E. Vazquez, *Chem. Commun.*, 2011, **47**, 10936.

[56] L. Yan, M. Lin, C. Zeng, Z. Chen, S. Zhang, X. Zhao, A. Wu, Y. Wang, L. Dai, J. Qu, M. Guo and Y. Liu, *J. Mater. Chem.*, 2012, **22**, 8367.

[57] I. Y. Jeon, H. J. Choi, M. Choi, J. M. Seo, S. M. Jung, M. J. Kim, S. Zhang, L. Zhang, Z. Xia, L. Dai, N. Park and J. B. Baek, *Sci. Rep.*, 2013, **3**, 1810-1818.

[58] X. Lü, J. Wu, T. Lin, D. Wan, F. Huang, X. Xie and M. Jiang, *J. Mater. Chem.*, 2011, **21**, 10685-10691.

[59] D. Deng, X. Pan, L. Yu, Y. Cui, Y. Jiang, J. Qi, W. X. Li, Q. Fu, X. Ma, Q. Xue, G. Sun and X. Bao, *Chem. Mater.*, 2011, **23**, 1188–1193.

[60] Z. Zuo, Z. Jiang and A. Manthiram, *J. Mater. Chem. A*, 2013, **1**, 13476-134881.

[61] D. Du, P. Li and J. Ouyang, *ACS Appl. Mater. Interfaces*, 2015, **7**, 26952–26958.

[62] A. Leela, M. Reddy, A. Srivastava, S. R. Gowda, H. Gullapalli, M. Dubey and P. M. Ajayan, *ACS Nano*, 2010, **4**, 6337–6342.

[63] S. Wang, L. Zhang, Z. Xia, A. Roy, D. W. Chang, J. B. Baek and L. Dai, *Angew. Chemie - Int. Ed.*, 2012, **51**, 4209–4212.

[64] L. Lai, J. R. Potts, D. Zhan, L. Wang, C. K. Poh, C. Tang, H. Gong, Z. Shen, J. Lin and R. S. Ruoff, *Energy Environ. Sci.*, 2012, **5**, 7936-7941.

[65] C. H. Choi, M. W. Chung, S. H. Park and S. I. Woo, *RSC Adv.*, 2013, **3**, 4246-6250.

[66] S. Stankovich, D. A. Dikin, R. D. Piner, K. A. Kohlhaas, A. Kleinhammes, Y. Jia, Y. Wu, S. B. T. Nguyen and R. S. Ruoff, *Carbon N. Y.*, 2007, **45**, 1558–1565.

[67] P. Wu, Z. Cai, Y. Gao, H. Zhang and C. Cai, *Chem. Commun.*, 2011, **47**, 11327-11332.

[68] D. Long, W. Li, L. Ling, J. Miyawaki, I. Mochida and S. H. Yoon, *Langmuir*, 2010, **26**, 16096–16102.

[69] H.-L. Guo, P. Su, X. Kang and S. K. Ning, *J. Mater. Chem. A*, 2013, **1**, 2248–2255.

[70] Y. Chang, G. Han, J. Yuan, D. Fu, F. Liu and S. Li, *J. Power Sources*, 2013, **238**, 492–500.

[71] Z. S. Wu, A. Winter, L. Chen, Y. Sun, A. Turchanin, X. Feng and K. Mllen, *Adv. Mater.*, 2012, **24**, 5130–5135.

[72] Y. Lu, S. T LoJheng, C. Lin, W. Zhang, J. Y. Lu, F. H. Liu, C. M. Tseng, Y. H. Lee, C. Liang and L. Li, *ACS Nano* 2013, **7**, 6522–6532.

[73] Zhao, L.; et al. *Science,* 2011, **333**, 999–1003.

[74] A. Lherbier, X. Blase, Y. Niquet, F. Triozon, S. Roche, *Phys. Rev. Lett.* 2008, **101**, 036808.

[75] C. Zhang, L. Fu, N. Liu, M. Liu, Y. Wang, Z. Liu, *Adv. Mater.* 2011, **23**, 1020–1024

[76] J. Zhang, C. Zhao, N. Liu *et al. Sci Rep,* 2016, **6,** 28330

[77] D. Usachov, O. Vilkov, A. Grüneis, D. Haberer, A. Fedorov, V. K. Adamchuk, A. B. Preobrajenski, P. Dudin⊥, A. Barinov⊥, M. Oehzelt, C. Laubschat, and D. V. Vyalikh, *Nano Lett.* 2011, **11**, 5401–5407.

[78] S. Seenithuraia , R. Pandyanb , S. V. Kumarc and M. Mahendran, Nano Hybrids, 2013**, 5,** 65-83.

[79] A. Jafari, A. Ramezani, and V.Fayaz, Canadian Journal of Physics, 2019, 9 , 97.

[80] A. Michota and J. Bukowska, *J. Raman Spectrosc.*, 2003, **34**, 21–25.

[81] K. Kneipp, *Phys. Today*, 2007, **60**, 40–46.

[82] X. C. Yu, J. Tao, Y. D. Shen, G. Z. Liang, T. Liu, Y. Z. Zhang and Q. J. Wang, *Nanoscale*, 2014, **6**, 9925–9929.

[83] P. H. C. Camargo, C. M. Cobley, M. Rycenga and Y. Xia, *Nanotechnology*, 2009, **20**, 434020-434029.

[84] Y. Liu, J. Zhou, B. Wang, T. Jiang, H. P. Ho, L. Petti and P. Mormile, *Phys. Chem. Chem. Phys.*, 2015, **17**, 6819–6826.
[85] Y. Li, X. Zhao, P. Zhang, J. Ning, J. Li, Z. Su and G. Wei, *J. Mater. Chem. C*, 2015, **3**, 4126–4133.
[86] S. Dutta, C. Ray, S. Sarkar, M. Pradhan, Y. Negishi and T. Pal, *ACS Appl. Mater. Interfaces,* 2013, **5**, 8724-8732.
[87] X. Kong and Q. Chen, *J. Mater. Chem.*, 2012, **22**, 15336–15341.
[88] G. Xu, J. Liu, Q. Wang, R. Hui, Z. Chen, V. A. Maroni and J. Wu, *Adv. Mater.*, 2012, **24**, 71-76.
[89] H. Hou, P. Wang, J. Zhang, C. Li and Y. Jin, *ACS Appl. Mater. Interfaces*, 2015, **7**, 18038–18045.
[90] C. Hu, J. Rong, J. Cui, Y. Yang, L. Yang, Y. Wang and Y. Liu, *Carbon N. Y.*, 2013, **51**, 255–264.
[91] K. Long, X. Luo, H. Nan, D. Du, W. Zhao, Z. Ni and T. Qiu, *J. Appl. Phys.*, 2013, **114**. 183520-183524.
[92] J. F. Li, J. R. Anema, T. Wandlowski and Z. Q. Tian, *Chem. Soc. Rev.*, 2015, **44**, 8399–8409.
[93] J. Wu, Y. Xu, P. Xu, Z. Pan, S. Chen, Q. Shen, L. Zhan, Y. Zhang and W. Ni, *Nanoscale*, 2015, **7**, 17529–17537.
[94] Y. Zhao, W. Zeng, Z. Tao, P. Xiong, Kong, M. S. Dresselhaus and X. Zhang, *Chem. Commun.*, 2015, **51**, 866–869.
[95] K. Ben Liew, W. R. W. Daud, M. Ghasemi, J. X. Leong, S. Su Lim and M. Ismail, *Int. J. Hydrogen Energy*, 2014, **39**, 4870–4883.
[96] F. Jaouen, E. Proietti, M. Lefevre, R. Chenitz, J. P. Dodelet, G. Wu, H. T. Chung, C. M. Johnston and P. Zelenay, *Energy Environ. Sci.*, 2011, **4**, 114-124.
[97] C. Song and J. Zhang, *PEM Fuel Cell Electrocatal. Catal. Layers*, 2008, **6**, 89–134.
[98] B. Wang, *J. Power Sources*, 2005, **152**, 1–15.
[99] M. Kaukonen, A. V. Krasheninnikov, E. Kauppinen and R. M. Nieminen, *ACS Catal.*, 2013, **3**, 159–165.
[100] M. Shao, Q. Chang, J. P. Dodelet and R. Chenitz, *Chem. Rev.*, 2016, **116**, 3594–3657.
[101] L. Zhang and Z. Xia, *J. Phys. Chem. C*, 2011, **115**, 11170–11176.
[102] Z. Wu, S. Yang, Y. Sun, K. Parvez, X. Feng and K. Mullen, *J. Am. Chem. Soc.*, 2012, **134**, 9082–9085.
[103] C. Zhang, N. Mahmood, H. Yin, F. Liu and Y. Hou, *Adv. Mater.*, 2013, **25**, 4932–4937.

[104] L. Chen, X. Cui, Y. Wang, M. Wang, R. Qiu, Z. Shu, L. Zhang, Z. Hua, F. Cui, C. Wei and J. Shi, *Dalt. Trans.*, 2014, **43**, 3420-3426.

[105] A. L. M. Reddy, A. Srivastava, S. R. Gowda, H. Gullapalli, M. Dubey and P. M. Ajayan, *ACS Nano*, 2010, **4**, 6337–6342.

[106] J. Liang, Y. Jiao, M. Jaroniec and S. Z. Qiao, *Angew. Chemie - Int. Ed.*, 2012, **51**, 11496–11500.

[107] R. Wang, D. C. Higgins, M. A. Hoque, D. Lee, F. Hassan and Z. Chen, *Sci. Rep.*, 2013, **3**, 2431-2440.

[108] D. Das, S. Kim, K.-R. Lee and A. K. Singh, *Phys. Chem. Chem. Phys.*, 2013, **15**, 15128-15136.

[109] Y. Liu, V. I. Artyukhov, M. Liu, A. R. Harutyunyan and B. I. Yakobson, *J. Phys. Chem. Lett.*, 2013, **4**, 1737–1742.

[110] D. H. Wu, Y. F. Li and Z. Zhou, *Theor. Chem. Acc.*, 2011, **130**, 209–213.

[111] S. Gao, Z. Ren, L. Wan, J. Zheng, P. Guo and Y. Zhou, *Appl. Surf. Sci.*, 2011, **257**, 7443–7446.

[112] Z. L. Wang, D. Xu, H. G. Wang, Z. Wu and X. B. Zhang, *ACS Nano*, 2013, **7**, 2422–2430.

[113] X. Wang, X. Cao, L. Bourgeois, H. Guan, S. Chen, Y. Zhong, D. M. Tang, H. Li, T. Zhai, L. Li, Y. Bando and D. Golberg, *Adv. Funct. Mater,* 2012, **22**, 2682–2690.

[114] L. Lai, J. Zhu, Z. Li, D. Y. W. Yu, S. Jiang, X. Cai, Q. Yan, Y. M. Lam, Z. Shen and J. Lin, *Nano Energy*, 2014, **3**, 134–143.

[115] D. Li, D. Shi, Z. Liu, H. Liu and Z. Guo, *J. Nanoparticle Res.*, 2013, **15**, 237-245.

[116] R. Lv, Q. Li, A. R. Botello-Méndez, T. Hayashi, B. Wang, A. Berkdemir, Q. Hao, A. L. Elías, R. Cruz-Silva, H. R. Gutiérrez, Y. A. Kim, H. Muramatsu, J. Zhu, M. Endo, H. Terrones, J. C. Charlier, M. Pan and M. Terrones, *Sci. Rep.*, 2012, **2**, 586-592.

[117] F. Niu, L. M. Tao, Y. C. Deng, Q. H. Wang and W. G. Song, *New J. Chem.*, 2014, **38**, 2269-2275.

[118] Y. Wang, Y. Shao, D. W. Matson, J. Li and Y. Lin, *ACS Nano*, 2010, **4**, 1790-1798.

[119] Y. Tian, F. Wang, Y. Liu, F. Pang and X. Zhang, *Electrochim. Acta*, 2014, **146**, 646–653.

[120] Y. Tian, Y. Liu, W. Wang, X. Zhang and W. Peng, *J. Nanoparticle Res.*, 2015, **17**, 1–9.

[121] J. Zhang, *PEM fuel cell electrocatalysts and catalyst layers: Fundamentals and applications*, 2008.

8

Coupling of Electrostatic Ion Cyclotron and Ion Acoustic Waves in a Five-Component Cometary Plasma

Manesh Michael[1], Sijo Sebastian[2], Shilpa S.[3], and Chandu Venugopal[4]

[1]Department of Physics, Bharata Mata College, India
[2]Department of Physics, St. Aloysius College, India
[3]International School of Photonics, Cochin University of Science and Technology, India
[4]School of Pure & Applied Physics, Mahatma Gandhi University, India

Abstract

The electrostatic ion cyclotron and ion acoustic waves are investigated in a five-component cometary plasma composed of hydrogen and positively and negatively charged oxygen ions and kappa defined slightly colder and hotter solar cometary electrons. A linear dispersion relation for the system is derived and studied numerically. It is found that at larger propagation angles, the separation between cyclotron and acoustic modes increases, which indicate a weak coupling between the two modes. For perpendicular propagation, the ion acoustic mode nearly vanishes and only the cyclotron mode exists. The influence of various parameters such as superthermality, densities of ions, charge numbers, and temperature ratios of pair ions on the dispersion characteristics is extensively investigated.

Keywords: Electrostatic ion cyclotron, ion acoustic waves, Five component cometary plasma

8.1 Introduction

The electrostatic ion cyclotron (EIC) and ion-acoustic (IA) modes are low-frequency waves, which exist in different plasma environments. EIC

DOI: 10.1201/9781003337102-8

waves have frequencies nearer to the cyclotron frequency and propagate perpendicular to the ambient magnetic field. The first experimental observations of EIC waves were carried out by D'Angelo and Motley [1, 2]. Mosier and Gurnett identified EIC waves in the upper layers of the ionosphere [3]. Various satellites such as ISEE-1, Viking, Polar, and FAST have unambiguously detected EIC waves in the auroral regions [4–7] ; these waves were also identified in the Earth's magnetosphere [8].

The low-frequency IA waves in multi-ion plasmas were also studied extensively by many authors in recent times [9–15] . Some experimental results on IA waves in plasma with negative ions were also reported by Song et al. [16, 17]. In these studies, they found that, as the negative ion concentration increased, the frequency of the IA waves increased, and the critical electric drift velocity needed for wave excitation decreased. A theoretical and experimental study on IA waves in plasma composed of negative dust grains was presented by Merlino et al. [18]. They found that the traces of dust grains reformed the properties of the IA waves through the quasi-neutrality condition.

Observations of IA waves in the solar wind during the first and second orbits of the Ulysses spacecraft were presented by Lin et al. [19]. IA waves were also identified in different regions of Earth's magnetosphere [20–22] . In most of these studies, both IA and EIC waves were studied independently. A study on the coupling between EIC and IA wave modes in a laboratory plasma was done by Ohnuma et al. [23]. The THEMIS satellite pointed out a possible coupling between IA and EIC modes in the magnetosphere of the Earth [8].

The traces of high energetic particles in the distribution tail forces plasmas existing in various space ambiences to deviate appreciably from the famous Maxwellian distribution. Vasyliunas first introduced a non-Maxwellian distribution from solar wind data [24]; this distribution, which was later named as the "kappa distribution," has been observed in many astrophysical and magnetospheric surroundings. The coupling of EIC and IA waves in solar wind composed of superthermal electrons, protons, and alpha particles has been studied recently by Sreeraj et al. [25].

A cometary plasma consists of new born heavier ions, hydrogen ions, and electrons with relative densities relying on their distances from the nucleus. Initially, positively charged oxygen ions were treated as the major heavier ion [26, 27]. However, Chaizy et al. reported the presence of negatively charged oxygen ions in the coma of comet Halley [28]. This important discovery enables one to treat a cometary plasma as a pair-ion plasma (O^+, O^-) with other lighter and heavier ions constituting the other ionic components of the plasma. Various studies have also revealed that a cometary environment has more than one component of electron [29, 30]. And, interestingly, the two species of electrons found at comet 67P/Churyumov–Gerasimenko were recently described by kappa distributions [31].

We, therefore, study the coupling of EIC and IA waves in a five-component cometary plasma consisting of hydrogen ions and positively and negatively charged oxygen ions and kappa modeled slightly colder and hot solar cometary electrons. A linear dispersion relation for the system is derived. It is found that at larger propagation angles, the separation between cyclotron and acoustic modes increases, which indicates a weaker coupling of the two modes at these angles.

8.2 Basic Equations and Theoretical Model

We discuss a magnetized five-component cometary plasma composed of hydrogen and positively and negatively charged heavier ions (O^+ and O^-) and kappa described slightly colder and hot solar cometary photo-electrons. The external magnetic field is applied in the z-axis ($B_0\hat{z}$). The charge neutrality of the multi-species plasma system is defined using the following equation:

$$n_{\mathrm{ce}0} + n_{\mathrm{se}0} + Z_1 n_{10} = n_{H0} + Z_2 n_{20},$$

where $n_{\mathrm{ce}0}$ and $n_{\mathrm{se}0}$ indicate the equilibrium densities of cometary and solar electrons, respectively, while n_{10}, n_{20}, and n_{H0} are the equilibrium densities of negatively charged oxygen (O^-) ions, positively charged oxygen (O^+) ions, and hydrogen ions, respectively. Z_1 and Z_2 represent the charge numbers of O^- and O^+ ions, respectively.

The kappa distribution of species "s" is given by

$$n_s = n_{s0}\left[1 - \frac{e\varphi}{k_B T_s(\kappa_s - \frac{3}{2})}\right]^{-\kappa_s + \frac{1}{2}}. \tag{8.1}$$

In eqn (8.1), $s = \mathrm{se}$ for solar electrons and $s = \mathrm{ce}$ for cometary photo-electrons. $n_s$indicates the density (with the subscript "0" denoting the equilibrium value), e the charge, T_s the temperature, and κ_s the spectral index for species "s." $k_B$is the Boltzmann's constant and φ is the potential.

The propagation characteristics of the ions are defined using the following fluid equations:

$$\frac{\partial n_j}{\partial t} + \nabla.(n_j v_j) = 0. \tag{8.2}$$

$$m_j n_j \left(\frac{\partial}{\partial t} + \overline{v}_j \cdot \nabla\right) \overline{v}_j = eZ_j n_j \left(-\nabla\phi + \frac{\overline{v}_j \times B_0}{c}\right) - \nabla P_j. \tag{8.3}$$

$$\nabla^2 \phi = -4\pi e \left(Z_2 n_2 + n_H - Z_1 n_1 - n_{\rm ce} - n_{\rm se}\right), \tag{8.4}$$

where v_j and m_j denote, respectively, the fluid velocity and mass of the *j*-species of ions (*j* =O^-, O^+, H^+). In eqn (8.3), the adiabatic equation of state for ions is $\frac{P_j}{P_{j0}} = \left(\frac{n_j}{n_{jo}}\right)^\gamma = N_j^\gamma$ where$P_{j0} = n_{j0} k_B T_j$ and $\gamma = \frac{N+2}{N}$ for an *N*-dimensional system. Here, we are considering a three-dimensional system and, hence, γ = 5/3.

Linearization and simplification of eqn (8.2)–(8.4) and the use of eqn (8.1) yield the following dispersion relation:

$$K - \frac{\omega_{p2}^2(\omega^2 - \Omega_2^2 \cos^2\theta)}{\omega^4 - \omega^2(\Omega_2^2 + \gamma_2 v_{t2}^2 k^2) + \gamma_2 v_{t2}^2 k^2 \Omega_2^2 \cos^2\theta} - \frac{\omega_{p1}^2(\omega^2 - \Omega_1^2 \cos^2\theta)}{\omega^4 - \omega^2(\Omega_1^2 + \gamma_1 v_{t1}^2 k^2) + \gamma_1 v_{t1}^2 k^2 \Omega_1^2 \cos^2\theta} - \frac{\omega_{\rm pH}^2(\omega^2 - \Omega_H^2 \cos^2\theta)}{\omega^4 - \omega^2(\Omega_H^2 + \gamma_H v_{tH}^2 k^2) + \gamma_H v_{tH}^2 k^2 \Omega_H^2 \cos^2\theta} = 0 \tag{8.5}$$

where $K = 1 + \frac{\omega_{\rm pce}^2}{k^2 v_{\rm tce}^2}\frac{(2\kappa_{\rm ce}-1)}{(2\kappa_{\rm ce}-3)} + \frac{\omega_{\rm pse}^2}{k^2 v_{\rm tse}^2}\frac{(2\kappa_{\rm se}-1)}{(2\kappa_{\rm se}-3)}$, $\omega_{pj} = \left(\frac{4\pi n_{j0} Z_j^2 e^2}{m_j}\right)^{1/2}$ is the plasma frequency of the *j*th species of ions (*j* =1 for O^- = 2 for O^+ and H for H^+), $\Omega_j = \frac{Z_j e B_0}{m_j c}$ is the corresponding cyclotron frequency of the ions, and $v_{ts} = \left(\frac{k_B T_S}{m_s}\right)^{1/2}$ is the thermal velocity of the species *s* (*s* = ce, se, H, 1, 2) with T_s as its corresponding temperature.

As a check on eqn (8.5), we notice that it changes to the corresponding relation in the work of Sreeraj et al. [25], when the second electron (se) and third ion component (O^-) are neglected.

8.3 Discussions

We now discuss the various limits of eqn (8.5). We first consider parallel propagation.

8.3.1 Parallel propagation

For parallel propagation, the propagation angle θ becomes 0° ($\cos\theta = 1$). The dispersion relation (8.5) then reduces to

$$K - \frac{\omega_{p2}^2}{\omega^2 - \gamma_2 v_{t2}^2 k^2} - \frac{\omega_{p1}^2}{\omega^2 - \gamma_1 v_{t1}^2 k^2} - \frac{\omega_{pH}^2}{\omega^2 - \gamma_H v_{tH}^2 k^2} = 0. \tag{8.6}$$

Eqn (8.6) is a dispersion relation of order 6 and represents the IA mode and IC mode; the IA mode is completely decoupled from the ion cyclotron mode under certain conditions. As a special case, we discuss the situation when $T_1 = T_2$ ($v_{t1} = v_{t2} = v_t$) and $\gamma_1 = \gamma_2 = \gamma$. Under these conditions, the dispersion relation becomes a quadratic equation in ω^2. The two solutions of the equation are obtained as

$$\omega_{\pm}^2 = \frac{B_1 \pm \sqrt{B_1^2 - 4KC_1}}{2K} \tag{8.7}$$

where

$$B_1 = Kk^2\left(\gamma v_t^2 + \gamma_H v_{tH}^2\right) + \omega_{p1}^2 + \omega_{p2}^2 + \omega_{\mathrm{pH}}^2$$

$$C_1 = (\omega_{p1}^2 + \omega_{p2}^2)\gamma_H v_{tH}^2 k^2 + \omega_{\mathrm{pH}}^2 \gamma v_t^2 k^2 + K\gamma\gamma_H v_t^2 v_{tH}^2 k^4$$

where ω_{+}^2 represents the fast acoustic mode and ω_{-}^2 represents the slow acoustic mode. It is found that in the fast acoustic mode, the plasma frequency of ions is modified by the acoustic contribution, and in the slow mode, it is driven by the acoustic contribution alone. It is also observed that solution (8.7) becomes imaginary for the condition $\omega_{p1}^2 + \omega_{p2}^2 + Kk^2\left(\gamma v_t^2 - \gamma_H v_{tH}^2\right) < 2\sqrt{K\gamma}k\omega_{\mathrm{pH}} v_t$. That is, the growth of instability of the wave modes occurs when the plasma frequency of the hydrogen ions dominates the sum of the plasma frequencies of the heavier ions.

8.3.2 Perpendicular propagation

For perpendicular propagation, θ is 90°, i.e., $\cos\theta = 0$. Therefore, the dispersion relation (8.5) becomes

$$K - \frac{\omega_{p2}^2}{\omega^2 - \Omega_2^2 - \gamma_2 v_{t2}^2 k^2} - \frac{\omega_{p1}^2}{\omega^2 - \Omega_1^2 - \gamma_1 v_{t1}^2 k^2} - \frac{\omega_{\mathrm{pH}}^2}{\omega^2 - \Omega_H^2 - \gamma_H v_{tH}^2 k^2} = 0. \tag{8.8}$$

Eqn (8.8) describes the ion cyclotron modes existing within the plasma. Here too, we discuss the special case when $T_1 = T_2$ ($v_{t1} = v_{t2} = v_t$), $\gamma_1 = \gamma_2 = \gamma$, and $Z_1 = Z_2$,i.e., $\Omega_1 = \Omega_2 = \Omega$. Under these conditions, eqn (8.8) has the following solutions:

$$\omega_{\pm}^2 = \frac{B_2 \pm \sqrt{B_2^2 - 4KC_2}}{2K} \tag{8.9}$$

where

$$B_2 = \omega_{p1}^2 + \omega_{p2}^2 + \omega_{\mathrm{pH}}^2 + K\left(\Omega^2 + \Omega_H^2 + k^2\gamma v_t^2 + k^2\gamma_H v_{tH}^2\right)$$
$$C_2 = \left[\omega_{p1}^2 + \omega_{p2}^2 + (\Omega^2 + \gamma v_t^2 k^2)K\right]\left(\Omega_H^2 + \gamma_H v_{tH}^2 k^2\right) + \omega_{\mathrm{pH}}^2(\Omega^2 + \gamma v_t^2 k^2)$$

where ω_+^2 represents the fast cyclotron mode and ω_-^2 represents the slow cyclotron mode. It is obvious that both the fast and slow modes are affected by the gyro frequencies of the ions. It is also observed that solution (8.9) becomes imaginary and the instability of the modes occurs for the condition

$$\omega_{p1}^2 + \omega_{p2}^2 + K\left[(\Omega^2 + \gamma k^2 v_t^2) - (\Omega_H^2 + \gamma_H k^2 v_{tH}^2)\right] < 2\omega_{\mathrm{pH}}\sqrt{K(\Omega^2 + \gamma k^2 v_t^2)}.$$

8.4 Results

We consider the parameters observed at comet Halley, namely a hydrogen density $n_{H0} = 4.95\ \mathrm{cm}^{-3}$, positively charged oxygen density $n_{20} = 0.5\ \mathrm{cm}^{-3}$, negatively charged oxygen density $n_{10} = 0.05\ \mathrm{cm}^{-3}$, hydrogen temperature $T_H = 8 \times 10^4$K, solar electron temperature $T_{\mathrm{se}} = 2 \times 10^5$ K, cometary electron temperature $T_{\mathrm{ce}} = 2 \times 10^4$ K, and oxygen temperatures $T_1 = T_2 = 1.16 \times 10^4$K. The background magnetic field is 8×10^{-5}G [28, 32].

For the numerical analysis, we normalize the dispersion relation (8.5) using the following parameters: the frequencies are normalized by the cyclotron frequency of negatively charged oxygen ions ($\frac{\omega}{\Omega}$), and wavenumber by the Larmor radius of the negatively charged oxygen ions ρ_1 given by the expression $\rho_1 = \frac{v_{t1}}{\Omega_1}$. EIC modes have a frequency greater than the cyclotron frequency of ions, and, hence, the normalized value of the frequency is greater than 1.

Dispersion relation (8.5) is an equation of order 12. Among the 12 roots, only two are feasible for the EIC mode and one for the IA mode for the

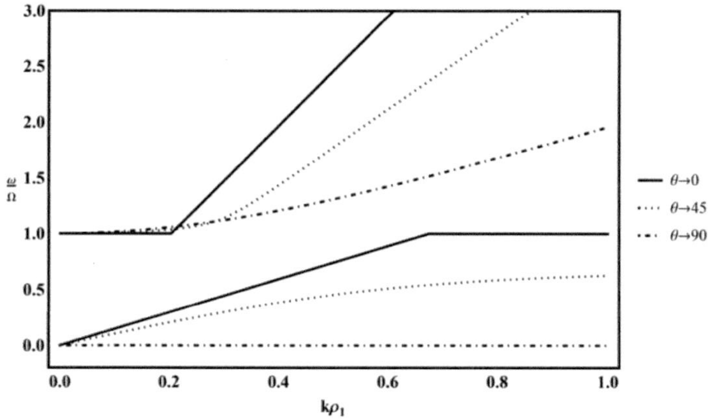

Figure 8.1 (a) Dispersion characteristics of EIC and IA modes for various angles of propagation.

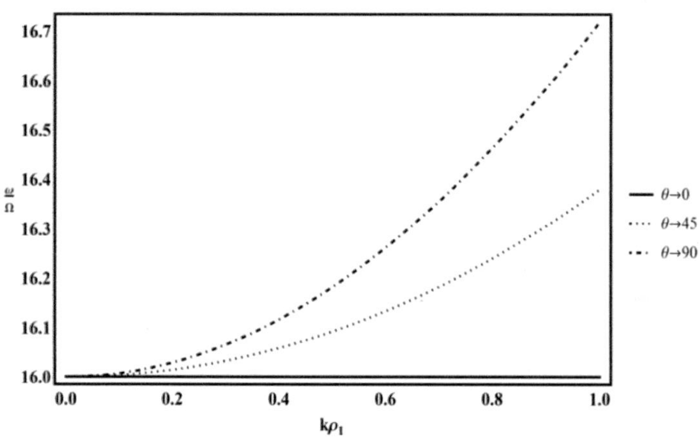

Figure 8.1 (b) Dispersion characteristics of EIC mode for various angles of propagation (for second possible solution).

parameters given above, and $\kappa_{ce} = \kappa_{se} = 2$. Figure 8.1(a) shows the dispersion characteristics of the EIC mode for the first possible root and IA mode as a function of the angle of propagation. It is evident from the figure that a rise in the angle of propagation widens the gap between the two modes; that is, the coupling weakens when the angle of propagation increases. The coupling nearly vanishes when the angle of propagation θ becomes 90°. Figure 8.1(b) represents the dispersion characteristic curve of the EIC mode obtained for a higher value of frequency (second feasible root) and corresponds to the hydrogen ion gyro frequency. It is evident that as the angle of propagation increases, the frequency of the EIC mode also increases. For parallel propagation, the frequency remains the same for all values of the normalized wavenumber.

The effect of the superthermality index κ (for simplicity, $\kappa_{ce} = \kappa_{se} = \kappa$) on the two wave modes is studied in Figure 8.2(a); the other parameters being the same as in Figure 8.1. It is obvious from the figure that for low values of the dimensionless wavenumber $k\rho_1$, the frequency of cyclotron mode is unaffected by the spectral indices κ. However, for higher $k\rho_1$,the frequency

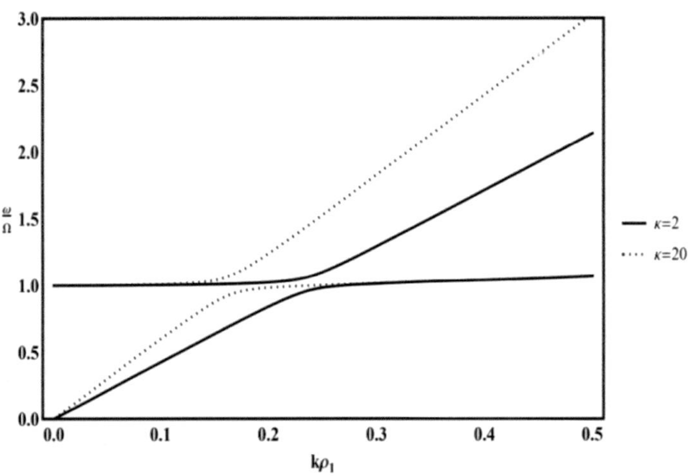

Figure 8.2 (a) Dispersion characteristics of EIC and IA modes for various spectral indices.

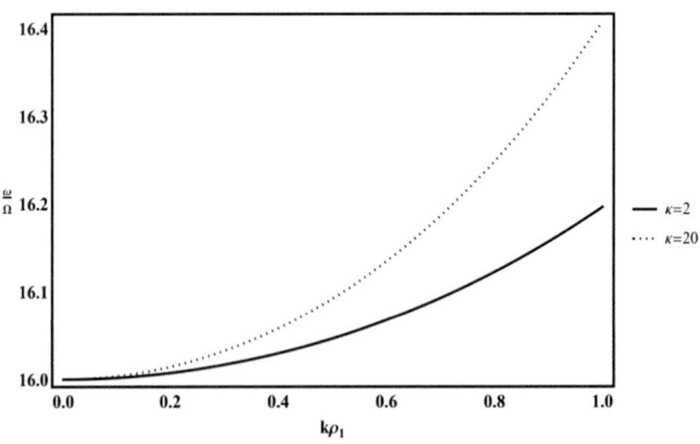

Figure 8.2 (b) Dispersion characteristics of EIC mode for various spectral indices (for the second possible solution).

of the ion cyclotron mode increases with an increase in the kappa values. In the case of the IA mode, there is a notable influence of kappa values on the frequency of the mode for low values of the wavenumber. In this region, the frequency of the IA wave increases with an increase of the spectral index κ. The highest frequencies for both the modes are obtained at larger values of κ, i.e., when the system approaches a Maxwellian distribution in agreement with the conclusions of Sulthana et al. [33]. It is also clear that when the plasma system approaches the Maxwellian limit ($\kappa \rightarrow 20$), the region of strongest coupling between two modes shifts toward higher wavelengths (lower k values). Figure 8.2(b) shows the characteristic curve of hydrogen ion EIC mode as a function of spectral indices κ. In this case too, a variation similar to that of Figure 8.2(a) is obtained.

The effect of positively charged oxygen ion densities is studied next. Figure 8.3 shows the variation of frequency of both the EIC and IA modes with dimensionless wavenumber as a function of O^+ ion density; the other parameters being $n_{10} = 0.05$ cm^{-3},$n_{H0} = 4.95$ cm^{-3},$\theta = 30°$,$T_1 = T_2 = 1.16 \times 10^4$K, $Z_1 = 1$, $Z_2 = 6$, and $\kappa_{ce} = \kappa_{se} = 3$. It is observed that the frequency of ion cyclotron mode is independent of the variation of positively charged oxygen ion densities for low values of $k\rho_1$; the frequency increases with an increase of positively charged oxygen ion densities for higher values of $k\rho_1$. In the case of the IA mode also, the frequency rises with an increase of O^+ ion density.

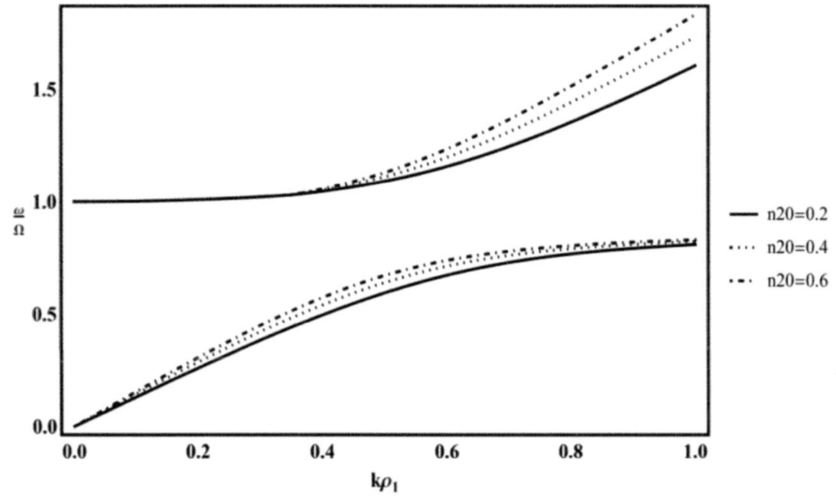

Figure 8.3 Dispersion characteristics of EIC and IA modes as a function of O^+ ion densities.

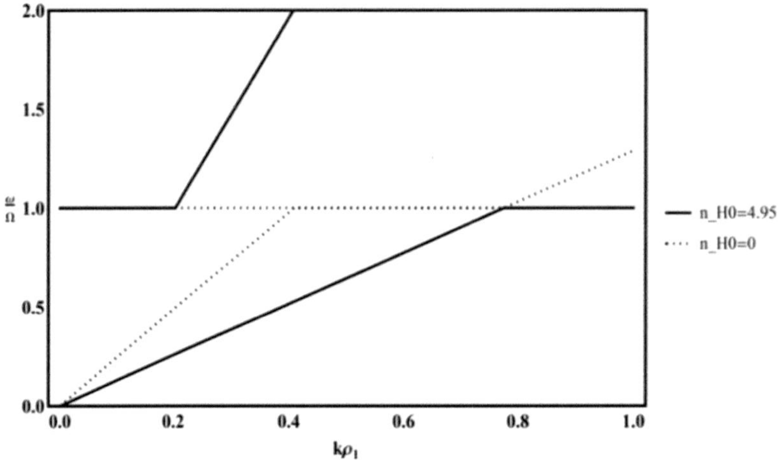

Figure 8.4 (a) Effect of hydrogen ions on dispersion characteristics of EIC and IA modes (for the first possible solution).

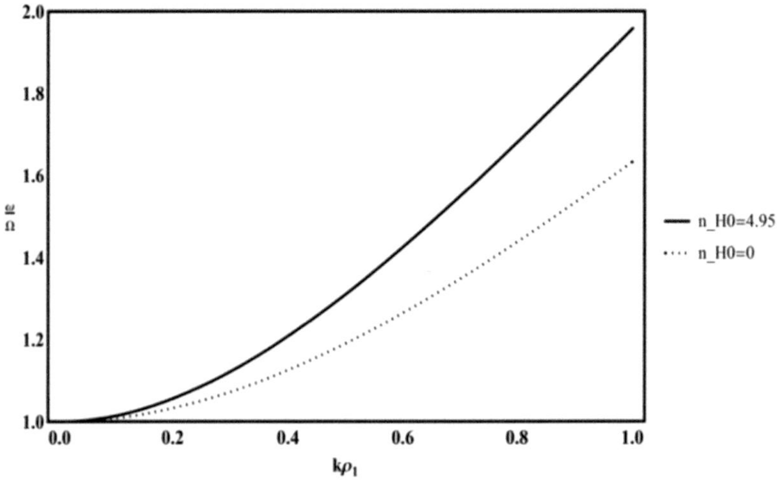

Figure 8.4 (b) Effect of hydrogen ions on dispersion characteristics of EIC mode (for the second possible solution).

In Figures 8.4(a) and 8.4(b), the influence of hydrogen ion densities on the frequencies of the wave modes is studied for parallel ($\theta = 0°$) and perpendicular ($\theta = 90°$) propagation. The other parameters for the figures are $n_{10} = 0.05\ \text{cm}^{-3}$, $n_{20} = 0.5\ \text{cm}^{-3}$, $T_1 = T_2 = 1.16 \times 10^4\text{K}$, $Z_1 = 1$, $Z_2 = 4$, and $\kappa_{ce} = \kappa_{se} = 2$. Figure 8.4 (a) represents the case when $\theta = 0°$. It

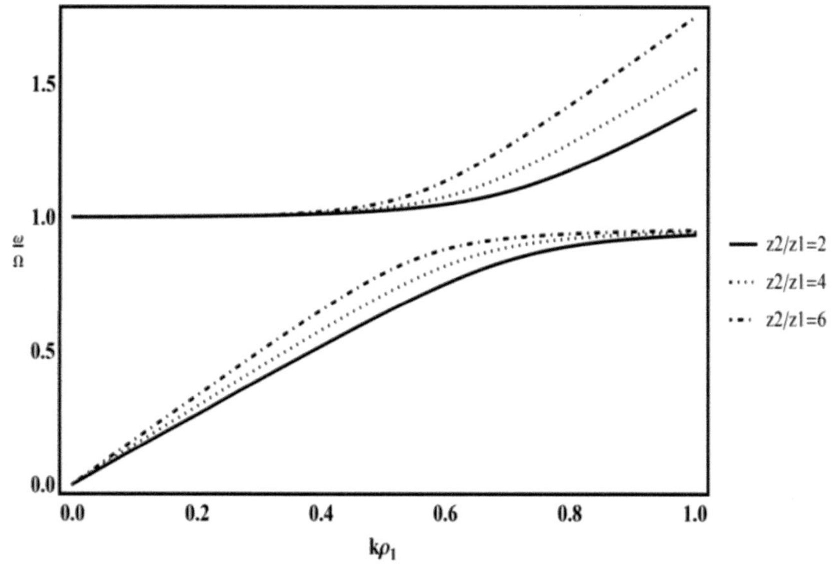

Figure 8.5 Dispersion characteristics of EIC and IA modes as a function of charge number ratio of oxygen ions.

is noticed that the presence of hydrogen ions weakens the coupling between the cyclotron and acoustic modes. In the absence of hydrogen ions, however, there exists a strong coupling in the region $0.4 \leq k\rho_1 \leq 0.8$. Figure 8.4(b) depicts the variation of frequency of cyclotron mode during $\theta = 90°$. It is very much evident that the presence of hydrogen ions enhances the frequency of the oxygen ion cyclotron mode.

In Figure 8.5, the effect of the variation of charge number ratio ($\frac{Z_2}{Z_1}$) of the pair ions on the frequency of the wave modes is studied; the parameters for the figure are$n_{10} = 0.05\ \text{cm}^{-3}$,$n_{20} = 0.3\ \text{cm}^{-3}$, $n_{H0} = 4.95\ \text{cm}^{-3}$, $T_1 = T_2 = 1.16 \times 10^4$K, $\theta = 15°$,and $\kappa_{ce} = \kappa_{se} = 3$. The curves indicate that as the charge number ratio increases, the region of the strongest coupling between two modes shifts toward the higher wavelength (lower *k* values) range. It is also evident from the figure that the frequency of both modes increases with an increase in the charge number ratio. In the case of ion cyclotron mode, the variation of frequency starts only after $k\rho_1 > 0.4$.

Finally, Figure 8.6 depicts the variation of frequencies of the two wave modes as a function of temperature ratio of oxygen ions ($\sigma = \frac{T_2}{T_1}$); the other parameters for the figure are $n_{10} = 0.05\ \text{cm}^{-3}$, $n_{20} = 0.5\ \text{cm}^{-3}$, $n_{H0} = 4.95\ \text{cm}^{-3}$, Z_1 = 1, Z_2 = 2, $\theta = 30°$,and $\kappa_{ce} = \kappa_{se} = 3$. It is obvious that

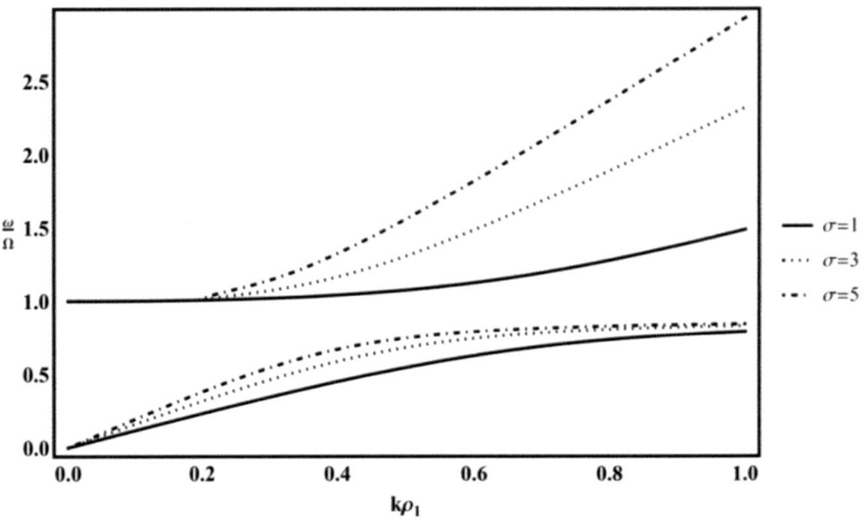

Figure 8.6 Dispersion characteristics of EIC and IA modes as a function of temperature ratio of oxygen ions.

frequencies of both ion cyclotron and IA modes increase with an increase of temperature ratios of ions. The coupling between the two modes also shifts toward lower $k\rho_1$ as σ increases.

8.5 Conclusions

We have, in this chapter, investigated the coupling between EIC and IA waves in a five-component cometary plasma composed of positively and negatively charged oxygen ions, hydrogen ions, and kappa defined hot solar electrons and slightly colder cometary electrons by deriving the linear dispersion relation for the system. The influence of the spectral indices kappa, temperatures of positively charged oxygen ions, and densities of positively charged oxygen and hydrogen ions on both EIC and IA modes has been studied. We find a weaker coupling between the two modes at higher values of propagation angles. For perpendicular propagation, the IA mode nearly vanishes and only the cyclotron mode exists.

It is also found that the influence of variation of spectral indices κ is not significant on the frequency of the cyclotron mode for low values of dimensionless wavenumber $k\rho_1$. However, for higher values, the frequency of the ion cyclotron mode increases with an increase in kappa values. The highest frequencies for both cyclotron modes are obtained for larger values of κ. Both hydrogen and oxygen ion densities also have a significant influence on the frequencies of IA and cyclotron modes. The effect of the temperature ratio of oxygen ions is appreciable on the frequencies of both the modes; it increases with an increase in the temperature ratios of these ions.

Acknowledgements

We gratefully acknowledge the financial assistance from Kerala State Council for Science, Technology & Environment, Thiruvananthapuram, Kerala, India (JRF for Shilpa S.).

References

[1] D'Angelo N, Motley RW. Electrostatic oscillations near the ion cyclotron frequency. Phys. Fluids 1962; 5: 633. DOI: 10.1063/1.1706672.

[2] Motley RW, D'Angelo N. Excitation of electrostatic plasma oscillations near the ion cyclotron frequency. Phys. Fluids 1963; 6: 296-299. Doi:10.1063/1.1706728.

[3] Mosier SR, Gurnett DA. Ionospheric observation of VLF electrostatic noise related to harmonics of the proton gyrofrequency. Nature 1969; 223: 605. Doi: 0.1038/223605a0.

[4] Cattell CA, Mozer FS, Roth I, Anderson RR, Elphic RC, Lennartsson W, Ungstrup E. ISEE 1 observations of electrostatic ion cyclotron waves in association with ion beams on auroral field lines from $\sim$ 2.5 to 4.5 Re. J. Geophys. Res. 1991; 96**:** 11421-11439. Doi. 10.1029/91JA00378.

[5] Andre M, Koskinen H, Gustafsson G, Lundin R. Ion waves and upgoing ion beams observed by the Viking satellite. Geophys. Res. Lett. 1987; 14**:** 463-466. Doi. 10.1029/GL014i004p00463.

[6] Mozer FS, Ergun R, Temerin M, Cattell C, Dombeck J, Wygant J. New features of time domain electric-field structures in the auroral acceleration region. Phys. Rev. Lett. 1997; 79**:** 1281. Doi. 10.1103/PhysRevLett.79.1281.

[7] Cattell C, Bergmann R, Sigsbee K, Carlson C, Chaston C, Ergun R, McFadden J, Mozer FS, Temerin M, Strangeway R, Elphic R. The association of electrostatic ion cyclotron waves, ion and electron beams and field-aligned currents: FAST observations of an auroral zone

crossing near midnight. Geophys. Res. Lett. 1998; 25: 2053-2056. Doi. 10.1029/98GL00834.

[8] Tang X, Cattell C, Lysak R, Wilson LB, Dai L, Thaller S. THEMIS observations of electrostatic ion cyclotron waves and associated ion heating near the Earth's dayside magnetopause. J. Geophys. Res. 2015; 120:3380-3392. Doi. 10.1002/2015JA020984.

[9] D'Angelo N. Low-frequency electrostatic waves in dusty plasmas. Planet. Space Sci. 1990; 38: 1143-1146. Doi. 10.1016/0032-0633.

[10] Shukla PK, Silin VP. Dust ion-acoustic wave. Phys. Scr. 1992; 45:508. Doi. 10.1088/0031-8949/45/5/015.

[11] Rosenberg M. Ion-and dust-acoustic instabilities in dusty plasmas. Planet. Space Sci. 1993; 41:229-233. Doi. 10.1016/0032-0633 (3)90062-7.

[12] D'Angelo N. Ion-acoustic waves in dusty plasmas. Planet. Space Sci. 1994; 42: 507-511. Doi. 10.1016/0032-0633 (94)90092-2.

[13] Ichiki R, Yoshimura S, Watanabe T, Nakamura Y, Kawai Y. Experimental observation of dominant propagation of the ion-acoustic slow mode in a negative ion plasma and its application. Phys. Plasmas 2002; 9: 4481-4487. Doi.10.1063/1.1515770.

[14] Nakamura Y, Bailung H, Ichiki R. Effects of a slow ion beam on ion-acoustic waves. Phys. Plasmas 2004; 11: 3795-3800. Doi. 10.1063/1.1764506.

[15] Sharma SC, Gahlot A. Ion beam driven ion-acoustic waves in a plasma cylinder with negative ions. Phys. Plasmas 2008; 15: 3705. Doi. 10.1063/1.2949708

[16] Song B, D'Angelo N, Merlino RL. Ion-acoustic waves in a plasma with negative ions. Phys. Fluids B 1991; 3: 284-287. Doi. 10.1063/1.859736

[17] Song B, Merlino RL, D'Angelo N. Potential relaxation instability and ion acoustic waves in a single-ended Q-machine plasma with negative ions. Phys. Lett. A 1991; 153:233-238. Doi. 10.1016/0375-9601(91)90800-N.

[18] Merlino RL, Barkan A, Thomson C, D'Angelo N. Laboratory studies of waves and instabilities in dusty plasmas. Phys. Plasmas 1998; 5:1607-1614. Doi. 10.1063/1.872828.

[19] Lin N, Kellogg PJ, MacDowall RJ, Gary SP. Ion acoustic waves in the heliosphere. Space Sci. Rev. 2001; 97:193-196. Doi. 10.1007/978-94-017-3230-7-29.

[20] Gurnett DA, Frank LA. Ion acoustic waves in the solar wind. J. Geophys. Res. 1978; 83: 58-74. Doi. 10.1029/JA083iA01p00058.

[21] Gurnett DA, Marsch E, Pilipp W, Schwenn R, Rosenbauer H. Ion acoustic waves and related plasma observations in the solar wind. J. Geophys. Res. 1979; 84:2029-2038. Doi. 10.1029/JA084iA05p02029.

[22] Backrud IM, Stenberg GM, Andre M, Morooka M, Hobara Y, Joko S, Rönnmark K, Cornilleau-Wehrlin N, Fazakerley A, Rème H. Cluster Observations and Theoretical Identification of Broadband waves in the Auroral Region. Ann. Geophys. 2005; 23:3739-3752. DOI : 10.5194/angeo-23-3739-2005.

[23] Ohnuma T, Miyake S, Watanabe T, Watari T, Sato T. Coupling between electrostatic ion cyclotron waves and ion acoustic waves. Phys.Rev.Lett.1973; 30:535. Doi.10.1103/PhysRevLett.30.535.

[24] Vasyliunas VM. Low-energy electrons on the day side of the magnetosphere. J. Geophys. Res. 1968; 73: 7519-7523. Doi. 10.1029/JA073i023p07519.

[25] Sreeraj T, Singh SV, Lakhina GS. Coupling of electrostatic ion cyclotron and ion acoustic waves in the solar wind. Phys. Plasmas 2016; 23:082901. Doi. 10.1063/1.4960657.

[26] Ipavich FM, Galvin AB, Gloeckler G, Hovestadt D, Klecker B, Scholer M. Comet Giacobini-Zinner: In situ observations of energetic heavy ions. Science 1996; 232: 366-369. DOI: 10.1126/science.232.4748.366.

[27] Coplan MA, Ogilvie KW, A'Hearn MF, Bochsler P, Geiss J. Ion composition and upstream solar wind observations at comet Giacobini-Zinner. J. Geophys. Res. 1987; 92: 39-46. Doi. 10.1029/JA092iA01p00039.

[28] Chaizy P, Reme H, Sauvaud JA, d'Uston C, Lin RP, Larson DE, Mitchell DL, Anderson KA, Carlson CW, Korth A, Mendis DA. Negative ions in the coma of comet Halley. Nature 1991; 349:393. Doi. 10.1038/349393a0

[29] Zwickl RD, Baker DN, Bame SJ, Feldman WC, Fuselier SA, Huebner WF, McComas DJ, Young DT. Three component plasma electron distribution in the intermediate ionized coma of comet Giacobini-Zinner. Geophys. Res. Lett. 1986; 13:401-404. Doi. 10.1029/GL013i004p00401.

[30] Bhardwaj A. On the solar EUV deposition in the inner comae of comets with large gas production rates. Geophys. Res. Lett. 2003; 30:24. Doi. 10.1029/2003GL018495.

[31] Broiles TW, Livadiotis G, Burch JL, Chae K, Clark G, Cravens TE, Davidson R, Eriksson A, Frahm RA, Fuselier SA, Goldstein J. Characterizing cometary electrons with kappa distributions. J. Geophys.Res.2016; 121:7407-7422. Doi. 10.1002/2016JA022972.

[32] Brinca AL, Tsurutani BT. Unusual characteristics of electromagnetic waves excited by cometary newborn ions with large perpendicular energies. Astron. Astrophys. 1987; 187:311-319. Doi. 10.1007/978-3-642-82971-0_57.

[33] Sultana S, Kourakis I, Saini NS, Hellberg MA Oblique electrostatic excitations in a magnetized plasma in the presence of excess superthermal electrons. Phys. Plasmas 2010; 17:032310. Doi. 10.1063/1.3322895.

9

Materials for Strongly Correlated Systems

M. S. Sajna[1], G. Vimal[2], N. V. Unnikrishnan[3], and V. P. Mahadevan Pillai[1]

[1]Department of Optoelectronics, University of Kerala, India
[2]St. Dominic's College Kanjirappally, India
[3]School of Pure & Applied Physics, Mahatma Gandhi University, India

Abstract

Strongly correlated materials, for which electron–electron interactions are significant, have fascinating properties, but a complete understanding of the underlying physical phenomenon is still under research. Controlling their complex behaviors, knowledge of distinct phases having considerably varying electronic as well as magnetic orders and transitions between the competing phases are presently the most exciting and challenging areas of material science. In recent years, a significant number of novel potential materials, especially 3d and 4f elements, have been revealed to their rich phases like superconductivity, charge density waves, and hard ferromagnetism. Alongside that, a strong interaction between electron correlation and spin−orbit interaction results in exotic phenomena with the emergence of novel quantum ground states. This chapter gives a quick overview of systems in which strong correlation impacts, as well as spin−orbit coupling, play a significant role in defining electronic, magnetic, optical, and thermodynamic properties. Within the scope of the chapter, the authors place emphasis on transition metal oxides, graphene, iron pnictides, rare-earth nickelates, iridates, and transition metal dichalcogenides. Since complete theoretical modeling is still under research, more attention is given to the progress and novelty of the materials and their intriguing properties.

Keywords: Correlated systems, Perovskite, Superconductors, Graphene, Rare-earth, Magnetoresistance, Transition Metals, Spin-Orbit interaction

DOI: 10.1201/9781003337102-9

9.1 Introduction

The strong correlation effect of electrons in condensed matter physics is an exciting area of research, and this chapter is aimed at catching some current advancements in the domain of highly correlated materials. Correlated materials frequently have notable characteristics and particular competing phases having drastically unique electronic and magnetic orders. The strong correlations of electrons may give rise to exciting low-temperature characteristics like heavy-fermion (HF) superconductivity, Kondo insulator, intermediate valence, or magnetic ordering with reduced magnetic moments. The new/advanced characteristics of the materials are mainly ranging from unusual high-temperature superconductivity (high T_c SC), colossal magnetoresistance (CMR), non-Fermi liquid phenomena, HF behaviors, quantum-phase transitions, etc. The advancements may lead us to reconsider the condensed matter aspects like electron–magnon/phonon interaction, Rashba effect, and the degree of freedom associated with the orbitals to gain interest to researchers. Interactions like electrons/atomic lattice, electrons/local magnetic moments, etc., can provide an electronic structure that is different from the calculations of non-interacting band structure, and those materials are likewise included under the category of highly correlated systems.

9.1.1 Theoretical aspects

For the conventional materials, the standard physical theories are adequate, but innovations of the new class "highly correlated materials" demand much more understanding and detailing of the existing theories in which the interactions between the electrons within metals have been regarded as negligible and were ignored. However, over the past decades, novel materials have been explored, and for explaining their mechanism, the strong interactions among the constituent electrons should be taken into account. In these materials, the energies of electron interaction are analogous with the electronic kinetic/translational energies; these electrons are termed as "strongly correlated." Methods like density functional theory (DFT) and local-density approximation are not appropriate for a better understanding of strongly correlated (SCE) compounds specifically at low temperatures since they are one-electron theories. According to Rau et al., there are various energy scales to consider strongly correlated materials: the atomic interactions of on-site Hubbard interaction U, Hund's coupling J_H, the spin–orbit coupling (SOC) λ, the crystal field Δ, and the electron kinetic energy described by hopping integral t [1]. In comparison with their 3D parent systems, one-

and two-dimensional systems exhibit quantum confinement and are highly correlated since the Coulomb interaction is more significant when electrons are on the surface.

9.1.2 Importance of correlated electron systems

Several new correlated electron systems have been adapted for solid-state devices over the past few decades. Examples include mobile phone network substations that integrate high-T_c cuprate superconductors and modern magnetic read head technologies that harness the Fe/Cr/Fe trilayer massive magnetoresistance (MR) properties. The advancement of novel materials with unique features is critical to advances, and the researchers are pursuing a broad range of interdisciplinary projects with incredible potential for advanced products in the field for industry, microelectronics, energy systems, medicine, aeronautics, and many other areas for future applications.

9.1.3 Objectives of the chapter

Ongoing research is mainly focused on developing and analyzing exotic materials with the objective of inventing novel systems with correlated electron quantum states. Research on fundamental physics of these novel materials as well as their device aspects has been conducted by several researchers worldwide. However, the investigation on new electronic quantum phenomena of the strongly correlated electronic systems is of elementary in nature. This chapter covers an insight of new knowledge on properties as well as applications of novel strongly correlated systems, including transition metal oxides, iron pnictides, graphene, chalcogenides, iridates, nickelates, borides, cerium monopnictides, etc.

Figure 9.1 gives some of the correlated electron materials as a function of on-site Hubbard interaction strength U with electron bandwidth D. The ratio varies with compounds and the given values are semi-qualitative. The interaction strength is symbolized by the color code in the figure, non-interacting by blue to very strongly interacting red.

9.2 Strongly Correlated Materials

With conventional theories of metals as well as insulators, behaviors of materials having strong electronic correlations cannot be comprehended since they are remarkably complex ones by competing interactions of spin, charge, lattice, as well as orbital degrees of freedom. This may further result in

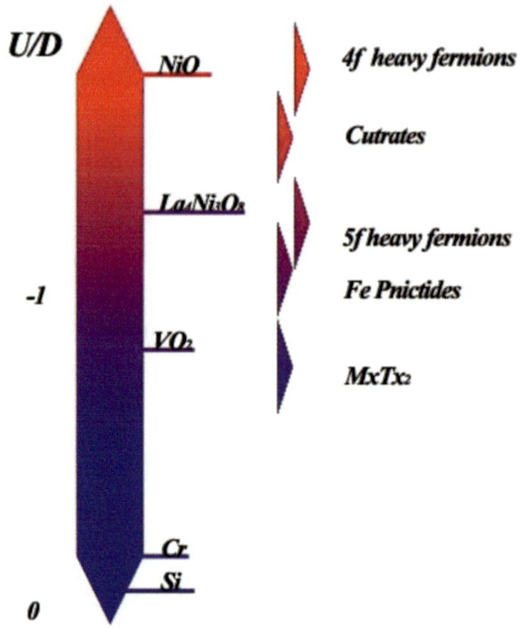

Figure 9.1 Strongly correlated systems as a function over the *U*/*D*.

high-temperature superconductivity, charge stripes in cuprates, pseudogap state, or even phase separation in case of certain manganites and cuprates. The novel oxide materials like manganites, vanadates, nickelates, etc., have strong electron–phonon interactions with highly correlated electrons. For a good insight of the same, theoretical as well as experimental investigation has to be carried out. A vital role has been played by computational and theoretical materials science to meet the quest for the design and development of novel materials and the realization of the improved properties of existing materials.

9.2.1 "Strongly correlated" transition metal oxides

The complex and intriguing characteristics of transition metal oxides derived from the closely correlated electronic degrees of freedom are catching research interest for a long time. The transition metal oxide compounds containing partially filled *d*-shells can exhibit spectacular variations from the single-electron band structure, and they have intriguing electronic properties including different types of magnetic, charge, and orbital ordering and unconventional high-temperature superconductivity. Since these materials have strong "lattice" effects, it is not possible to give either the kinetic or potential energy in simple electron gas forms. The interaction energies turn out to be extensively larger than the kinetic energies; therefore, the compounds are coming under the category "strongly correlated systems." However, the spin–orbit interactions in 3d transition metal oxides like high-temperature cuprates, manganites, and vanadium oxides are relatively small compared to 4f and 5d materials, whereas the atomic interactions, crystal field, and the kinetic terms are significant. These oxides generally exhibit crystal structures that are similar to ABO_3 perovskite and are shown in Figure 9.2 (a).

The transition metal occupies the B-site surrounded by (and bonded to) six oxygen ions, so as to form a B-site centered octahedron. La_2CuO_4 exhibits a "Ruddlesden-Popper" type structure, which is a modified version of the perovskite structure. Transition metal oxides like Sr_2FeMoO_6 exhibit a double perovskite structure as in Figure 9.2 (b). In all the above cases, the transition metal-oxygen coordination is octahedral or tetrahedral, which is an essential condition.

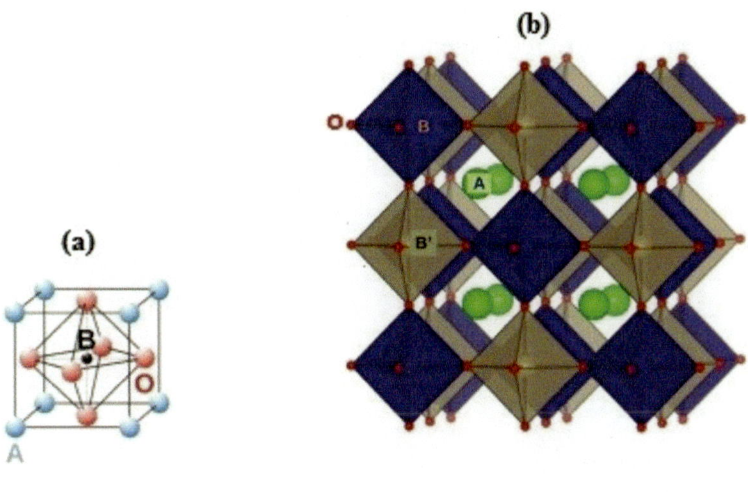

Perovskite: ABO_3 **Double Perovskite: $A_2BB'O_6$**

Figure 9.2 (a) The perovskite and (b) double perovskite structures (reused from [2]).

9.2.1.1 Cuprates

The cuprate (copper oxide) superconductor, with its rich phases and exceptionally high transition temperature, is one of the spectacular examples of complex solid phenomena. The superconducting property of cuprates is discovered in the mid-1980s by Bednorz and Muller. The cuprate superconductors are made up of oxygen and copper layers, and, perhaps, it is the most well-known among all the strongly correlated materials.

From characterization techniques, it was revealed that there is a pseudogap in density of states below a particular temperature and the materials undergo a transition into d-wave superconducting state. Onset of the pseudogap and its magnitude varies significantly with that of the nanometer distance range [3]. Angle-resolved photoemission spectroscopy (ARPES) is the characterization technique to acquire insight into multibody physics, which accounts for the cuprate superconductor-rich phase diagram shown in Figure 9.3.

The layered structure of cuprate family consisted of CuO_2 planes and a charge reservoir layer, which mainly determines the carrier concentration. The super conductivity arises on doping as the insulating phase of the parent compound vanishes.

Soon after the discovery of the first cuprate superconductor $La_{2-x}Ba_xCuO_4$ (LBCO) with T_c = 30 K in 1986, the transition temperature has increased by discovering $La_{2-x}Sr_xCuO_4$ (LSCO) up to 40 K. The key issue in the initial stage is to figure out the brittle nature of these materials and pave the way to create them into various practical shapes like wires

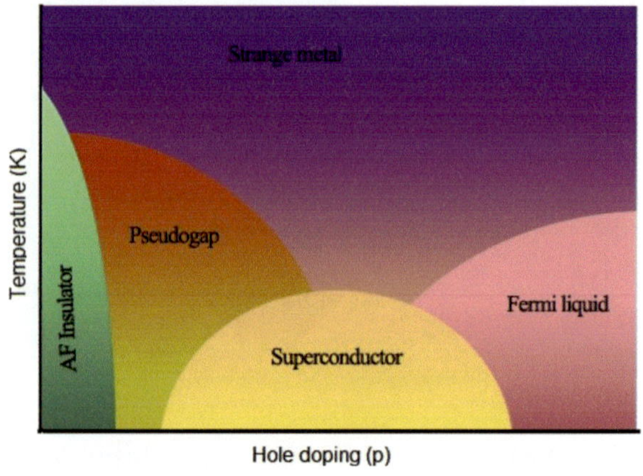

Figure 9.3 Schematic phase diagram of high-T_c cuprate superconductors.

or tapes. Various compounds were investigated to overcome the issue and to increase the transition temperature. Bismuth strontium calcium copper oxide (BSCCO) is one of many answers to the problem. Depending on the chemical composition, BSCCO have different forms having T_c well above liquid nitrogen temperature. For example, $Bi_2Sr_2Ca_2Cu_3O_x$ (BSCCO-2223) and $Bi_2Sr_2CaCu_2O_x$ (BSCCO-2212) were discovered with T_c = 110 and 96 K, respectively. Now a family of compounds which belongs to BSCCO was investigated and used in some very high-tech transmission cables, transformers, motors, and generators. MagLab scientists and engineers, in collaboration with industry partners, have done extensive research and development on cuprates and achieved a transition temperature of about 92 K for Yttrium barium copper oxide (YBCO). YBCO is nowadays used in a wide range of applications in the field of power transmission cables, motors, generators, and microwave filters for mobile phone towers. The above cuprates are regarded as layer-type compounds in the cuprate family, and the T_c of compounds perhaps increased with the number of CuO_2 layers in the unit cell (UC). The layered crystal structures of the LSCO, YBCO, and BSCCO family of cuprates are depicted in Figure 9.4.

The explosion of research works in the last three decades ensuing exploration of more than 200 cuprate superconductors belongs to seven families: lanthanum barium copper oxide ($La_{2-x}Ba_xCuO_4$), YBCO ($YBa_2Cu_3O_7$ and $YBa_2Cu_4O_8$), BSCCO ($Bi_2Sr_2Ca_{n-1}Cu_nO_{2n+4}$), thallium barium calcium copper oxide ($Tl_2Ba_2Ca_{n-1}CunO_{2n+4}$ and $Tl_2Ba_2Ca_{n-1}CunO_{2n+3}$), and

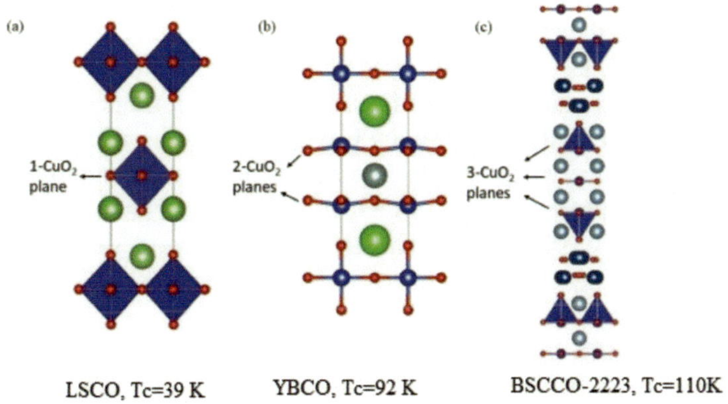

Figure 9.4 Crystal structure of (a) $La_{2-x}Sr_xCuO_4$ (LSCO), (b) $YBa_2Cu_3O_{7-\delta}$ (YBCO), and (c) $Bi_2Sr_2Ca_2Cu_3O_x$ (BSCCO-2223).

mercury barium calcium copper oxide ($HgBa_2Ca_{n-1}Cu_nO_{2n+2}$) with T_c up to 134 K in $HgBa_2Ca_2Cu_3O_9$ at ambient pressure and 164 K under 30 GPa.

9.2.1.2 Manganates

Another important transition metal oxide with perovskite structure is manganite compounds where spin, charge, and orbital degrees of freedom are at play simultaneously. These materials exhibit a rich phase diagram influenced by a variety of parameters like Mott-type strong d electron Coulomb interactions, Jahn–Teller distortions, orbital and charge ordering, and ferromagnetic and antiferromagnetic (AF) ordering, which are mainly emerged via doping. The end members of series, $RMnO_3$ (R – rare earth atom) and $BMnO_3$ (B – Ca or Sr), are AF insulators. The material gains significant attraction due to the broad range of symmetry breaking and ordering phenomena, which provides extensive possibilities for researchers to design functional applications.

The fascinating characteristics of perovskite manganites are the CMR, which have a crucial role in the development of magnetic memory, sensors, and switching gadgets. The giant CMR observed in perovskite or double perovskite manganites typically originates from the double-exchange interaction between the Mn^{3+} and the Mn^{4+} ions [4]. Stimulated from the widespread application, investigations on the new materials with large MR is in the forefront of the experimental research, and, consequently, researchers evidenced that parameters like oxygen vacancies, composition, and B-site cation ordering are key factors to control the feature. However, one of the crucial obstacles that still remain for the technological applications is the requirement of high magnetic field. To overcome this, substitution of different dopants, fabrication of thin films, multilayer films, core-shell, heterostructures, etc., were reported [5, 6]. Recently, Das et al. illustrated the significant enhancement of MR of $La_{1-x}Ca_xMnO_3$ (x = 0.52, 0.54, and 0.55) with a reduction of particle size in nanometer scale [7]. Besides CMR, lanthanum calcium manganite (LCMO) is intensively explored for their influences on the high-temperature superconducting properties of YBCO. It is evidenced that YBCO/LCMO heterointerface may stabilize incommensurate antiferromagnetism or triplet superconductivity by enhancing charged density waves [8].

9.2.1.3 Vanadium oxides

Vanadium dioxides (VO_2) have partially occupied d-orbitals, which prompt the electron–electron interactions along with strong couplings among electronic degrees of freedom. Recent steps in thin-film synthesis and recent efforts to execute vanadium-based oxides for viable applications that expand existing innovations outperform confinements of traditional materials. There exist some systems for which optical responses are obtained for thermally induced changes termed as “thermochromic” systems [9].

VO_2 has a first-order phase transition at 67 °C from a high temperature, rutile metallic phase to the low temperature, monoclinic insulating phase. The electronic conductivity reduces by four orders as the VO_2 cooled through the transition. This considerable variation in the electronic as well as optical properties at the vicinity of room temperature boosts the possibility of material for various applications. For instance, the optical switching linked with the metal–insulator transition (MIT) of vanadium dioxide (VO_2) has been recommended extending from anti-laser shields to “intelligent” windows, which can selectively block heat radiations in warm weather [10].

The mechanism of insulator-to-metal transition (IMT) is driven principally by structural change due to electron–phonon interactions or by electron–electron interactions, which is a topic of conflict. The former leads to a Peierls insulator, whereas the latter results in a Mott insulator. Crystal structures of VO_2 with metal, intermediate, and insulating phases with corresponding band structures are depicted in Figure 9.5. The near-field infrared

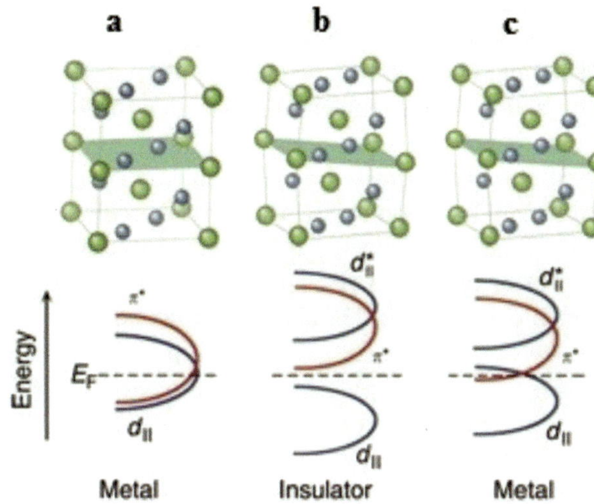

Figure 9.5 Crystal structures of VO_2 phases with the corresponding band structures. (a) The metal rutile phase. (b) The insulating monoclinic phase. (c) Intermediates of the metal-like monoclinic phase with the proposed band diagram (reused from [11]).

studies on VO_2 in the IMT regime reveals that it is phase-separated into nanoscale metallic "puddles" in the insulating host. It has a novel collective electronic state with different optical masses in the metallic puddles that vary from the macroscopic rutile metallic phase of VO_2. The varying optical mass proposed that electron–electron interactions determine the IMT in VO_2 and the IMT is primarily a Mott transition [12].

It is inferred that as the infused electrons were adequate, the rutile metal state could be settled up to room temperature. Whereas the electron density in a metal-like monoclinic portion was unable to suppress the Peierls bending of the monoclinic lattice and is attributable to less infused charge, disentangling the orbital-specific Mott correlation for the void opening of the VO_2 MIT.

9.2.2 Iron-based superconductors

9.2.2.1 Iron pnictide

Iron pnictides or ferropnictides, relatively newcomers in the high T_c family, are composed of layers of iron and a pnictide, for example, arsenic or phosphorus. Since iron was considered to be irreconcilable with superconductivity, the discovery of fluorine-doped LaFeAsO by Hideo Hosono and co-workers [13] in 2008 with T_c = 26 K came as astonishing information to the condensed matter researchers. With the advent of this, high-temperature superconductivity in materials other than copper-based compounds starts to be investigated widely. From then onwards, research was ongoing for the advancement in this field by raising T_c values of LaFeAs(O, F) to 43 K by the applying pressure and then to a temperature as high as 55 K by replacing La by other rare-earth elements [14, 15].

Some essential structural classes of ferro superconductors (FeSCs) are shown in Figure 9.6, all based on a typical layered structure, which is having a planar layer of Fe atoms connected by tetrahedrally coordinated chalcogen (S, Se, and Te) or pnictogen (P and As) anions arranged in a stacked series divided by "blocking layers" of alkali, alkaline-earth, or rare-earth and oxygen/fluorine.

Figure 9.6 gives the general crystal structures of iron-pnictide superconductors reported so far. Figure 9.6 (a) is shortened as "1111" structure, which is RFeAsO with ZrCuSiAs-type structure having space group P4/nmm. In the case of RTPnO compounds like $LaFeAs(O_{1-x}F_x)$, RO layers and FeAs layers are stacked along the c-axis (R – rare earths such as Ce, Pr, and Nd; T – late-transition metal; Pn – pnictogen). Either replacing 3% oxygen by fluorine [$RFeAs(O_{1-x}F_x)$] or by introducing oxygen deficiency ($RFeAsO_{1-\delta}$),

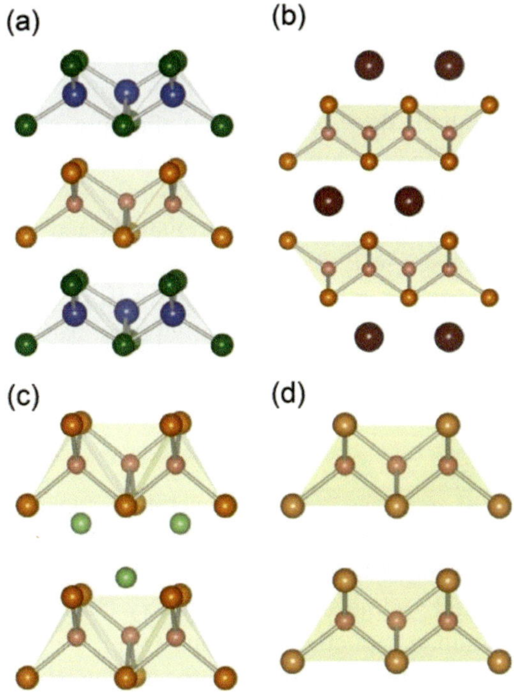

Figure 9.6 Crystal structures of (a) LaFeAsO (referred to as "1111" in the article), (b) BaFe2As2 ("122"), (c) LiFeAs ("111"), and (d) FeSe ("11") (reused from [16]).

superconductivity emerges [17]. For $SmFeAs(O_{1-x}F_x)$, the highest T_c = 55 K was reported [18]. Superconductivity has also been observed for A = Ca at x = 0:1 with transition temperature T_c = 22 and 4 K for A = Sr with x = 0:125, respectively [19, 20]. In Figure 9.6 (b), an iron-pnictide family with $ThCr_2Si_2$-type structure is shown and the structure is termed as "122" and is very well-known in "HF" compounds, and the first discovered HF superconducting compound is $CeCu_2Si_2$ [21]. Figure 9.6 (c) gives the third structure "111," and it was reported in superconducting LiFeAs and the highest T_c = 18 K is reported for $Li_{0.6}Fe_2As_2$ [22]. The LiFeAs crystallizes into tetragonal structure Cu_2Sb-type with the FeAs layer having an average iron valence of +2 similar to that of "1111" or "122" parent compounds. Figure 9.6 (d) shows α-FeSe compounds with the α-PbO-type structure, and its superconductivity has been reported at 8 K. This particular structure, abbreviated as "11" structure, has an exact planar crystal sublattice identical to the FeAs layer in the above three structures [23].

Iron-based superconductors (IBSCS) have been found to display a switchover between orbital, spin, and degrees of freedom of the lattice, drastically influencing their low-energy electronic properties comprising conductivity. Even though the exact blending mechanism stays unidentified, few interactions, both in the orbital and in the spin channel, have been recommended to intervene the superconducting pairing. In the group of iron pnictides, it was reported that optimally doped NaFeCoAs displays the absolute most grounded selective electronic correlations and was demonstrated by the optical spectroscopy (OS), *ab initio* band structure, ARPES, and Eliashberg calculations. The in-built complexity of IBSCS arises from several partially filled Fe-3d orbitals that concurrently play a part in the low-energy quasi-particle dynamics and are strongly influenced by the Hund's coupling correlations [24].

9.2.2.2 FeSe films

In 2012, Wang et al. reported a high-temperature superconductivity with a transition temperature of around 77 K by analyzing its extensive energy gap in one UC thick single-layer FeSe system film grown on a Se-etched SrTiO (001) substrate by molecular beam epitaxy (MBE), which possesses one of the highest superconducting transition temperatures in FeSCs [25]. This not just shows a ground-breaking way for finding new superconductors or on account of raising T_c; yet, in addition, it offers a very much characterized analysis of the mechanism of HT_c superconductivity in a variety of materials.

Single-layer FeSe is a vital candidate because it has one of the highest superconducting T_c amongst the Fe-based superconductors. As compared with the other Fe-based superconductors, the FeSe films have a fairly simple Fermi surface having two symmetry-related electron pockets; without accounting the minus sign issue, by the magnetic fluctuations, the inner pocket scattering can be analyzed. The mechanism of superconductivity is suggested from the fluctuations arising from the orbital ordering and/or nearer AF or nematic phases. Since Fe-pnictide/Fe-chalcogenides are high T_c superconductors, it is manifested that there should be a principle interaction mechanism, in general, for the Fe family because of their high T_c.

9.2.3 Graphene

Strong electron correlations result in discrete exciton behaviors in carbon nano tubes (CNTs), which were first predicted in 1997 by Ando et al. [26]. Graphene bilayers possessed interacting electron behaviors like superconductivity, magnetism, etc. Recent reports evidenced the presence of correlated flat band in a graphene bilayer framework [27], which results from the superlattice modulation of two graphene sheets twisted in the moiré structure by a certain angle close to "magic angle." At zero magnetic field, the correlated flat-band indicates half-filling insulating phases, which acts as a correlated insulator.

While doping, an electrically tunable superconductivity that resembles high T_c SC in cuprates can be seen. These distinct features of bilayer graphene twisted at the magic angle pave a way for interesting many-body quantum phases in the 2D platform composed of pure carbon and with no magnetic field. Simple accessibility of flat bands, electrical tenability, and the bandwidth tunability through the twist angle may open up to more appealing correlated systems, like quantum spin liquids, correlated topological insulators (TIs), etc. The thread brought a breakthrough in the year 2018 that while working with a dual layer of graphene, a group at the MIT, Cambridge noticed that a minute relative twist among the layers caused insulating as well as superconductive states that were entirely absent in a single sheet of graphene [27]. They found a similarity between cuprate high-temperature superconductors as well as double layers of graphene. Twisted bilayer graphene (TBG) is an intriguing quantum material with regard to its very own privilege since it gives access to strongly correlated electron phenomena experimentally and their ongoing improvements in the field of "graphene twistronics" very quickly. Graphene shows a strong correlation in the presence of intense perpendicular magnetic field, which allows electrons to orbit in tight circles causing π electron localization. This localization of π electrons leads to the quantum hall effect, one of the characteristics of strong correlation in graphene systems at low temperature [28].

At low perpendicular magnetic fields, the disorder-induced MR effect is quadratic while linear at higher fields. This result is technologically impressive, particularly in a two-dimensional (2D) material like graphene, because of the potential applications in a 2D magnetic sensor with nanoscale resolution. A novel giant negative MR of 10,000% at room temperature for a small magnetic field of 2 kilogauss is achieved in layered graphene recently [29]. Since graphene has a lot of potential with its excellent electrical and thermal conductivity, research works were rapidly growing to explore new insights of graphene systems that may help for faster, next-generation electronics like quantum computing technologies.

9.2.4 Transition metal dichalcogenides

Transition metal dichalcogenides (TMDCs) with formula MX_2 (M = transition metal and X = chalcogen) is another potential group that deserves much attention, owing to their distinct features such as high T_c superconductivity, interesting ferromagnetic properties, and charge density waves. TMDCs constitute a class of compounds with layers of transition metals, such as Mo or W, and chalcogens, such as S, Se, or Te ions, which forms stacked X-M-X layers. An illustration showing a three-dimensional structural view of MX_2 is given in Figure 9.7.

In the domain of layer-structured materials, graphene is, as of yet, the most attractive topic due to its very strong electrical and thermal conductivity. TMDCs maintain various benefits over graphene, which makes them an alluring area of research. While graphene is a zero-gap semiconductor, which requires actuating a gap to make it a viable transistor, TMDC can alter the electrical properties by way of simply altering the transition metal, the chalcogen, or both. More than 60 MX_2 frameworks have been known for quite a long time and several polytypes can be shaped for given M and X blends by stacking faults due to the inherent weakness of the interlayer Van der Walls forces. Hence, TMDC compounds exhibit a wide range of physical properties and material spans from insulators (HfS_2), semiconductors (MoS_2; MX_2, M = Pt, Pd and X = S, Se), semi-metals (WTe_2 and TcS_2), and metals (NbS_2 and VSe_2). Henceforth, there have been impressive endeavors to incorporate excellent TMDCs, for primary investigations as well as for integration in future level devices. Moreover, TMDC systems uphold many of the desirable properties of graphene, such as layered structure for fabricating thin film devices and the capability to intercalate atoms between the layers.

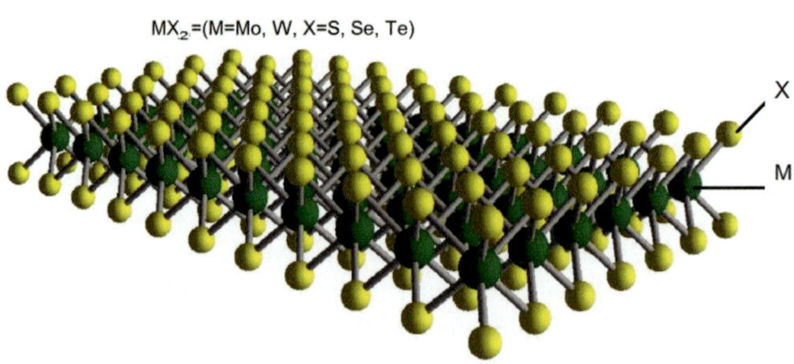

Figure 9.7 Stacked layer structure of transition metal dichalcogenides (reused from [30]).

Finally, TMDCs are easy to synthesis in high yield with quality. Hence, because of the variety of the properties and the accessibility of more cost-effective manufacturing methods, TMDCs have become one of the hot topics in material sciences.

One of the promising candidates that belong to the TMDC is MoS_2 because of the existence of an electronic bandgap, which makes these structures a potential choice for fabricating nanoscale electronic devices [31]. Apart from graphene, the potential applications of MoS_2 are spintronics and valleytronics devices. The strong intrinsic SOC of monolayer MoS_2 gives rise to splitting of valence bands with opposite spins at the two valleys as required by means of time-reversal symmetry. Exciting the samples with circularly polarized light, controlling of valley population can be accomplished. This impact ought to be even more pronounced for MX_2 materials that exhibit stronger spin–orbit splitting in the monolayers, i.e., for WS_2 and WSe_2. Single-layer MoS_2 is considered as an analog of graphene in the field of fabricating interband tunnel field-effect transistors (FETs) with lower energy consumption [32] and other devices that require transparent thin films of semiconductors. Recently, Mak et al. [33] studied the optical activities of the monolayer and recommended its use for photostable markers and sensors in nanoscale. Apart from this, a wide range of applications had also been reported for other molybdenum dichalcogenides such as $MoSe_2$, $MoTe_2$, etc. [34, 35].

However, the flagship of 2D TMDCs is MoS_2; tungsten dichalcogenides gained a lot of scientific attention due to the large size of tungsten that may change the two-dimensional nature. Moreover, compared to MoS_2, its abundance, low toxicity, and price are also favoring the rapid growth of scientific research, despite the fact that the greater part of the characteristics are the same as that of molybdenum dichalcogenides. Subsequently, an extensive variety of applications were accounted in the field of electronics and optoelectronics. New studies proposed that WSe_2 quantum dots can be utilized for integrated solid-state quantum photonics and quantum information processing [36] and that WTe_2 have had remarkable magnetic properties with its large uniaxial positive MR [37].

Another fascinating property of the TMDC is the unconventional superconductivity, which is rather different from standard s-wave cooper pairs in the famous BCS theory. Experimental observations demonstrated the presence of "Ising" superconductivity in monolayer TMDCs, such as MoS_2 [38] and $NbSe_2$ [39], in which the in-plane upper critical field $H_{c2,\parallel}$ is far away from the paramagnetic limit. Zhao detailed a fascinating phenomenon

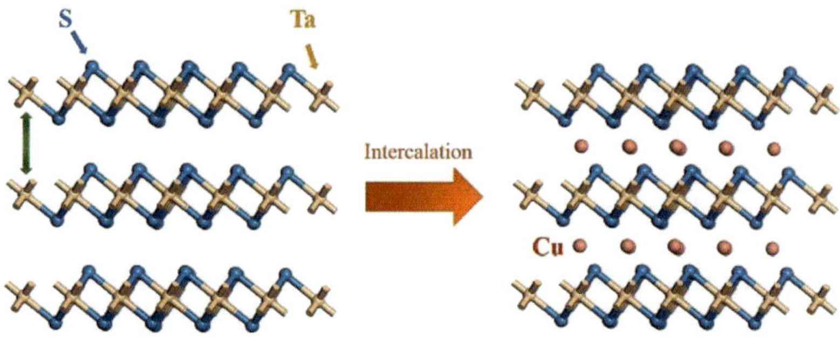

Figure 9.8 Crystal structure of TaS_2 and the intercalation of Cu occupied at the Van der Waals positions in between the TaS_2 layers (reused from [44]).

of $NbSe_2$ that, unlike conventional superconductors, the transition temperatures of $NbSe_2$ monolayers are substantially increased by the disorder [40]. The reason for significant enhancement possibly arises from multifractality of electron wave functions as predicted by various theoretical models. Intercalated TMDCs, a new level of complexity accomplished through the intercalation of various complexes T in between the layers to form T_xMX_2 as depicted in Figure 9.8, exhibit intriguing physical properties relying upon the nature of the intercalant T. For instance, the magnetic and transport properties of the TMDCs surprisingly modified to offer more versatility by intercalation of Cr, Mn, Fe, Co, or Ni [41, 42]. Versatile properties that have potential applications such as sharp magnetic domain switching by the intercalation of Cu or Fe in TaS_2, superconductivity induced by the Cu intercalation in $TiSe_2$, the emergence of unusual phases such as Mott insulating phase, nearly commensurate charge-density-wave (NCCDW) phase, superconductivity, commensurate charge-density-wave (CCDW) phase, and normal metallic phase by the intercalation of Se in TaS_2, are the few examples of windows opened by intercalated TMDCs [43].

9.2.5 Iridium oxides

Pyrochlore iridate $R_2Ir_2O_7$, where R is generally a rare-earth element, pulled a lot of enthusiasm in the last decade because of their richness of phases such as topological Mott insulator, chiral spin liquids, Weyl semi-metal (WSM), and axion insulator [45]. Recent studies uncover that a large portion of the members of pyrochlore iridate family exhibit an all-in-all-out (AIAO) magnetic ordering where all the magnetic moments point toward or outwards from the center of the Ir tetrahydra. [46]. Much scientific attention is devoted to $Nd_2Ir_2O_7$, which offers an ideal platform to explore remarkable features like anomalous domain-wall conductance, magnetic field-induced insulator–semi-metal transition, etc. [47]. Further opportunities were opened by the exploration of quadratic band touching dispersion of $Pr_2Ir_2O_7$ and topological semi-metal and Fermi-arc surface in $Y_2Ir_2O_7$ [45, 48]. However, a lot of theoretical and empirical shreds of evidence came in the last decade, emphasizing a strong correlation of the pyrochlore iridates; research works on quantum ground states that would emerge within the sight of interactions in these systems are still in progress. The following sections give a concise outline of the structure and the most significant updates of the correlative phenomena.

Prevalence and significance of the pyrochlore iridates are certainly due to the interplay of comparable energy scales, for example, on-site Coulomb repulsion, crystal field splitting, and SOC. The common motif in these materials is the pyrochlore-type structure ($A_2B_20_7$), well known as a geometrically frustrated lattice, made out of two nested corner-sharing tetrahedra of each R ion and Ir one, as displaced from one another by a half UC [50], as depicted in Figure 9.9. The 5d orbitals of the iridium are spatially extendable,

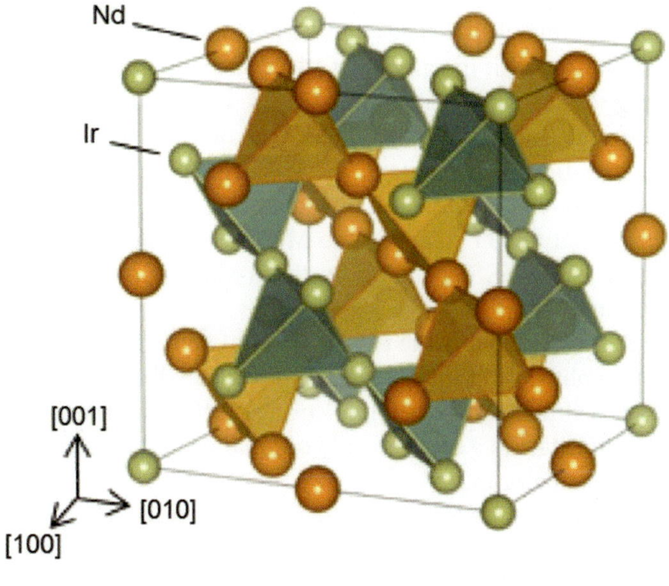

Figure 9.9 Pyrochlore lattice structure of $Nd_2Ir_2O_7$ showing the Ir tetrahedra (reused from [49]).

and, hence, the size of the rare-earth atom also affects the effective bandwidth of 5d electrons via Ir–O–Ir bond angle, thereby tuning adequate strength of the interaction. As a consequence, variations in several quantum phases have occurred as the gradual replacement of rare-earth ions. For instance, the MIT temperature decreases and tends to zero between Nd and Pr as the RE^{3+} ionic radius increases [51]. It is also found that at the paramagnetic phase, the electrical conductivity has a direct dependence with the ionic radius of the rare earth atom. In addition to that, their electronic as well as magnetic properties can also be tuned by thermodynamic variables such as pressure and temperature, and, thus, pyrochlore iridate is regarded as an ideal system to learn about a variety of transport and magnetic properties.

9.2.5.1 Topological insulator

A strong spin–orbit interaction plays the key factor behind TI phases in transition metal oxides, and pyrochlore iridate is considered as one of the ideal systems for the study of various topological phases. Stability of such phases in these materials depends on the trigonal crystal-field effect emerged by a local distortion of IrO_6 octahedra, and that is different for distinct *A*-site ion in $A_2Ir_2O_7$. Hence, by appropriate substitutions at the *A*-site, the pyrochlore iridate offers to control the relative strength of the crystal-field effect and, thus, the stability of the topological and magnetic phases such as a TI, AIAO, and related AF states and various kinds of topological semi-metals. Hence, they exhibit metallic as well as insulating ground states both at low temperatures and finite-temperature MITs. High-pressure experiments on pyrochlore iridate may reveal that depending on the choice of the *A*-site ion or upon application of hydrostatic pressure, the TI state and the related Mott insulating state can have distinctive magnetic ordering patterns. Also, it is well-studied that the TI state and the Mott insulating state may exhibit distinct magnetic ordering, depending on the *A*-site ion or the presence of hydrostatic pressure.

Recent studies revealed that the trigonal crystal field effect could destabilize the insulating phase of the system, leading to a metallic ground state. The reverse is also possible that a sufficiently strong electron interaction can turn the metallic ground state to a magnetically ordered insulator or a more subtle form of the Mott insulator [52].

AIAO spin ordering is another exciting phenomenon of pyrochlore iridates that have attracted the burgeoning interest of the scientific world. In the AIAO phase, as shown in Figure 9.10, the four magnetic moments on vertices of every first tetrahydron are directed toward the center (all-in), while that of

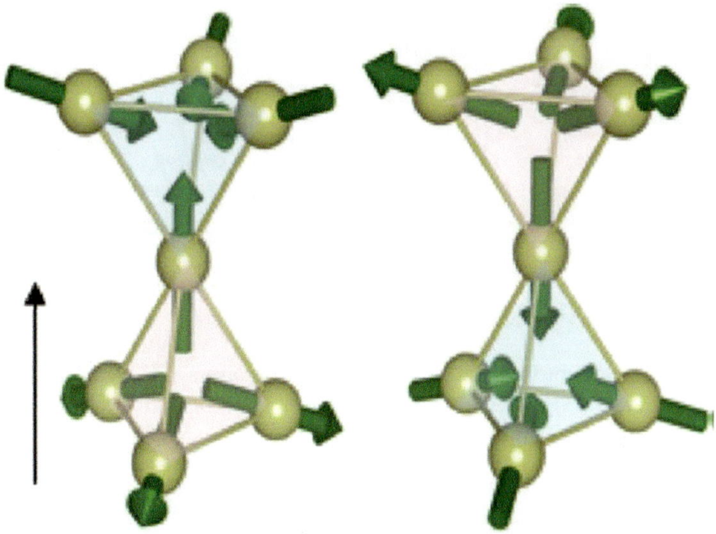

Figure 9.10 AIAO phase of pyrochlore iridate structure in which the magnetic moments are indicated by arrows (reused from [49]).

the other tetrahydron are pointed outwards (all out), and the direction gets interchanged on time reversal.

On time reversal, many magnetic phenomena like the magnetostriction, magnetocapacitance, and Voigt magneto-optic effect of pyrochlore iridates become tunable with the external magnetic field. Also, AIAO spin ordering is considered to be one of the prerequisites for the origin of the WSM phase, a nontrivial topological phenomenon in which the conduction band and valance band cross at Weyl nodes, in pyrochlore iridates. These are the earliest compounds that are proposed to realize a magnetic WSM by theoretical estimations. It was originally proposed by Wan et al. in $Y_2Ir_2O_7$, but, later, a large number of investigations have come out with the strong indications of WSM phase in pyrochlore iridates [45]. Yang et al. [53] and other researchers have evidenced the pressure-triggered anomalous Hall effect (AHE) and a field-induced charge-density-wave (CDW), which are unique features of the WSM phase, but in the absence of a direct experimental evidence, the topic is still under intense discussion.

9.2.5.2 Honeycomb iridates

The A_2IrO_3 (A = Li, Na, and K) family of layered honeycomb iridates build by layers of A atoms alternating with the AIr_2O_6 network stacked along the

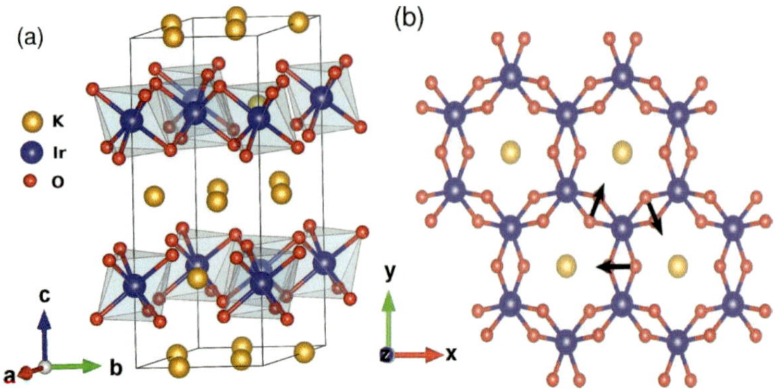

Figure 9.11 Crystal structure of K_2IrO_3. (a) Alternating layers of K atoms alternating with IrO_6 octahedra. (b) Top view of the honeycomb structure of K_2IrO_3 (reused from [56]).

c-axis. K atoms occupy voids between edge-sharing IrO_6 octahedra, which form a honeycomb lattice as shown in Figure 9.11.

As in the case of pyrochlore iridates, spin–orbit interactions and electron correlations have comparable strength, and an interplay between the two can lead to unusual physics. Despite the enormous theoretical interest, from the magnetic, thermal, and transport measurements, Singh et al. have reported that the Mott insulator phase of Na_2IrO_3 and Li_2IrO_3 lies close to the spin-liquid regime around the Kitaev limit [54]. In the strong coupling limit, Jackeli and Khaliullin suggested that the structure can possibly realize the Kitaev model and its spin-liquid ground state [55].

Another potential candidate in the iridate family, which deserves much attention, is electron-doped Sr_2IrO_4, and recent studies proposed that these materials may realize high T_c superconductivity similar to the cuprates [57]. Even though the complete understandings of the d-wave-like excitation gap in Sr_2IrO_4 is not done, the material offers an excellent platform to study the analogy of the *d*-wave gap with the pseudogap present in the cuprates.

9.2.6 Rare-earth nickelates

Another fascinating system in the current decade is rare-earth nickelates with a perovskite structure for fabricating a novel oxide interface and heterostructures. This family of material exhibits a rich-phase diagram that comprises charge ordering, strong electron–lattice coupling, sharp MITs, and a long-period magnetically ordered state, owing to their strong correlation of electrons. For instance, at high temperatures, they have a paramagnetic phase but exhibit an AF insulating nature at a lower temperature with a considerable conductivity variation at MI transition temperature (TMI). These systems have additionally been regarded as a new class of "bad metals" in the sense that their resistivity escalates above the semi-classical Mott-Ioffe-Regel limit and does not saturate at high temperatures. Significant covalent effects that are instigated by means of strong overlaps between O-2p and Ni-3d states characterize the electronic structure of rare-earth nickelates. Subsequently, external stimuli, such as temperature, chemical, or hydrostatic pressure, can modify the electronic bandwidth and influence the MIT [58]. These nickelates exhibit strong magnetically induced ferroelectric polarizations, which is two orders of magnitude larger than one of the typical multiferroics, for example, $TbMnO_3$ [59]. Recent investigations on $RNiO_3$ films predict that, through systematic engineering, they can behave similar to high T_c cuprate superconductors [60]. Rare-earth nitrates have a great deal of interest in solid oxide fuel cell (SOFC) applications as an electrolyte due to their ionic conductivity and low activation energy [61].

Perovskite nickelate $RNiO_3$, where R is a trivalent rare earth (except $LaNiO_3$), exhibits MIT as a function of temperature. The transition is from their high-temperature metallic orthorhombic phase with space group Pbnm to the low-temperature insulating monoclinic phase with space group *P*21/*n*. The phase diagram displayed in Figure 9.12 briefed some of the physical properties of the rare-earth nickelates.

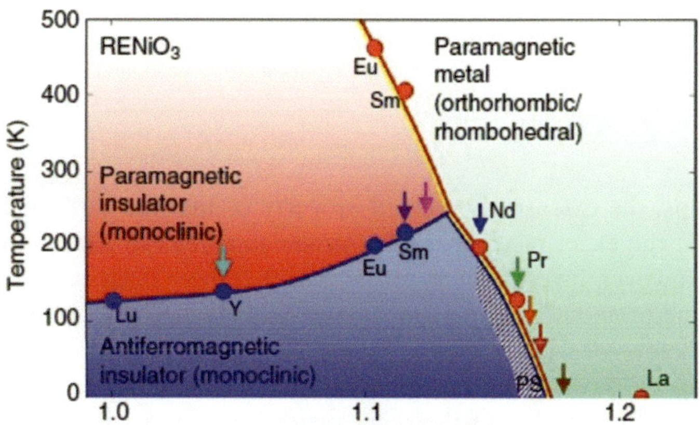

Figure 9.12 Phase diagram of the rare-earth nickelates with ionic radius of the rare-earth ion in the horizontal axis and the temperature taken along vertical axis (reused from [62]).

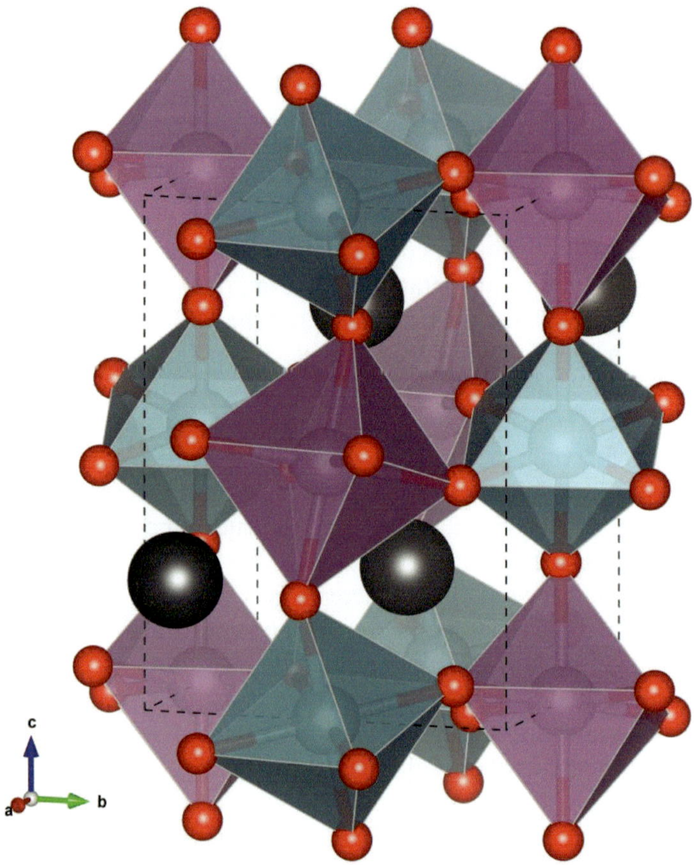

Figure 9.13 Monoclinic phase of $LuNiO_3$ at 60 K, showing LB octahedra (purple) and SB octahedra (green). Lu atoms and O atoms are represented by black and red spheres (reused from [64]).

Perhaps, the extensively studied property of the rare-earth nickelates is their MIT; the exact mechanism that drives this unusual MIT is still under debate. However, investigations recognized that there is an expansion of the UC volume of about 0.2% accompanying MIT [63]. The *P21/n* structure of $LuNiO_3$ below MIT is depicted in Figure 9.13.

During the structural transition, a volume of the half of the NiO_6 octahedra increases and that of the other octahydra reduces by altering Ni–O bond lengths, which can be assumed as a cooperative breathing distortion of the oxygen octahedra resulting in a bond disproportionated (BD) state with an alternating long bond (LB) and short bond (SB) octahedra. Latest investigations propose that the presence of LB and SB bonds are critical for the exhibitance of the MIT [65]. Varignon et al. [66] reported a complete phase diagram of rare-earth nickelates from first principles and successfully described their experimental electronic and magnetic ground state. The investigation also explained dependence of the MIT with the ionic radius of RE and predicted second-order to the first-order transition for R = Pr and Nd based on their theoretical phase diagram.

The insulating phase of the rare-earth nickelates becomes antiferromagnetically ordered at $T < T_{Néel}$, and the large ordering is generally explained by the super-exchange interaction between the 2p (oxygen) and the partially filled d-orbitals of the nickel [67]. A small excitation energy (1 eV) promotes the transfer of 2p electrons into the unoccupied d electron states. Henceforth, the nickelates can be treated as self-doped Mott insulators. The coupling oxygen-derived states and the localized correlated d states are accountable for the self-doping mechanism in the nickelates and thus unusual AF ordering.

Apart from the MI phase transition and the magnetic structure, a significant number of the later investigations are set apart by the exploration of the potential application of nickelates in a variety of fields. Cao et al. [68] reported the effectiveness of $LaNiO_3$ as an anode in high performance supercapacitors. $SmNiO_3$ particularly explored broadly and found applications in many of the fields, such as optoelectronics, catalysis, resistance switching, magneto transport [69], and so forth. In the current decade, there is an enormous quest in the development of high-yield heterostructures such as thin films, ultra-thin films, and multilayers or superlattices of rare-earth nickelate due to the vast number of new phenomena emerged by the mixture of epitaxial strain, quantum confinement, and interfacial charge transfer. Recently, Prajapati et al. [70] evidenced the mosaic structure-controlled M–I transition and conductivity dynamics of $Pr_{0.5}Sm_{0.5}NiO_3$ thin films and proposed the utilization of digital-to-analog type terahertz modulation controlled by the mosaicity of the substrate template of the thin films. In short, a large panel of applications has been identified for rare-earth nickelate thin films, even though many are at the level of ideas. Since the electronic, magnetic, and optical properties can be tuned by the stoichiometry, electron doping, defect engineering, etc., rare-earth nickelate gained considerable scientific attention. Nowadays, scientists are also looking to the development of single crystals of rare-earth nickelates expecting to open a large window of functionalities.

9.2.7 Borides

Magnesium diboride (MgB_2) was categorized under HT_c superconductor in 2001. It is actually placed somewhere in the middle of a low-temperature and high-temperature superconductor. Cooper pairs are the mechanism that causes superconductivity at low temperatures, whereas MgB_2 keeps superconducting at higher temperatures (critical temperature is about 39 K). One of the main advantages of MgB_2 is that it is abundant, low-cost, and light-weight. It is confirmed by the scientists at CERN that, at low temperatures, it is a fantastic conductor, and the results suggest that it could be used for long-distance power transportation.

9.2.8 Cerium monopnictides

Cerium monopnictides CeX (X = Bi and Sb) crystallized in the NaCl structure is another important material that shows a strong electron-correlated phenomenon. In particular, they are the ideal playground for investigating the interplay of magnetism and topology. At low temperatures, they exhibit a wide range of magnetically ordered phases. For instance, CeBi undergoes a transition from the paramagnetic phase to an AF type-I phase (AFM) and then to another AFM type-IA phase at 25 and 13 K, respectively [71]. The most studied material of the cerium pnictide family, CeSb, undergoes a transition from a paramagnetic phase to an antiferroparamagnetic (AFP) phase and then to AFM type-IA phase when the temperature reduces to 16 K and further to 8 K [72]. These low-temperature phases in this kind of system were investigated widely and attributed to the change of Wigner crystal liquid by the strong magnetic-polaron lattice at low temperatures. The phase transitions in magnetic behavior are also reflected in transport properties like temperature dependence of the resistivities of these materials.

Extreme magnetoresistance (XMR), observed in rare-earth monopnictides, in particular, CeSb, deserves attention due to its technological aspects. A large magnetic response of resistivity of about $1.6 \times 10^6\%$ at a magnetic field of 9 T is reported for CeSb, which is much larger compared to XMR observed in other rare-earth monopnictides. Investigations show that Ce orbital state is partly responsible for the XMR and suggested that the unique combination of orbital inversion and type-I magnetic ordering in CeSb triggered the large response [73]. Many intriguing possibilities and phenomena were explored recently as the progress of the research on cerium monopnictides and, very recently, orbital-flop induced MR [74] anisotropy was reported for CeSb, which can shed light into more exotic properties and applications.

9.3 Conclusion

In this chapter, we focused on the materials in which electronic, magnetic, thermodynamic, and optical properties are significantly influenced by the electron–electron correlations and the spin–orbit interactions that are insignificant in the case of ordinary materials. This correlative effect introduced a rich panoply of competing phases in these materials and, subsequently, novel and distinct properties having significant fundamental and technological interest emerged by controlling phase transitions through fluctuating the parameters doping concentration, pressure, and magnetic field. With regard to superconductivity, by glancing through the advancement of high T_c superconductors, we can see that all the exploration was focused on the cuprate family for a considerable period of time; however, as of late, there developed other correlated materials like pnictides and perovskites with exciting possibilities. For instance, T_c above 50 K is accounted for $Sm[O_{1-x}F_x]$ FeAs, which can be varied by changing the rare-earth counterpart. In any case, both theoretical and experimental investigations are to be done to realize the intriguing properties of these materials. Similarly, graphene is, by a long shot, the most studied two-dimensional crystal; however, different TMDCs like MoS_2, $MoSe_2$, WS_2, and WSe_2 attracted great consideration for their prevalent electronic and spintronics applications. Recent experimental evidence of MoS_2 recommends that the spin, orbit, and valley properties of these TMDCs are quite distinctive and may prompt to yet unforeseen applications. On the other hand, substantial theoretical and experimental development has been accomplished on account of the iridate family of materials that exhibit various exotic phases due to the nontrivial interplay of spin–orbit interaction and electron correlation. A wide scientific attention is still needed for the proper synthesis of strongly correlated materials and to interface with different frameworks, which may uncover additional opportunities to fulfill an innovative requirement for smart materials with high sensitivity and selectivity to external stimuli like temperature, pressure, and electric and magnetic fields. Moreover, further advanced computational techniques are necessary for the whole depiction, formulation of design principles, and the manufacturing of new material systems with prerequisite properties. However, advances in the algorithm developments for computation and theoretical developments such as cluster dynamical mean-field theory might assist to overcome the challenges in the future.

Acknowledgements

One of the authors, Sajna M. S., acknowledges UGC for Dr. D.S. Kothari Post-Doctoral Fellowship (F.4–2/2006 (BSR)/PH/18–19/0037, dated February 6 2019).

References

[1] J.G. Rau, E.K. Lee, H.Y. Kee, Spin-Orbit Physics Giving Rise to Novel Phases in Correlated Systems: Iridates and Related Materials, Annual Rev. Cond. Mat. Phy. 7 (1) (2016) 195-221. doi.org/10.1146/annurev-conmatphys-031115-011319

[2] Tanusri Saha-Dasgupta, Double perovskites with 3d and 4d/5d transition metals: compounds with promises, Mater. Res. Express 7 (2020) 014003 1-11, https://doi.org/10.1088/2053-1591/ab6293

[3] K.K. Gomes, A. N. Pasupathy, A. Pushp, S. Ono, A. Yazdani, Visualizing pair formation on the atomic scale in the high- T_c superconductor $Bi_2Sr_2CaCu_2O_{8+\delta}$, Nature, 447 (2007) 569–572. doi:10.1038/nature05881

[4] Y. Moritomo, A. Asamitsu, H. Kuwahara, Y. Tokura, Giant magnetoresistance of manganese oxides with a layered perovskite structure, Nature, 380(1996) 141–144. https://doi.org/10.1038/380141a0

[5] K. Das, R. Rawat, B. Satpati, I. Das, Giant enhancement of magnetoresistance in core-shell ferromagnetic charge ordered nanostructures, Appl. Phys. Lett. 103 (2013) 202406/1–202406/5. https://doi.org/10.1063/1.4830376

[6] A. Kumar, D.D.Sarma, Effect of "dipolar-biasing" on the tunability of tunneling magnetoresistance in transition metal oxide systems. Appl. Phys. Lett. 100 (2012) 262407/1–262407/4. https://doi.org/10.1063/1.4731206

[7] K. Das, P. Dasgupta, A. Poddar, I. Das, Significant enhancement of magnetoresistance with the reduction of particle size in nanometer scale, Scie. Rep. 6 (2016) 20351/1-20351/12. | DOI: 10.1038/srep20351

[8] A. Frano, S.B. Canosa, E. Schierle, Y. Lu, M. Wu, M. Bluschke, M. Minola, G. Christiani, H.U. Habermeier, G. Logvenov, Y.

Wang, P.A. van Aken, E. Benckiser, E.Weschke, M.L. Tacon, B. Keimer, Long-range charge-density-wave proximity effect at cuprate/manganate interfaces, Nature Mater. 15 (2016) 831-834. https://doi.org/10.1038./nmat4682.

[9] Y. Cui, Y. Ke, C. Liu, Z. Chen, N. Wang, L. Zhang, Y. Zhou, S. Wang, Y. Gao, Y. Long, Thermochromic VO_2 for Energy-Efficient Smart Windows, Joule, 2, 9, (2018) 1707-1746, https://doi.org/10.1016/j.joule.2018.06.018.

[10] Tomczak, J.M. and Biermann, S. (2009), Optical properties of correlated materials – Or why intelligent windows may look dirty. phys. stat. sol. (b), 246: 1996-2005. doi:10.1002/pssb.200945231

[11] Z. Li, J. Wu, Z. Hu, Y. Lin, Q. Chen, Y. Guo, Y. Liu,Y. Zhao, J. Peng, W. Chu, C. Wu, Y. Xie. Imaging metal-like monoclinic phase stabilized by surface coordination effect in vanadium dioxide nanobeam., Nat. Commun. **8**, 15561 (2017). DOI: 10.1038/ncomms15561

[12] M.M. Qazilbash, M. Brehm, B.-G. Chae, P.C. Ho, G.O. Andreev, B.-J. Kim, S.J. Yun, A.V. Balatsky, M.B. Maple, F. Keilmann, H.-T. Kim, D.N. Basov, Mott Transition in VO_2 Revealed by Infrared Spectroscopy and Nano-Imaging, Science 318, 5857(2007)1750-1753.DOI: 10.1126/science.1150124.

[13] Y. Kamihara, T. Watanabe, M. Hirano, & Hosono, H. Iron-based layered superconductor La[$O_{1-x}F_x$]FeAs(x =0.05-0.12) with T_c=26 K.J. Am. Chem. Soc. 130 (2008) 3296-3297. https://doi.org/10.1021/ja800073m.

[14] H. Takahashi, K. Igawa, K. Arii, Y. Kamihara, M. Hirano, H. Hosono, Superconductivity at 43 K in an iron-based layered compound LaO(1-x)F(x)FeAs. *Nature*. 2008;453(7193):376-378. doi:10.1038/nature06972

[15] Z.-A. Ren, J. Yang, W. Lu, W. Yi, X.-L. Shen, Z.-C. Li, G.-C. Che, X.-L. Dong, L.-L. Sun, F. Zhou, and Z.-X. Zhao: Superconductivity in the iron-based F-doped layered quaternary compound Nd[O_{1-x} F_x]FeAs. Europhys. Lett. 82 (2008) 57002; https://doi.org/10.1209/0295-5075/82/57002

[16] K. Ishida, Y. Nakai, H. Hosono, To What Extent Iron-Pnictide New Superconductors Have Been Clarified: A Progress Report, Journal of the Physical Society of Japan, 78, 062001 (2009) 10.1143/JPSJ.78.062001.

[17] H. Kito, H. Eisaki, A. Iyo, Superconductivity at 54 K in F-Free $NdFeAsO_{1-y}$, J. Phys. Soc. Jpn. 77 (2008) 063707. https://doi.org/10.1143/JPSJ.77.063707.

[18] Z.-A. Ren, W. Lu, J. Yang, W. Yi, X.-L. Shen, Z.-C. Li, G.-C. Che, X.-L. Dong, L.-L. Sun, F. Zhou, and Z.-X. Zhao: Superconductivity at 55 K in Iron-Based F-Doped Layered Quaternary Compound Sm[O1-xFx] FeAs, Chin. Phys. Lett. 25 (2008) 2215.https://doi.org/10.1088/0256-307X/25/6/080

[19] S. Matsuishi, Y. Inoue, T. Nomura, H. Yanagi, M. Hirano, and H. Hosono, Superconductivity Induced by Co-Doping in Quaternary Fluoroarsenide CaFeAsF, J. Am. Chem. Soc. 130 (2008) 14428-14429. https://doi.org/10.1021/ja806357j

[20] S. Matsuishi, Y. Inoue, T. Nomura, M. Hirano, and H. Hosono, Cobalt-Substitution-Induced Superconductivity in a New Compound with ZrCuSiAs-Type Structure, SrFeAsF, J. Phys. Soc. Jpn. 77 (2008) 113709. https://doi.org/10.1143/JPSJ.77.113709

[21] F. Steglich, J. Aarts, C. D. Bredl, W. Lieke, D. Meschede, W. Franz, H. Schaĺfer, Superconductivity in the Presence of Strong Pauli Paramagnetism: $CeCu_2Si_2$, Phys. Rev. Lett. 43 (1979) 1892. https://doi.org/10.1103/PhysRevLett.43.1892

[22] X.C. Wang, Q. Liu, Y. Lv, W. Gao, L.X. Yang, R.C. Yu, F.Y. Li, C. Jin, The superconductivity at 18 K in LiFeAs system. Solid. State. Commun. 148 (2008) 538. https://doi.org/10.1016/j.ssc.2008.09.057

[23] F.C. Hsu, J.Y. Luo, K.W. Yeh, T.K. Chen, T.W. Huang, P. M. Wu, Y.C. Lee, Y.L. Huang, Y.Y. Chu, D. C. Yan, and M.K. Wu:Proc. Natl. Acad. Sci. U.S.A. 105 (2008) 14262.

[24] L.d Medici, G. Giovannetti, M. Capone, Selective Mott Physics as a Key to Iron Superconductors. Phys. Rev. Lett. 112 (2014) 177001/1-177001/5. https://doi.org/10.1103/PhysRevLett.112.177001

[25] W.Q. Yan, L. Zhi, Z.W. Hao, Z.Z. Cheng, Z.J. Song, L. Wei, D. Hao, O. Yun-Bo, D. Peng, C. Kai, W. Jing, S. Can-Li, H. Ke, J. Jin-Feng, J. Shuai-Hua, W. Ya-Yu, W. Li-Li, C. Xi, M. Xu-Cun, X. Qi-Kun, Interface-Induced High-Temperature Superconductivity in Single Unit-Cell FeSe Films on $SrTiO_3$; Chinese. Phys. Lett. 29 (3) (2012) 037402/1- 037402/4. https://doi.org/10.1088/0256-307X/29/3/037402

[26] T. Ando, Excitons in carbon nanotubes, J.Phys.Soc. Jpn, 66 (1997) 1066–1073. https://doi.org/10.1143/JPSJ.66.1066.

[27] Y. Cao, V. Fatemi, A. Demir, S. Fang, S.L. Tomarken, J.Y. Luo, J.D. Sanchez-Yamagishi, K. Watanabe, T. Taniguchi, E. Kaxiras, R.C. Ashoori, P.J. Herrero, Unconventional superconductivity in magic-angle graphene superlattices, Nature, 556 (2018) 43-50. DOI: 10.1038/nature26160.

[28] K. Bolotin, F. Gharhari, M. Schulman, H. Stormer, P. Kim, Observation of the fractional quantum Hall effect in graphene Nature 462 (2009) 196– 199. https://doi.org/10.1038/nature08582.

[29] S.C. Bodepudi, A. P. Singh, S. Pramanik, Giant Current-Perpendicular-to-Plane Magnetoresistance in Multilayer Graphene as Grown on Nickel, Nano Lett., 14, 5 (2014) 2233-2241. DOI:10.1021/nl4030853

[30] K. Matsuda, Optical Properties of Atomically Thin Layered Transition Metal Dichalchogenide, J. Phys. Soc. Jpn., 84 (2015) 121009. https://doi.org/10.7566/JPSJ.84.121009

[31] Q.H Wang, K Kalantar-Zadeh, A. Kis, J.N. Coleman, M.S. Strano, Electronics and optoelectronics of two-dimensional transition metal dichalcogenides, Nature Nanotechnology, 7 (2012) 699-712. https://doi.org/10.1038/nnano.2012.193

[32] B. Radisavljevic, A. Radenovic, J. Brivio, V. Giacometti, A. Kis, Single-layer MoS_2 transistors, Nature Nanotechnology, 6 (2011).147-150. https://doi.org/10.1038/nnano.2010.279.

[33] K.F. Mak, C. Lee, J. Hone, J. Shan, T.F. Heinz, Atomically thin MoS_2: A new direct-gap semiconductor, 105 (2010) 136805/1-136805/4 DOI: 10.1103/PhysRevLett.105.136805

[34] A.E. Ekhari, Molybdenum diselenide ($MoSe_2$) for energy storage, catalysis, and optoelectronics, Appl. Mater. Today, 8 (2017) 1–17, DOI: 10.1016/J.Apmt.2017.01.006.

[35] E. Revolinsky, D. Beerntsen, Electrical properties of The $MoTe_2$–WTe_2 and $MoSe_2$–WSe_2 Systems, J. Appl. Phys., 35 (1964) 2086-2089, DOI: 10.1063/1.1702795.

[36] R. Hanson, D.D. Awschalom, Coherent manipulation of single spins in semiconductors, Nature, 453 (2008) 1043–1049, DOI: 10.1038/Nature 07129.

[37] M.N. Ali, J. Xiong, S. Flynn, J. Tao, Q.D. Gibson, L.M. Schoop, T. Liang, N. Haldolaarachchige, M. Hirschberger, N.P. Ong, R.J. Cava, Large, Non- Saturating Magnetoresistance In WTe_2, Nature, 514 (2014) 205–208, DOI: 10.1038/Nature13763

[38] J.M. Lu, O. Zheliuk, I. Leermakers, N.F. Yuan, U. Zeitler, K.T. Law, J. Ye, Evidence For Two-Dimensional Ising Superconductivity In Gated MoS_2, Science 350 (2015) 1353-1357. 10.1126/Science.Aab2277

[39] X. Xi, Z. Wang, W. Zhao, J.-H. Park, K. T. Law, H. Berger, L. Forró, J. Shan, K. F. Mak, Ising Pairing in Superconducting $NbSe_2$ Atomic Layers, Nat. Phys. 12 (2015) 139-144. DOI: 10.1038/nphys3538

[40] K. Zhao, H. Lin, X. Xiao, W. Huang, W. Yao, M. Yan, Y. Xing, Q. Zhang, Z-X Li, S. Hoshino, J. Wang, S Zho, L. Gu, M.S. Bahramy, H. Yao, N. Nagaosa, Q-K Xue, K.T. Law, X. Chen, S-H Ji, Disorder-Induced Multifractal Superconductivity In Monolayer Niobium Dichalcogenides, Nat. Phys. 15 (2019) 904-910 027001. 10.1038/S41567-019-0570-0

[41] H. Narita, H. Ikuta, H. Hinode, T. Uchida., T. Ohtani, M. Wakihara, Preparation and Physical Properties of FexTaS_2 ($0.15 \leq x \geq 0.50$) Compounds, J. Solid State Chem., 108 (1994)148-151, https://doi.org/10.1006/jssc.1994.1022.

[42] J. A. Wilson, A.D. Yoffe, The transition metal dichalcogenides discussion and interpretation of the observed optical, electrical and structural properties, Adv. Phys. (1969), 18, 193-335. https://doi.org/10.1080/00018736900101307

[43] Y. Liu, R. Ang, W.J. Lu, W.H. Song, L.J. Li, Superconductivity Induced By Se-Doping In Layered Charge-Density-Wave System 1T-Tas_2-Xs, Appl. Phy. Lett. 102 (2013) 192602/1-192602/5. 10.1063/1.4805003

[44] R. Liu, C. Wang, Y. Li, Y. Xie, Q. Chen, Z. Chen and Q. Liu, intercalating copper into layered TaS_2 van der Waals gaps, RSC Adv., 2017, 7, 46699, DOI: 10.1039/C7RA08630J

[45] X. Wan, A.M. Turner, A. Vishwanath, S. Y. Savrasov, Topological semimetal and Fermi-arc surface states in the electronic structure of pyrochlore iridates, Phy. Rev. B, 83 (20) (2011) 205101/1-205101/9, doi:10.1103/ PhysRevB.83.205101

[46] H. Sagayama, D. Uematsu, T. Arima, K. Sugimoto, J.J. Ishikawa, E. O'Farrell, S. Nakatsuji, Determination of long-range all-in-all-out ordering of Ir^{4+} moments in a pyrochlore iridate $Eu_2Ir_2O_7$ by resonant x-ray diffraction, Phys. Rev. B 87 (2013)100403/1-10403/4. DOI: 10.1103/PhysRevB.87.100403

[47] K. Ueda, J. Fujioka, Y. Takahashi, T. Suzuki, S. Ishiwata, Y. Taguchi, M. Kawasaki, Y. Tokura. Anomalous domain-wall conductance in pyrochlore-type $Nd_2Ir_2O_7$ on the verge of the metal-insulator transition, Phys. Rev. B 89 (2014) 075127/1-075127/5. DOI: 10.1103/PhysRevB.89.075127

[48] T. Kondo, M. Nakayama, R. Chen, J.J. Ishikawa, E.G. Moon, T. Yamamoto, Y. Ota, W. Malaeb, H. Kanai, Y. Nakashima, Y. Ishida, R. Yoshida, H. Yamamoto, M. Matsunami, S. Kimura, N. Inami, K. Ono, H. Kumigashira, S. Nakatsuji, L. Balents, S. Shin, Quadratic Fermi

node in a 3D strongly correlated semimetal, Nat. Commun. 6 (2015) 10042/1-10042/12. DOI: 10.1038/ncomms10042.

[49] K. Ueda, R. Kaneko, H. Ishizuka, J. Fujioka, N. Nagaosa,Y.Tokura. Spontaneous Hall effect in the Weyl semimetal candidate of all-in all-out pyrochlore iridate. *Nat Commun* **9,** 3032 (2018). https://doi.org/10.1038/s41467-018-05530-9

[50] B.J. Kim, H. Jin, S.J. Moon, J.Y. Kim, B.G. Park, C.S. Leem, J. Yu, T.W. Noh, C. Kim, S.J. Oh, J.-H. Park, V. Durairaj, G. Cao, E. Rotenberg, Novel Jeff ij 1=2 Mott State Induced by Relativistic Spin-Orbit Coupling in Sr_2IrO_4, PRL 101 (2008) 076402/1-076402/4. DOI: 10.1103/PhysRevLett.101.076402

[51] K. Matsuhira, M. Wakeshima, Y. Hinatsu, S. Takagi, Metal–Insulator Transitions in Pyrochlore Oxides $Ln_2Ir_2O_7$. J. Phys. Soc. Jpn., 80(9) (2011) 094701/1-094701/7. doi:10.1143/jpsj.80.094701

[52] B.J. Yang, Y.B. Kim, Topological insulators and metal-insulator transition in the pyrochlore iridates, Phys. Rev. B, 82 (2010) 085111/1-085111/11, DOI: 10.1103/PhysRevB.82.085111

[53] K.Y. Yang, Y.-M. Lu, Y. Ran, Quantum Hall effects in a Weyl semimetal: Possible application in pyrochlore iridates, Phys. Rev. B 84 (2011) 075129/1-075129/11 DOI: 10.1103/PhysRevB.84.075129

[54] Y. Singh, P. Gegenwart, Antiferromagnetic Mott insulating state in single crystals of the honeycomb lattice material Na_2IrO_3, Phys. Rev. B 82 (2010) 064412/1-064412/7, DOI: 10.1103/PhysRevB.82.064412

[55] G. Jackeli, G. Khaliullin, Mott Insulators in the Strong Spin-Orbit Coupling Limit: From Heisenberg to a Quantum Compass and Kitaev Models, Phys. Rev. Lett. 102 (2009) 017205/1-017205/4. DOI: 10.1103/PhysRevLett.102.017205

[56] R.Yadav, S.Nishimoto, M.Richter, J. Brink, R. Ray, Large off-diagonal exchange couplings and spin liquid states in C_3-symmetric iridates Phys. Rev. B 100(2019)144422 1-10

[57] Y.K. Kim, N.H. Sung, J.D. Denlinger, B. J. Kim, Observation of a d-wave gap in electron-doped Sr_2IrO_4, Nature.Phys., 12 (2016) 37-41. doi.org/10.1038/nphys3503

[58] J.B. Torrance, P. Lacorre, A.I. Nazzal, E.J. Ansaldo, C. Niedermayer, Systematic study of insulator-metal transitions in perovskites RNiO3 (R = Pr, Nd, Sm, Eu) due to closing of charge-transfer gap. Phys. Rev. B 45 (1992) 8209–8212. Doi:10.1103/PhysRevB.45.8209

[59] T. Kimura, T. Goto, H. Shintani, K. Ishizaka, T. Arima, Y. Tokura, Magnetic control of ferroelectric polarization. Nature, 426 (6962) 55–58. doi:10.1038/nature02018

[60] J. Chaloupka, G. Khaliullin, Orbital order and possible superconductivity in $LaNiO_3/LaMO_3$ superlattices. Phys. Rev. Lett. 100 (2008) 016404/1-016404/4. doi:10.1103/PhysRevLett.100.016404

[61] Y. Zhou1, X. Guan, H. Zhou, K. Ramadoss, S. Adam, H. Liu, S. Lee, J. Shi, M. Tsuchiya, D.D. Fong, S. Ramanathan, Strongly correlated perovskite fuel cells, Nature 534 9 (2016) 231-234. doi:10.1038/nature17653

[62] B.A. Frandsen, L.Liu, Y.J.Uemura, Volume-wise destruction of the antiferromagnetic Mott insulating state through quantum tuning. Nat Commun 7, 12519 (2016). https://doi.org/10.1038/ncomms12519

[63] J.A. Alonso, J.L. García-Muñoz, M.T. Fernández-Díaz, M.A.G. Aranda, M. J. Martínez-Lope, M. T. Casais, Charge Disproportionation in $RNiO_3$ Perovskites: Simultaneous Metal-Insulator and Structural Transition in$YNiO_3$. Phy. Rev. Lett., 82(19) (1999) 3871–3874. doi:10.1103/PhysRevLett.82.3871

[64] A. Hampel, C. Ederer, Interplay between breathing mode distortion and magnetic order in rare-earth nickelates $RNiO_3$ within DFT+U, Phys. Rev. B 96 (2017)165130 1-12

[65] S. Catalano, M. Gibert, J. Fowlie, J. Íñiguez, J.-M. Triscone, J. Kreisel, Rare-earth nickelates $RNiO_3$: thin films and heterostructures. Reports on Progress in Physics, 81(4) (2018) 046501. doi:10.1088/1361-6633/aaa37a

[66] J. Varignon, M. N. Grisolia, J. Íñiguez, A. Barthélémy, M. Bibes, Complete phase diagram of rare-earth nickelates from firstprinciples, npj Quant. Mater. 2, 21 (2017) 1-9, 10.1038/s41535-017-0024-9

[67] J. Zaanen, G.A. Sawatzky, J. W. Allen, Band gaps and electronic structure of transition-metal compounds. Phy. Rev. Lett., 55(4) (1985) 418–421. doi:10.1103/PhysRevLett.55.418

[68] Y. Cao, B. Lin, Y. Sun, H. Yang, X. Zhang, Sr-doped Lanthanum Nickelate Nanofibers for High Energy Density Supercapacitors. Electrochim. Acta, 174 (2015) 41–50. doi:10.1016/j.electacta.2015.05.131

[69] S.D. Ha, J. Shi, Y. Meroz, L. Mahadevan, S. Ramanathan, Neuromimetic Circuits with Synaptic Devices Based on Strongly Correlated Electron Systems, Phys. Rev. Applied 2 (6) (2014) 064003/1-064003/11. doi:10.1103/physrevapplied.2.064003

[70] G.L. Prajapati, S. Das, D.S. Rana, Digital to analog type Terahertz modulation controlled by mosaicity of the substrate template in rare-earth nickelate thin films. ACS Appl. Mater & Interfaces. 11, 36 (2019) 33109-33115. doi:10.1021/acsami.9b09362

[71] H. Kumigashira, S.-H. Yang, T. Yokoya, A. Chainani, T. Takahashi, A. Uesawa, T. Suzuki, O. Sakai, Y. Kaneta, High-resolution angle-resolved photoemission spectroscopy of CeBi, Phys. Rev. B 54 (1996) 9341-9345. 10.1103/PhysRevB.54.9341

[72] A. Takayama, S. Souma, T. Sato, T. Arakane, T. Takahashi, Magnetic phase transition of CeSb studied by low-energy angle-resolved photoemission spectroscopy, J. Phys. Soc. Jpn 78 (2009) 073702. 10.1143/jpsj.78.073702

[73] L. Ye, T. Suzuki, C.R. Wicker, J.G. Checkelsky, Extreme magnetoresistance in magnetic rare-earth monopnictides. Phys. Rev. B 97 (2018) 081108/1-081108/8. 10.1103/PhysRevB.97.081108

[74] J. Xu, F. Wu, J. Bao , F. Han, Z-L Xiao, I. Marti , Y-Y. Lyu, Y-L. Wang, D. Y. Chung , M. Li, W. Zhang, J. E. Pearson, J. S. Jiang , M. G. Kanatzidis,W-K.Kwok, Orbital-flop Induced Magnetoresistance Anisotropy in Rare Earth Monopnictide CeSb, Nat. Commun. 10 (2019) 2875/1-2875/8. 10.1038/s41467-019-10624-z

10

Synthesis of Selected Strongly Correlated Multiferroic Systems

Anshida Mayeen[1], Praveena M. G.[2], Kala M. S.[2], Shemim S. S.[3], and Nandakumar Kalarikkal[1,4]

[1]School of Pure and Applied Physics, Mahatma Gandhi University, India
[2]Department of Physics and Centre for Research, St. Teresa's College, India
[3]Thangal Kunju Musaliar College of Engineering, India
[4]International and Interuniversity Centre for Nanoscience and Nanotechnology, Mahatma Gandhi University, India

Abstract

Strongly correlated systems belong to the class of heavy fermion compounds that consist of insulators, electronic material, and materials, which exhibit exceptional electronic and magnetic properties. Due to the vast array of potential applications, a significant amount of research has been conducted in this field. Among them, a remarkable feature of these strongly correlated materials is their ability to exhibit excellent magnetic and transport properties. Because of this property, these materials can be considered as functional materials. Amid these strongly correlated systems, multiferroic systems are gaining huge attention. Multiferroic materials are an energizing region of research since they simultaneously display ferromagnetism and ferroelectric polarization and inter-coupling between them. This chapter deals with the synthesis of selected strongly multiferroic systems.

Keywords: Nanomultiferroics, Wet Chemical Synthesis, Composite multiferroics

DOI: 10.1201/9781003337102-10

10.1 Introduction

Strongly correlated perovskite oxides having the capability to exhibit a wide range of functionalities within the same system simultaneously are gaining huge interest among the researchers. These perovskite materials exhibit excellent magnetic, ferroelectric, magneto resistive, and superconductive behaviors. Out of these materials, multiferroics are considered as a most relevant group because it possesses both ferromagnetism and ferroelectricity and a magnetoelectric coupling in between. In these materials, an external electric field can switch magnetization and an external magnetic field can adjust electric polarization. Therefore, multiferroics can also be specified as magnetic ferroelectrics. These magnetoelectric materials are widely used in data storage, spintronic, sensors, and microelectronics applications [1–3].

In multiferroics, magnetoelectric coupling occurs because of the following facts:

- Multiferroics in which the ferroelectricity is mainly due to the hybridization, covalence, or structural effects.
- Multiferroics in which the ferroelectricity arises due to other electronic mechanism or correlation effects [4].

Multiferroic materials can generally be divided into two categories: single-phase and composite.

For single-phase multiferroic systems, it is challenging to achieve room-temperature magnetoelectric coupling, while in composite systems, it is effortless to achieve a high degree of magnetoelectric coupling. Single-phase multiferroics contain the anti-ferromagnetic relaxor ferroelectrics like $Pb(Fe_{1/2}Nb_{1/2})O_3$ (PFN) and $Pb(Fe_{1/2}W_{1/3})O_3$ (PFW). The list includes orthorhombic manganites such as $REMnO_3$ and $REMn_2O_5$, where RE is a rare-earth element, $LiCoPO_4$ and $BiFeO_3$, and its solid solutions with $BaTiO_3$ [4–6].

Composite multiferroic materials are hybrid multicomponent material in which one is ferroelectric and the other will be ferromagnetic. The cross-coupling interaction between these two components, that is, ferromagnetic filler coupling with ferroelectric filler, results in magnetoelectric coupling [6]. These composite multiferroic materials can be categorized into polymer-based and ceramic-based multiferroics. This chapter provides an overview of the synthesis methods for single-phase and composite nano-multiferroics based on ceramics, including the sol$^-$gel method, hydrothermal, and pulse laser deposition. The details of these methods are discussed further.

10.2 Synthesis of Multiferroics

This chapter examines the two main approaches for synthesizing multiferroics, bottom-up and top-down methods, and provides examples for each method in detail.

10.2.1 Bottom-up approach

In the bottom-up approach, the desired material is built up from the bottom either by atom by atom, molecule by molecule, or cluster by cluster. In this method, properties such as shape, surface area, pore size, stoichiometry, etc., can be controlled. Zero-dimensional (nanoparticles), one-dimensional particles such as nanowires, nanotubes, and nanorods, and two dimensions such as thin film can be prepared with the aid of the bottom-up approach [7]. The bottom-up technique is used to develop the desired material from the underlying material. This technique is more critical for the fabrication of nanomaterials. The bottom-up approach always offers a higher chance to get nanomaterials with lesser defects with uniform chemical composition and with enhanced short-range and long-range ordering [8]. The different bottom-up processes involve:

⇒ self-assembly of nanomaterials;
⇒ sol^{-}gel method;
⇒ electrodeposition;
⇒ physical vapor deposition;
⇒ pulsed laser deposition.

The bottom-up approach can be again divided into the vapor phase and liquid phase. Vapor phase includes pyrolysis and inert gas condensation; on the other hand, liquid phase includes solvothermal, sol^{-}gel synthesis, etc. [7–11].

10.2.1.1 Co-precipitation method

In the co-precipitation method, metal cations from a medium were usually precipitated as hydroxides, carbonates, oxalates, etc. The precipitate obtained from the reaction is calcined at a higher temperature to yield the final desired powder. In particular reaction, mixing takes place at the atomic scale; hence, the calcination temperature required is shallow. This method helps to prepare particles with smaller size. Certain synthesis parameters should be controlled in order to control the reaction such as concentration, pH, temperature, stirring speed, etc. Although this method has certain advantages, this method cannot be employed in the preparation of high purity final stoichiometric phase, and this method is not effective, when the reactants have different solubility as well as different precipitation rate [12, 13]. The synthesis of $BiFeO_3$ and $BaTiO_3$ using co-precipitation method is discussed in the following.

10.2.1.1.1 Synthesis of bismuth ferrite ($BiFeO_3$) by the co-precipitation method

The co-precipitation method entails the separation of a solid containing a mixture of ionic species from a liquid solution. In this method, all cations precipitate simultaneously, resulting in the formation of a mixed-metal precursor. In this method, for the preparation of $BiFeO_3$, bismuth nitrate and iron nitrate were used as precursors and ammonium hydroxide to obtain a hydroxide precipitate. Here, the calcination temperature ranges from 550 to 750 °C to obtain phase-pure $BiFeO_3$. The main disadvantage of this method is controlling the particle size and size distribution. If the precipitation is fast or uncontrollable, it results in larger particle size. A modified precipitation method, involving the addition of ammonia and nitrate solutions to maintain a constant pH level, can be used to overcome the limitations of the co-precipitation method, as seen in the synthesis of bismuth ferrite nanoparticles [13, 14].

10.2.1.1.2 Synthesis of barium titanate ($BaTiO_3$) by the co-precipitation method

The oxalate route is less preferred in the preparation of $BaTiO_3$ because it is tough to optimize the synthesis condition since Ti and Ba precipitate simultaneously. In the oxalate route, titanium is precipitated as titanyl oxalate in the presence of alcohol at a pH value less than or equal to 2. However, barium is precipitated as BaC_2O_4 at a pH greater than 4. Between the pH range 2–4, titanium forms soluble ionic species while $TiO(C_2O_4)_{22}$ leads to a distortion (Ba:Ti ratio) during the simultaneous precipitation. So, the synthesis of barium titanate by the co-precipitation method is less considered due to the above difficulty.

10.2.1.2 Combustion method

The combustion method is a complex series of chemical reactions that take place between a fuel and an oxidant, and it is characterized by the generation of heat, light, or both, in the form of a glow/flames. The combustion method is based on rapid thermal degradation of reactants in the presence of oxygen, which is effectively used for the creation of desired metal oxide nanoparticles.

As the name implies, the combustion reaction is a series of flaming and explosive reactions, and the rate of combustion is determined by the nature of the reaction, which can include decomposition and deflagration [17]. Ceramic materials are often synthesized through smoldering and flaming reactions.

10.2.1.2.1 Synthesis of bismuth ferrite ($BiFeO_3$) by the combustion method

The synthesis of bismuth ferrite ($BiFeO_3$) is achieved by using bismuth nitrate and iron nitrate as oxidants and citric acid and ammonium nitrate as fuels. The specifics of the reaction are explained below:

$$2Bi(NO_3)_3\text{ (aqueous)} + 2Fe(NO_3)_3\text{(aqueous)} + 5C_6H_8O_7\text{ (aqueous)} + 15NH_4NO_3\text{(aqueous)} \rightarrow 2BiFeO_3\text{(solid)} + 15N_2\text{ (gas)} + 50H_2O\text{ (liquid)} + 30CO_2\text{ (gas)} \ldots\ldots\ldots\ldots\ldots \quad (10.1)$$

This equation works based on rocket fuel chemistry. As per this, the valencies of carbon, hydrogen, nitrogen, and oxygen are +4, +1, 0, and −2, respectively. The valency of nitrogen is considered as zero because, during combustion, nitrogen is converted to molecular nitrogen. Valencies of the bismuth and iron depend upon the metal ions in the compound and will be +3 and +3, respectively [18, 19].

10.2.1.2.2 Synthesis of barium titanate ($BaTiO_3$) by the combustion method

Ba $(NO_3)_2$, TiO_2, NH_4NO_3, and reducing agents were used for the preparation of $BaTiO_3$ nanoparticles preparation. An aqueous solution of the precursors was prepared in the first stage. The above suspension has to be homogenized with a micro ball mill for an hour and these suspensions are transfer into a heat-resistant beaker and placed in a muffle furnace preheated up to 500 °C. Reactive charge undergoes fast boiling during heating, which results in the formation of foam where the intense exothermal reaction occurred, leading to a fluffy powder formation. The powder obtained was calcined at 800 °C for 30 minutes, resulting in the formation of $BaTiO_3$ particles, which is adequately grounded in an agate mortar, which results in the formation of fine $BaTiO_3$ powder [20, 21].

10.2.1.3 Sonochemical method

Chemical reactions (in reactants) that take place under the influence of ultra-frequency sound results in the formation of desired nanostructures, which is called the sonochemical method. Due to its simplicity and low cost of equipment, the sonochemical method has garnered significant interest among researchers. The sonochemical method possesses certain advantages over other synthesis methods; striking features of the sonochemical method are the uniform particle size of the particles obtained, increased reaction speed, higher surface energy, high purity samples, etc. In the sonochemical method, complex oxides can be formed easily at room temperature, followed by calcination at moderate temperature; here, an acoustic activation produced during the irradiation of liquids along with ultrasound generates transient local conditions, which decomposes the metal precursors to nano-sized particles. Particle size and particle shape can also be controlled by varying ultrasound powder leads [23].

10.2.1.3.1 Synthesis of $BaTiO_3$ by using the sonochemical method

In the sonochemical synthesis of $BaTiO_3$, high-purity barium chloride dehydrates ($BaCl_2$ů$2H_2O$; 99%) and titanium (IV) chloride (TiCl4; 99%) of appropriate proportions were used as precursors.

The chemical reaction that takes place during the preparation of $BaTiO_3$ by sonochemical method is shown below:

$$BaCl_2 \text{ (aqueous)} + TiCl_4(l) + 6NaOH \text{ (aqueous)} \rightarrow BaTiO_3 \text{ (solid)} + 6NaCl \text{ (aqueous)} + 3H_2O \ldots\ldots\ldots\ldots\ldots \quad (10.2)$$

Here, a stoichiometric amount of precursors (barium chloride and titanium chloride) were first dissolved in 100 ml of distilled water. Instantly, a small volume of sodium hydroxide solution was added to the above solution under constant magnetic stirring. This addition of alkaline solution is to maintain the pH of the system and provide a strongly alkaline environment during the reaction time. After that, the solution was exposed to ultrasound (high intense) radiation for a given time. Here, the sonication was carried out without cooling. Therefore, the temperature of the solution will be increased as a result of sonication and reach up to 80 °C. After the completion of sonication, the solution was cooled to room temperature and results in the formation of a precipitate, which is washed separately with an ammonia solution until the complete removal of the chloride. The high pH provided by ammonia (10.5) greatly reduces the loss of barium ions from the surface of $BaTiO_3$ nanoparticles. This powder was dried in an oven at 100 °C for several hours, resulting in the formation of $BaTiO_3$ particles [23, 24].

10.2.1.4 Hydrothermal method

A hydrothermal process is a reaction that occurs between reactants in a sealed vessel containing solvents, where the temperature and pressure of the solvent can be elevated to the critical point through simultaneous heating. When water is used as a solvent, the term hydrothermal is used. The hydrothermal method is considered a highly effective and promising technique for the production of nanoparticles with desired morphology, crystallinity, and adjustable surface properties. Nanoparticles created through this method possess exceptional optical, electronic, magnetic, mechanical, and chemical characteristics, as a result of the quantum confinement effect and the high surface-to-volume ratio. For preparing different shapes like cube, sphere, tube, etc., the hydrothermal method is more preferred. Nanomaterials can be produced in large amounts and with relatively low cost, with high crystallinity, and with well-controlled dimensionality, which are the significant advantages of this method [25].

10.2.1.4.1 Synthesis of $BaTiO_3$ hydrothermal method

Hydrothermal synthesis of crystalline $BaTiO_3$ has gained wide acceptance in recent times. Hydrothermal synthesis of $BaTiO_3$ is a straightforward and one-step process, and also the precursors are less expensive; this is why the synthesis of $BaTiO_3$ by the hydrothermal method is more preferred by the researchers. $Ba(OH)_2$ or $BaCl_2$, $Ba(NO_3)_2$ and TiO_2, $TiCl_4$ or gels of Ba-Ti acetate mixtures were used as precursors. Barium titanate particles prepared by this method will be fine and uniform in size. The particle size can be regulated by altering specific synthesis parameters such as temperature, reaction time, pressure on the ionic equilibrium in the hydrothermal reaction medium, and pH. The pH of the reaction can be controlled by adding an adequate amount of KOH (potassium hydroxide) or NaOH (sodium hydroxide). Addition of KOH and NaOH will prevent HCL formation if chloride precursors are used for $BaTiO_3$ synthesis. The procedure for the synthesis of $BaTiO_3$ using chloride precursors is illustrated below.

A stoichiometric amount (1 M) of $BaCl_2.2H_2O$ with 0.6 mol of $TiCl_4$ was added to de-ionized water, and 1%–3% of NaOH was added to the mixture of reactants to neutralize the acidic formation and to maintain the pH value at the specific limit [26]. After the addition of NaOH, the solution was stirred for 30 minutes, which results in the formation of a colloidal sol. The final volume was set at 90 ml by adding distilled water to the sol, and after that, the sol was transferred to a sealed Teflon-lined autoclave vessel (100 ml) and heated at 240 °C for 20 hours. The final precipitate obtained has to be washed, centrifuged and dried at 80 °C in vacuum oven for 12 hours. The $BaTiO_3$ powder obtained by this method will be fine and of uniform particle size [27, 28].

10.2.1.5 Sol‾gel method

The sol‾gel method involves a lot of chemistry, which is the main factor for the formation of the required material. Certain synthesis parameters like pH of the sol, concentration of the positive ion, temperature, etc., control the sol‾gel chemistry and the final product. The sol‾gel method takes place in various i.e., it is not a single-step process like the hydrothermal method. In the sol‾gel method, metal oxides, glasses, and ceramics were prepared by hydrolyzing a chemical precursor to sol, and, sequentially, the sol is converted to gel, which on drying or heating results in the formation of the powder. After further calcination, crystalline powder will be formed.

Various steps involved in the sol‾gel method are as follows:

1. Semi-hydrolysis of the metal alkoxide enables the formation of reactive monomers

Step 1 – Partial hydrolysis M (OR) *n* + mH2O ——... M (OR) *n-m* (OH) *m* + ROH.

2. Polycondensation of the monomers, which results in the formation of oligomers in the form of colloids.

2M (OR) *n-m* (OH) *m* ——... (OR) *n*-mM-O-(OR) *n-m* + HOH

23OR

(OR) n-1M-(OR) + (HO)-M (OR) n-1 ——-... (OR) n-1M-O-M (OR) n-1 + ROH

3. Further hydrolysis to be continued to encourage the formation of a 3D matrix, called a gel, through polymerization and crosslinking: three-dimensional cross-linking between (OR) n-mM-O-(OR) n-m molecules [28–30].

As the polymerization and cross-linking take place, the viscosity of the sol increases until the formation of gel. The sol‾gel process results in products with varying structural and electrical properties, which are influenced by factors such as the type of precursor solution used as well as conditions during deposition like pH, temperature, and reaction time [28].

10.2.1.5.1 Synthesis of $BaTiO_3$ by the sol⁻gel method

There are basically two strategies adopted for the synthesis of $BaTiO_3$ particles through the sol⁻gel method:

a) acetic acid gel method;
b) stearic acid gel method.

Acetic acid gel method for the preparation of $BaTiO_3$ nanoparticles:

The starting materials used for the preparation of $BaTiO_3$ by this method were barium acetate, tetrabutyl titanate, isopropyl alcohol, and glacial acetic acid. Tetrabutyl titanate was dissolved in isopropyl alcohol (0.1 mol) at room temperature, and glacial acetic acid (0.3 mol) was gradually stirred into the solution to form titanyl acylate. To this solution, 0.1 mol of barium acetate that was dissolved in distilled water was added and mixed well, and acetic acid was added dropwise in order to control the pH in the range 3–4. The resulting solution is kept under vigorous magnetic stirring for half an hour to form a transparent sol, and the sol is kept at 90 °C until the gel formed. Thereafter, the gel is dried properly at 120 °C for 12 hours in an oven and calcined in different temperatures to form barium titanate nanoparticles. The average particle size of the resultant powder was to be around 50–80 nm [28–30] .

Stearic acid gel method for the preparation of $BaTiO_3$ nanoparticles:

The reagents used for the preparation of $BaTiO_3$ nanoparticles by stearic-acid-based sol⁻gel methods were barium stearate, tetrabutyl titanate, and stearic acid. Stearic acid was melted at 73 °C, and barium stearate was added to form a yellow, transparent solution. Tetrabutyl titanate was then added in stoichiometric amounts and stirred until a brown sol was formed. The sol was dried for 12 hours to form a gel, which was calcined at different temperatures to obtain $BaTiO_3$ nanoparticles. The average particle size of the $BaTiO_3$ particles was determined to be between 23 and 50 nm [28–31] .

10.2.1.5.2 Synthesis of bismuth ferrite ($BiFeO_3$) by the sol⁻gel method

For the synthesis of bismuth ferrite nanoparticles through the sol⁻gelgel method, bismuth nitrate and iron (111) nitrate nonahydrate were used as the precursors. These precursors were separately dissolved in glacial acetic acid with the stoichiometric ratio 1:1. When the solution becomes completely transparent, ethylene alcohol was added to the solution under continuous magnetic stirring. After stirring for half an hour, the solution was mixed correctly. The final sol will be transparent blackish red. A stable sol was obtained by maintaining the precursor solution at 0.05–0.2 M by adding acetic acid and ethylene alcohol. The precursor solution was dried at 40 °C for a week to obtain bismuth ferrite in gel form. The dried gel was grounded, employing an agate mortar. The grounded powder was calcined at a higher temperature for the formation of high purity crystalline nanoparticles [32, 33].

10.2.1.6 Pechini method (modified sol⁻gel)

The Pechini method was named after the author of the original patent, who builds some modification in sol⁻gelgel chemistry and introduced small molecule chelating ligands. The initial stage is the formation of a homogeneous solution of metal/citrate complexes; after that, this solution is converted into a covalent polymeric network to entrap these metal ions. The key advantage of this method is the ability to regulate the thermal decomposition of the organic matrix, which enhances control over the growth of the ceramic product. The Pechini method is based on a primary reaction of transesterification between citrate and ethylene glycol, resulting in a homogeneous precursor solution containing metal-citrate complexes. Here, the solution should be heated at a specific temperature to promote polyesterification between the citric acid and ethylene glycol. Initially, a metal salt is dissolved in a citric acid solution, and then ethylene glycol is added to create an extended covalent network. The sample is heated in a furnace to eliminate the organic matrix after the polymer network has formed, resulting in the creation of a ceramic product. In the Pechini method, multiple metals can be evenly dispersed throughout the network [34–38] .

10.2.1.6.1 Synthesis of $BiFeO_3$ by the Pechini method

High purity metal nitrate solutions were used as precursors here. Synthesis of bismuth ferrite by this method needs more care because if we add bismuth nitrate directly in water, bismuth cations will strongly hydrolyze to form insoluble bismuth oxynitrate ($Bi_2O_3N_2O_52H_2O$) and if acid is added to stabilize these bismuth cations, it will hinder the formation of a metal–citrate complex network. So bismuth nitrate is directly added to ethylene glycol and citric acid is dissolved in distilled water; after that, both solutions were mixed properly. The reactants were mixed in the required proportions for 3g of $Bi(NO_3)_3.5H_2O$, 1.26 g of citric acid is taken, and 50% extra amount of ethylene glycol is added with the total volume of the water taken in order to dissolve the citric acid properly. Now the bismuth cations will be stabilized in the presence of ethylene glycol since it contains two alcoholic hydroxyl functional groups having a strong affinity toward metal ions. Ferric nitrate was also dissolved in the solvent mixture containing citric acid and ethylene glycol. After some time, both solutions will be mixed well and stirred vigorously with the aid of a magnetic stirrer and the temperature was set at 120 °C. When the solution becomes concentrated, its color changes to orange and NO_2 gas is produced due to the decomposition of nitrate ions. After the above reaction, the temperature of the solution is raised at 190 °C in order to promote polymerization, hence speeding up the metal complex formation. As time passes, the solution gets condensed; it becomes highly viscous and changes its color to light brown showing that $BiFeO_3$ containing gel is formed without any turbidity and precipitation.

Further heating is needed to remove organic impurity such as unreacted ethylene glycol from the sophisticated polymeric gel, so that the brown solid is heated at 350 °C. The resulting form is a dark brown fluffy ash, which is appropriately grinded; the obtained powder needs two more stages of heating at 600 °C for 3 hours to form phase-pure $BiFeO_3$ [35, 36].

10.2.1.6.2 Synthesis of barium titanate ($BaTiO_3$) by the Pechini method

The steps involved in the synthesis of $BaTiO_3$ nanoparticles are discussed here. Titanium tetra isopropoxide was first dissolved in citric acid aqueous solution to form titanium citrate solution; a proper amount of ethylene glycol is added to the above solution. The as-prepared solution was heated at 60 °C with constant stirring for 15 minutes. Simultaneously, barium acetate is dissolved in an aqueous solution of citric acid. The solution is heated at 90 °C for several minutes; when the solution becomes transparent, ethylene glycol is added to the solution. The molar ratio of citric acid and ethylene glycol should be maintained at 1:4, for both citrate solutions. When both solutions become viscous, mix them thoroughly and let it solidify and become a dark brown glassy resin; at that time, raise the temperature to 120–140 °C. After that, the sample is placed in an oven at 200 °C for 1 hour for the organic part to decompose. The sample undergoes different stages of heating in a furnace with a heating rate of 120 °C per hour; at first, the resin is heated at 300 °C for 4 hours and then the resin becomes a solid black mass. Then the sample was calcined at 500 °C for 4 hours, 700 °C for 3 hours, and 750 °C for 2 hours. The agglomerates obtained were crushed by milling, and cubic barium titanate is obtained by this method. Tetragonal-shaped barium titanate can be prepared by heating the above-calcined powder at 1200 °C for 2 hours [37, 38].

10.2.2 Top-down approach

In the top-down approach, a bulk material is taken, and it is treated according to the desired shape. The process involves utilizing external tools to precisely cut, mill, and form the material into the desired size and shape through microfabrication techniques. Micropatterning techniques such as photolithography and inkjet belong to this category. Some examples are given below:

→ laser ablation;
→ high energy milling;
→ chemical mechanical milling;
→ electro-explosion;
→ sputtering;
→ vapor-phase condensation;
→ solid-state reaction.

There are different procedures that exist for the preparation of multiferroics. They are solid-state reaction, mechanochemical, chemical synthesis, colloidal processing, and template grain growth [39, 40].

10.2.2.1 Solid-State Reaction

The most common and straightforward method for the production of multiferroic powder is a solid-state method. Oxide or carbonate precursors were generally used, and this precursor was continuously grinded for a long time. After grinding, high-temperature sintering is required to obtain the exact crystal phase. Particles obtained after the solid-state reaction is almost in the micro range, and the powder will lack purity. This method is traditionally used for the synthesis of multiferroic materials. For nano-sized multiferroics synthesis, the solid-state reaction method is less preferred due to its limitation, but the method is relatively inexpensive. The main limitations of this method include non-uniformity in the size and shape of the prepared particles, the chance of multiphase character, and lack of reproducibility. Combustion-based high-temperature solid-state reactions result in the production of larger particles [41].

10.2.2.1.1 Synthesis of $BiFeO_3$ using the solid-state method

For the solid-state synthesis of bismuth ferrite, Bi_2O_3 and Fe_2O_3 were used as reactants in a 1:1 (equal molar) ratio, and they are heated at 850 °C to enable interdiffusion of the positively charged ions. The initiation of chemical reaction, characterized by a slow rate of kinetics, occurs when the ion diffusion takes place. The reaction will take place at the point of contact between the component reactants and will continue sequentially by interdiffusion over the product phase. In a solid-state reaction, the initial reaction will be quicker, but the further reaction will be slower. Once the pure $BiFeO_3$ is formed, the samples were appropriately grinded to obtain $BiFeO_3$ fine powder. For further characterizations like magnetoelectric, magnetic, and dielectric measurements, $BiFeO_3$ pellets were made from these fine powders with the aid of a pelletizer.

The challenge in obtaining phase-pure $BiFeO_3$ particles lies in the narrow temperature range for phase stabilization during preparation. The chances of getting impurity phases are high; $Bi_2Fe_4O_9$, $Bi_{25}FeO_{40}$ are the different impurity phases, which can be formed during the synthesis process. If the proper $BiFeO_3$ phase is not obtained, then it will be complicated to achieve a ferroelectric loop because of leakage [42, 43].

10.2.2.1.2 Synthesis of $BaTiO_3$ using the solid-state method

For preparing $BaTiO_3$, the reactants used were TiO_2 and $BaCO_3$. During the solid-state synthesis of $BaTiO_3$, the reactants were mixed thoroughly in order to get the particles with a smaller size, uniformity without much agglomeration. After proper mixing, the reactants were heated at a high temperature to enable the formation of $BaTiO_3$. The following is a description of the expected chemical reaction between barium carbonate and titanium dioxide during barium carbonate formation:

$$BaCO_3 \rightarrow BaO + CO_2 \ldots\ldots\ldots\ldots \quad (10.3)$$

$$BaO + TiO_2 \rightarrow BaTiO_3 \ldots\ldots\ldots\ldots \quad (10.4)$$

The reaction takes place swiftly at the point where the reactants come into contact [44, 45].

10.2.2.2 Mechanical activation method

The mechanochemical method is based on frequent welding, distortion, and rupture of a combination of reactants. Here, the chemical reaction takes place at the interfaces of the nano-sized grains produced during constant milling. As a result, the reaction that requires high-temperature calcination enables the formation of the desired product, which can be limited to a lower temperature along with the ball milling. Here, the reaction will be in a stable state manner or self-propagating combustive manner. In the mechanochemical method, the combustion process can be avoided by setting the milling parameter properly to reduce collision energy and by adding non-reactive diluents to the powder mixture to reduce the reaction rate. By optimizing suitable parameters for the reaction such as stoichiometry of the reactants, milling conditions, and mechanochemical processing, well-dispersed nanocrystalline particles can be obtained. Moreover, this method enables the formation of agglomeration-free nanopowder [46–48] .

10.2.2.2.1 Synthesis of bismuth ferrite by mechanical activation

In the mechanochemical method, the chemical reactivity between the reactants is comparatively high. Another peculiarity of this method is that a comparatively low-temperature calcination is required here. The oxides Bi_2O_3 and Fe_2O_3 were mixed well and successively crushed for about 100 hours in a high energy planetary ball mill. Milling results in mixing of the reactants in the atomic scale. Thus, a uniform amorphous mix is formed. Calcination at 700 °C gives $BiFeO_3$ phase, but sometimes the prepared $BiFeO_3$ does not show the signature of the ferroelectric phase. A small fraction of other phases can also be present along with the pure $BiFeO_3$ phase [46, 47].

10.2.2.2.2 Synthesis of $BaTiO_3$ using mechanical activation method

Here, $BaCO_3$ and TiO_2 were used as the precursors, and they are mechanically activated for a few hours. $BaCO_3$ and TiO_2 were mixed in a 1:1 stoichiometric ratio to produce $BaTiO_3$, and the mixture was subjected to 4 hours of homogenization via high-energy ball milling at room temperature after mechanical activation. After mixing, the powder needs to be calcined at a higher temperature approximately at 900 °C for 2 hours. For the removal of carbonate from the resultant powder, the powder is subjected to washing with dilute formic acid; after this procedure phase, pure $BaTiO_3$ is obtained. We can also prepare $BaTiO_3$ using BaO as a precursor [48].

10.3 Need for Multiferroic Composite Systems and Core–Shell Structures

There exists a small number of single-phase multiferroic materials, such as $BiFeO_3$, $YMnO_3$, $BiMnO_3$, etc., because of mutual exclusivity (ferroelectricity requires an empty "d" orbital, while ferromagnetism requires partially filled "d" orbitals), and also the extent of magnetoelectric coupling in these systems is weak. It is tough to achieve magnetoelectric coupling at room temperature. Due to the above limitations, for device applications, multiferroic composite systems with high magnetoelectric coupling were preferred. Combinations of ferroelectric oxides with magnetic oxides such as nickel ferrite and cobalt ferrite were selected to create high magnetoelectric coupling by sintering both materials at high temperatures. However, sometimes such type of composite system possesses low magnetoelectric (ME) effect than the expected value, which is due to the inherent preparation problems such as atomic interfacial interdiffusion difficulties, thermal expansion mismatching, reaction problems, etc. [49, 50].

In ferrite-based 0–3 ceramic composite system, the low magnetoelectric effect occurs due to ferrite content, which is either conductive or semiconductive. Thus, introducing ferrite into the ferroelectric oxides separately leads to the reduction in the dielectric properties of the whole composite system and creates leakage problems. One of the best methods to overcome these limitations of the composite system and to create high magnetoelectric coupling in these materials is to disperse a high concentration of ferrite particles inside a piezoelectric ceramic matrix. An effective pathway is to create a core–shell structure of the ferrite–ferroelectric system, in which the ferrite system as the core and the piezoelectric layer forms a shell; thus, to a greater extent, we can prevent ferrite from being in direct contact during sintering. Nevertheless, it is very difficult to prepare a suitable structure for ferrite-piezoelectric core–shell particles [50].

10.4 Synthesis of Multiferroic Composite Structures

10.4.1 Synthesis of $BiFeO_3$–$NaNbO_3$ multiferroic composite by the Pechini method

The reactants such as ammonium niobate oxalate, bismuth nitrate, ferric nitrate, and sodium nitrate were used for the preparation of $Bi_{1-x}Na_xFe_{1-x}Nb_xO_3$ nanocomposites. The precursors were dissolved in the aqueous solution of citric acid. The molar ratio of the reactants is maintained at 1:1, concerning the metal cation. The pH of the solution needs to be set at 5, and the solution is stirred well. When the solution becomes transparent, the temperature is set at 100 °C to promote the gel formation; once the gel is formed, it should be heated at 500 °C to get ceramic powder. Again, the powder is heated at 850 °C for 1 hour in order to get phase-pure bismuth ferrite powder [51].

10.4.2 Synthesis of $Ni_{1-x}Co_xFe_2O_4 - Ba_{1-x}Sr_xTiO_3$ multiferroic composite by the sol-gel method

Ferric nitrate, cobalt nitrate, and nickel nitrate were used as the starting materials for the synthesis of $Ni_{1-x}Co_xFe_2O_4$ (NCF). Nickel nitrate, cobalt nitrate, and ferric nitrate were added to a mixture of ethanol and distilled water in the necessary molar ratios, with ethanol and water serving as solvents. The above solution was stirred well using a magnetic stirrer for 8 hours and the sol obtained was oven dried at 60 °C for 24 hours; this results in the formation of a brown colored gel, which was thoroughly dried at 120 °C for 12 hours before being calcined at temperatures ranging from 800 to 1000 °C for 6 hours. Single-phase NCF was formed as a result of calcination.

Synthesis of $Ba_{1-x}Sr_xTiO_3$ (BST) was carried out by using the solid-state reaction method. The precursors, barium carbonate, strontium carbonate, and titanium oxides, were mixed thoroughly in distilled water with the use of zirconia balls for one day to obtain a slurry. This slurry was then dried at 120 °C for 24 hours in an oven, before being finely ground with an agate mortar. Finally, the mixture was calcined at 1050 °C for 6 hours.

A multiferroic composite system was prepared by mixing nickel-cobalt ferrite and barium strontium titanate powders in the required composition. Proper mixing was carried out by using the ball milling technique; for 12 hours, acetone is used as the liquid for proper mixing. After proper mixing, the powder is heated at a high temperature; during this process, NCF and BST react correctly to form the desired composite system. If the system does not dry properly, one more heating is required, and the powder has to be heated at 1050 °C for 6 hours [52].

10.4.3 Synthesis of multiferroic core–shell structures

The core–shell multiferroic nanoparticle preparation method is a three-step process:

- preparing ferrite nanoparticles (core);
- preparing the solution of the ferroelectric layer (layer);
- adding the ferrite particle into the sol of the ferroelectric layer.

The resulting gel is dried and sintered at a high temperature and grounded properly using an agate mortar.

For example, the process involved in the preparation of Fe_2O_3@$PbTiO_3$ nanoparticles is described below.

10.4.3.1 Preparation of Fe_2O_3 nanoparticles

For the particular synthesis, preparation of 2 mmol concentration of ferrite (core) particles, $FeCl_3$ů$6H_2O$, and 0.3 g of NaOH was dissolved in 40 ml of ethylene glycol (EG) and stirred for some time. The solution was transferred into a Teflon-lined stainless-steel autoclave of volume 50 ml, and the solution was heated at a temperature of 200 °C for 10 hours. The black product formed after the reaction was washed with ethanol and deionized water several times and then dried in vacuum oven at 60 °C for 4 hours properly [48].

10.4.3.1.1 Synthesis of the final core–shell structure of Fe_2O_3@$PbTiO_3$

To synthesize Fe_2O_3@$PbTiO_3$ nanoparticles, 0.6 g of ferrite particles were dispersed in 5 ml of polyethylene glycol-400 with the help of sonication. Simultaneously, 2 mmol of $Ti(SO_4)_2$ was dissolved in 150 ml of distilled water. The two solutions were mixed and stirred adequately for 12 hours at room temperature to obtain a uniform coating of Ti-hydroxide. After washing in an alkaline solution (1 mol/L KOH), these particles underwent aging, followed by dispersion in 40 ml of deionized water and the addition of lead nitrate [$Pb(NO_3)_2$] and 0.4 g of KOH. The above mixture was transferred into sealed Teflon-lined stainless-steel autoclave and kept at 160 °C for 4 hours. The powder was subjected to heat treatment at 600°C for 2 hours, followed by thorough grinding to effectively synthesize the Fe_2O_3@$PbTiO_3$ core-shell structure [53].

10.4.3.2 Synthesis of $CoFe_2O_4$@ $BaTiO_3$ multiferroic core–shell nanoparticles

$CoFe_2O_4$@ $BaTiO_3$ (CFO@BTO) core–shell particles were prepared by combining precipitation and the organosol method. Initially, cobalt ferrite nanoparticles were prepared by the co-precipitation method of average particle size around 40 nm. The CFO@BTO particles were prepared through a two-step process, starting with the coating of hydrophobic cobalt ferrite nanoparticles with oleic acid and oleylamine, forming a ferrofluid in toluene. This coating bound carboxylic acid groups to the core material surface, promoting proper dispersion of the ferrite material. The next step involved transferring the cobalt ferrite nanoparticles in toluene solution to an organosol of $BaTiO_3$.

The synthesis procedure is as follows: $CoFe_2O_4$ nanopowder was added to toluene and heated at 80 °C with stirring to form a stable ferrofluid suspension containing 30% oleic acid and oleylamine. The suspension was mixed mechanically for 30 minutes, followed by 12 hours of rotating milling to maintain fine, surface-coated ferrite particles. The particles were then centrifuged at 7000 rpm, washed to remove excess surfactant and unmodified particles, and re-suspended in toluene to form a stable ferrofluid.

For the synthesis of the core–shell structure, the organosol method was employed. The chemicals used for the synthesis of the $BaTiO_3$ shell particles are Ba $(C_2H_3O_2)_2$, titanium (IV) *n*-isopropoxide, oleic acid, and toluene. Here, barium acetate and titanium isopropoxide were appropriately mixed and oleic acid and toluene were added to the above suspension to form a yellowish organosol. A suitable amount of ferrofluid was added to the sol having a content of 50 wt.% of the cobalt ferrite particles in the final powder; at this stage, tetramethyl ammonium hydroxide (TMAH) was added to the sol at 90 °C in order to promote gel formation. When the gel is formed, it was dried for 12 hours and calcined at a higher temperature at 800 °C for 8 minutes; the resultant sample was washed with 0.2 mol of acetic acid to dissolve the impurities such as barium carbonate. If the addition of TMAH is omitted, then the sample formed will be only a composite and not a core–shell material [54, 55].

10.5 Conclusion

This chapter discusses different synthesis methods of selected multiferroic systems. Synthesis of both single-phase ceramics-based multiferroics and selected composite multiferroics was discussed in detail. Different fine and potential methods for the synthesis of multiferroic materials involves sol¯gelgel method, electrodeposition, physical vapor deposition, pulsed laser deposition, laser ablation, high energy milling, chemical mechanical milling, electro-explosion, sputtering, and solid-state reaction. It is possible to prepare both nano-sized and micro-sized multiferroic materials by using these methods. Based on the requirements such as particle size, morphology, thickness, etc., we can adopt the different synthesis strategies. However, among the different methods, the sol¯gelgel method, the hydrothermal method, and solid-state methods were widely preferred for the synthesis of multiferroic materials.

References

[1] Prellier, W., Singh, M.P. and Murugavel, P., 2005. The single-phase multiferroic oxides: from bulk to thin film. *Journal of Physics: Condensed Matter, 17*(30), p. R803.

[2] Martin, L.W., Chu, Y.H. and Ramesh, R.J.M.S., 2010. Advances in the growth and characterization of magnetic, ferroelectric, and multiferroic oxide thin films. *Materials Science and Engineering: R: Reports, 68*(4-6), pp.89-133.

[3] Martin, L.W. and Ramesh, R., 2012. Multiferroic and magnetoelectric heterostructures. *Acta Materialia, 60*(6-7), pp.2449-2470.

[4] Izyumskaya, N., Alivov, Y. and Morkoc, H., 2009. Oxides, oxides, and more oxides: high-κ oxides, ferroelectrics, ferromagnetics, and multiferroics. *Critical Reviews in Solid State and Materials Sciences, 34*(3-4), pp.89-179.

[5] Moure, C. and Peña, O., 2015. Recent advances in perovskites: Processing and properties. *Progress in Solid State Chemistry, 43*(4), pp.123-148.

[6] Dagotto, E. and Tokura, Y., 2008. Strongly correlated electronic materials: present and future. *MRS bulletin, 33*(11), pp.1037-1045.

[7] Ren, S., 2009. *Bottom-up multiferroic nanostructures* (Doctoral dissertation).

[8] Yang, Y., Priya, S., Wang, Y.U., Li, J.F. and Viehland, D., 2009. Solid-state synthesis of perovskite-spinel nanocomposites. *Journal of Materials Chemistry, 19*(28), pp.4998-5002.

[9] Safi, R. and Shokrollahi, H., 2012. Physics, chemistry and synthesis methods of nanostructured bismuth ferrite (BiFeO3) as a ferroelectromagnetic material. *Progress in Solid State Chemistry, 40*(1-2), pp.6-15.

[10] Stojanovic, B.D., Dzunuzovic, A.S. and Ilic, N.I., 2018. Review of methods for the preparation of magnetic metal oxides. In *Magnetic, Ferroelectric, and Multiferroic Metal Oxides* (pp. 333-359). Elsevier.

[11] Bobruk, M., Brylewska, K., Durczak, K., Wojciechowski, K., Adamczyk, A. and Brylewski, T., 2017. Synthesis of manganese-cobalt spinel via wet chemistry methods and its properties. *Ceramics International, 43*(17), pp.15597-15609.

[12] Mouzon, J., 2005. *Synthesis of Yb: Y2O3 nanoparticles and fabrication of transparent polycrystalline yttria ceramics* (Doctoral dissertation, Luleå tekniska universitet).

[13] Ivanov, S., 2012. Multiferroic complex metal oxides: Main features of preparation, structure, and properties. In *Science and Technology of Atomic, Molecular, Condensed Matter & Biological Systems* (Vol. 2, pp. 163-238). Elsevier.

[14] Rendón-Angeles, J.C., Matamoros-Veloza, Z., Montoya-Cisneros, K.L., Cuevas, J.L. and Yanagisawa, K., 2016. Synthesis of Perovskite Oxides by Hydrothermal Processing–From Thermodynamic Modelling to Practical Processing Approaches. *Perovskite Materials: Synthesis, Characterisation, Properties, and Applications*, p.27.

[15] Vijatović, M.M., Bobić, J.D. and Stojanović, B.D., 2008. History and challenges of barium titanate: Part II. *Science of Sintering*, *40*(3), pp.235-244.

[16] Yan, T., Liu, X.L., Wang, N.R. and Chen, J.F., 2005. Synthesis of monodispersed barium titanate nanocrytals—hydrothermal recrystallization of BaTiO3 nanospheres. *Journal of crystal growth*, *281*(2-4), pp.669-677.

[17] Han, Y., Li, S., Wang, X. and Chen, X., 2004. Synthesis and sintering of nanocrystalline hydroxyapatite powders by citric acid sol–gel combustion method. *Materials Research Bulletin*, *39*(1), pp.25-32.

[18] Farhadi, S. and Zaidi, M., 2009. Bismuth ferrite (BiFeO3) nanopowder prepared by sucrose-assisted combustion method: a novel and reusable heterogeneous catalyst for acetylation of amines, alcohols and phenols under solvent-free conditions. *Journal of Molecular Catalysis A: Chemical*, *299*(1-2), pp.18-25.

[19] Azam, A., Jawad, A., Ahmed, A.S., Chaman, M. and Naqvi, A.H., 2011. Structural, optical and transport properties of Al3+ doped BiFeO3 nanopowder synthesized by solution combustion method. *Journal of Alloys and Compounds*, *509*(6), pp.2909-2913.

[20] Takatori, K., Tani, T., Watanabe, N. and Kamiya, N., 1999. Preparation and characterization of nano-structured ceramic powders synthesized by emulsion combustion method. *Journal of Nanoparticle Research*, *1*(2), pp.197-204.

[21] Aghayan, M., Zak, A.K., Behdani, M. and Hashim, A.M., 2014. Sol–gel combustion synthesis of Zr-doped BaTiO3 nanopowders and ceramics: Dielectric and ferroelectric studies. *Ceramics International*, *40*(10), pp.16141-16146.

[22] Yadav, R.S., Mishra, P. and Pandey, A.C., 2008. Growth mechanism and optical property of ZnO nanoparticles synthesized by sonochemical method. *Ultrasonics sonochemistry*, *15*(5), pp.863-868.

[23] Xu, M., Lu, Y.N., Liu, Y.F., Shi, S.Z., Qian, T.S. and Lu, D.Y., 2006. Sonochemical synthesis of monosized spherical $BaTiO_3$ particles. *Powder technology*, *161*(3), pp.185-189.

[24] Vuttivong, S., Niemcharoen, S., Seeharaj, P., Vittayakorn, W.C. and Vittayakorn, N., 2013. Sonochemical synthesis of spherical $BaTiO_3$ nanoparticles. *Ferroelectrics*, *457*(1), pp.44-52.

[25] Prado-Gonjal, J., Villafuerte-Castrejón, M.E., Fuentes, L. and Moran, E., 2009. Microwave–hydrothermal synthesis of the multiferroic $BiFeO_3$. *Materials Research Bulletin*, *44*(8), pp.1734-1737.

[26] Polsongkram, D., Chamninok, P., Pukird, S., Chow, L., Lupan, O., Chai, G., Khallaf, H., Park, S. and Schulte, A., 2008. Effect of synthesis conditions on the growth of ZnO nanorods via hydrothermal method. *Physica B: Condensed Matter*, *403*(19-20), pp.3713-3717.

[27] Wang, W., Cao, L., Liu, W., Su, G. and Zhang, W., 2013. Low-temperature synthesis of $BaTiO_3$ powders by the sol–gel-hydrothermal method. *Ceramics International*, *39*(6), pp.7127-7134.

[28] Lu, S.W., Lee, B.I., Wang, Z.L. and Samuels, W.D., 2000. Hydrothermal synthesis and structural characterization of $BaTiO_3$ nanocrystals. *Journal of Crystal Growth*, *219*(3), pp.269-276.

[29] Farhadi-Khouzani, M., Fereshteh, Z., Loghman-Estarki, M.R. and Razavi, R.S., 2012. Different morphologies of ZnO nanostructures via polymeric complex sol–gel method: synthesis and characterization. *Journal of sol-gel science and technology*, *64*(1), pp.193-199.

[30] Hernandez, B.A., Chang, K.S., Fisher, E.R. and Dorhout, P.K., 2002. Sol– gel template synthesis and characterization of $BaTiO_3$ and $PbTiO_3$ nanotubes. *Chemistry of Materials*, *14*(2), pp.480-482.

[31] Wang, W., Cao, L., Liu, W., Su, G. and Zhang, W., 2013. Low-temperature synthesis of $BaTiO_3$ powders by the sol–gel-hydrothermal method. *Ceramics International*, *39*(6), pp.7127-7134.

[32] Xu, J.H., Ke, H., Jia, D.C., Wang, W. and Zhou, Y., 2009. Low-temperature synthesis of $BiFeO_3$ nanopowders via a sol–gel method. *Journal of Alloys and Compounds*, *472*(1-2), pp.473-477.

[33] Lin, Z., Cai, W., Jiang, W., Fu, C., Li, C. and Song, Y., 2013. Effects of annealing temperature on the microstructure, optical, ferroelectric and photovoltaic properties of $BiFeO_3$ thin films prepared by sol–gel method. *Ceramics International*, *39*(8), pp.8729-8736.

[34] Lin, Y.J., Chang, Y.H., Yang, W.D. and Tsai, B.S., 2006. Synthesis and characterization of ilmenite $NiTiO_3$ and $CoTiO_3$ prepared by a modified Pechini method. *Journal of non-crystalline solids*, *352*(8), pp.789-794.

[35] Wei, J. and Xue, D., 2008. Low-temperature synthesis of BiFeO3 nanoparticles by ethylenediaminetetraacetic acid complexing sol–gel process. *Materials Research Bulletin*, *43*(12), pp.3368-3373.

[36] Silva, J., Reyes, A., Esparza, H., Camacho, H. and Fuentes, L., 2011. BiFeO3: a review on synthesis, doping and crystal structure. *Integrated Ferroelectrics*, *126*(1), pp.47-59.

[37] Lazarevica, Z., Romcevica, N., Vijatovicb, M., Paunovica, N., Romcevica, M., Stojanovicb, B. and Dohcevic-Mitrovica, Z., 2008, September. Characterization of barium titanate ceramic powders by Raman spectroscopy. In *Proceedings of the Tenth Annual Conference of the Materials Research Society of Serbia*.

[38] Mayeen, A., Kala, M.S., Jayalakshmy, M.S., Thomas, S., Rouxel, D., Philip, J., Bhowmik, R.N. and Kalarikkal, N., 2018. Dopamine functionalization of BaTiO 3: an effective strategy for the enhancement of electrical, magnetoelectric and thermal properties of $BaTiO_3$-PVDF-TrFE nanocomposites. *Dalton Transactions*, *47*(6), pp.2039-2051.

[39] Zhou, H., 2009. Synthesis, characterization, and morphology-dependent investigation of zero-dimensional and one-dimensional formulations of various classes of nanomaterials (Doctoral dissertation, State University of New York at Stony Brook).

[40] Basith, M.A., Ngo, D.T., Quader, A., Rahman, M.A., Sinha, B.L., Ahmmad, B., Hirose, F. and Mølhave, K., 2014. Simple top-down preparation of magnetic $Bi_{0.9}Gd_{0.1}Fe_{1-x}Ti_xO_3$ nanoparticles by ultrasonication of multiferroic bulk material. *Nanoscale*, *6*(23), pp.14336-14342.

[41] Sun, Z.P., Liu, L., Zhang, L. and Jia, D.Z., 2006. Rapid synthesis of ZnO nano-rods by one-step, room-temperature, solid-state reaction and their gas-sensing properties. *Nanotechnology*, *17*(9), p.2266.

[42] Valant, M., Axelsson, A.K. and Alford, N., 2007. Peculiarities of a solid-state synthesis of multiferroic polycrystalline $BiFeO_3$. *Chemistry of Materials*, *19*(22), pp.5431-5436.

[43] Bernardo, M.S., Jardiel, T., Peiteado, M., Caballero, A.C. and Villegas, M., 2011. Reaction pathways in the solid state synthesis of multiferroic $BiFeO_3$. *Journal of the European Ceramic Society*, *31*(16), pp.3047-3053.

[44] Simon-Seveyrat, L., Hajjaji, A., Emziane, Y., Guiffard, B. and Guyomar, D., 2007. Re-investigation of synthesis of BaTiO3 by conventional solid-state reaction and oxalate coprecipitation route for piezoelectric applications. *Ceramics international*, *33*(1), pp.35-40.

[45] Hennings, D.F., Schreinemacher, B.S. and Schreinemacher, H., 2001. Solid-state preparation of $BaTiO_3$-based dielectrics, using ultrafine raw materials. *Journal of the American Ceramic Society*, *84*(12), pp.2777-2782.

[46] Maurya, D., Thota, H., Nalwa, K.S. and Garg, A., 2009. BiFeO3 ceramics synthesized by mechanical activation assisted versus conventional solid-state-reaction process: A comparative study. *Journal of alloys and compounds*, *477*(1-2), pp.780-784.

[47] Layek, S., Verma, H.C. and Garg, A., 2015. Enhancement in magnetic properties of Ba-doped $BiFeO_3$ ceramics by mechanical activation. *Journal of Alloys and Compounds*, *651*, pp.294-301.

[48] Gomez-Yañez, C., Benitez, C. and Balmori-Ramirez, H., 2000. Mechanical activation of the synthesis reaction of $BaTiO_3$ from a mixture of BaCO3 and TiO2 powders. *Ceramics International*, *26*(3), pp.271-277.

[49] Liu, R., Zhao, Y., Huang, R., Zhao, Y. and Zhou, H., 2010. Multiferroic ferrite/perovskite oxide core/shell nanostructures. *Journal of Materials Chemistry*, *20*(47), pp.10665-10670.

[50] Woldu, T., Raneesh, B., Hazra, B.K., Srinath, S., Saravanan, P., Reddy, M.R. and Kalarikkal, N., 2017. A comparative study on structural, dielectric and multiferroic properties of $CaFe_2O_4/BaTiO_3$ core–shell and mixed composites. *Journal of Alloys and Compounds*, *691*, pp.644-652.

[51] Ummer, R.P., Sreekanth, P., Raneesh, B., Philip, R., Rouxel, D., Thomas, S. and Kalarikkal, N., 2015. Electric, magnetic and optical limiting (short pulse and ultrafast) studies in phase pure (1− x) BiFeO 3–x NaNbO 3 multiferroic nanocomposite synthesized by the pechini method. *RSC Advances*, *5*(82), pp.67157-67164.

[52] Pahuja, P., Sharma, R., Singh, V. and Tandon, R.P., 2015. Novel method of synthesis of multiferroic nickel cobalt ferrite–barium strontium titanate composite system. *International Journal of Applied Ceramic Technology*, *12*, pp.E156-E163.

[53] Liu, R., Zhao, Y., Huang, R., Zhao, Y. and Zhou, H., 2010. Multiferroic ferrite/perovskite oxide core/shell nanostructures. *Journal of Materials Chemistry*, *20*(47), pp.10665-10670.

[54] Corral-Flores, V., Bueno-Baques, D. and Ziolo, R.F., 2010. Synthesis and characterization of novel $CoFe_2O_4$–$BaTiO_3$ multiferroic core–shell-type nanostructures. *Acta Materialia*, *58*(3), pp.764-769.

[55] Etier, M., Gao, Y., Shvartsman, V.V., Elsukova, A., Landers, J., Wende, H. and Lupascu, D.C., 2012. Cobalt ferrite/barium titanate core/shell nanoparticles. *Ferroelectrics*, *438*(1), pp.115-122.

11

An Introduction to Strongly Correlated Multiferroic Core–Shell Nanoparticles: Synthesis, Characterization, and Applications

Praveena M. G. and Kala M. S.

Department of Physics and Centre for Research, St. Teresa's College Affiliated to Mahatma Gandhi University, India

Abstract

Strongly correlated systems contain materials that exhibit exceptional electronic and magnetic properties. Among the various strongly correlated systems, multiferroic core–shell systems are attracting the scientific world. The strongly correlated multiferroic core–shell nanoparticles are widely studied because of their multifunctional properties, such as remarkable magnetic and electric properties. Multiferroics are materials that can have the two properties of ferroelectricity and ferromagnetism. Electron(fermion) behavior plays a key role in the material properties. Multiferroic core–shell nanoparticle has got significant attention in the scientific community because of their multifunctional applications. The materials have a magnetic/electric core with an electric/magnetic shell so that they are capable of manipulating electric polarization with an applied magnetic field and vice versa. The multiferroic core–shell nanoparticles are widely used for the spintronic, biological, and catalytic applications. The chapter covers the general aspects and brief description of the classification of multiferroics, core–shell nanoparticles, and their synthesis, characterization, and applications.

Keywords: strongly correlated multiferroics, core–shells, synthesis, applications

11.1 Introduction

The highly correlated perovskite oxides with multifunctional properties are gaining the attention of the research community due to the ferroelectric, magnetic, and magnetoresistive properties. Strongly correlated multiferroic materials with ferroelectric and ferromagnetic properties are one of the most studied highly correlated perovskite systems [21–23]. Nanomaterials, graphene and carbon nanotubes, smart and advanced functional materials, spintronic materials, bio-materials, and multiferroic materials are just a few of the hopeful solid-state physics areas of today that promise to bring about another wave of technological advancements and economic impact akin to the modern silicon revolution of the 1950s. The overwhelming interest in multiferroic materials in recent decades has been sparked by their alluring physical characteristics and enormous potential for practical applications. Multiferroics are substances that can possess both ferromagnetism and ferroelectricity. The behavior of electrons (fermions) is crucial to the characteristics of materials (Tokura, 2011).

11.1.1 Multiferroics

Multiferroic materials, which typically occur in at least one of the following order states including electric, magnetic, or piezo-elastic phases simultaneously, are a special category of strongly correlated fermion compound materials. In this manner, well-known materials, for example, ferromagnets showing solid piezo-attractive properties, and besides piezo-ferroelectrics are ostensibly a remarkable class of magneto- or electro-elastic coupled multiferroics. The link between the ferro-phases and the piezo-elastic properties in the two cases makes these materials essential for cutting-edge applications since it allows for the rapid control of ferroelectric and ferromagnetic properties using mechanical stress that is delivered externally. There are also opposite effects, whereby externally induced magnetic and electric fields can cause a mechanical strain in a material to change the substance's shape.

In Figure 11.1, the baffling connection between polarization–electric field–strain, magnetization–field–strain, and stress–strain–polarization–charge is graphically depicted.

While the concurrence of Ferro-arrange and piezo-elastic properties can offer ascent to intriguing piezo-electric and piezo-magnetic couplings, there are two extra multiferroic conceivable states, each showing cross-couplings between the order states. These different sorts of multiferroics are: (i)

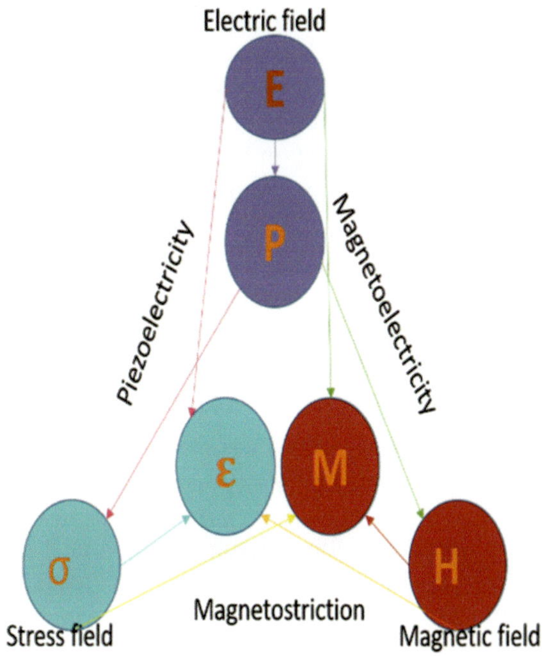

Figure 11.1 Possible cross-couplings in multiferroics. *E* – electric field; *P* – electric polarization; *s* – applied mechanical stress; *e* – strain; *H* – magnetic field; *M* – magnetization.

magneto-electric multiferroic, in which the electric and magnetic order states exist together; and (ii) elasto-magneto-electric multiferroic, in which every one of the three arrange states is available. It is imperative to refer to that, by electric and magnetic order state, any kind of ferroic ordering is seen, including ferri-magnetic, anti-ferromagnetic, ferromagnetic, paramagnetic, super-paramagnetic, and their electrical counterparts, if applicable.

In any case, what makes multiferroic materials scientifically and technologically captivating is not just their capacity to show various order states at the same time, in specifically, the effects that can result from cross-coupling between the order states. A noteworthy additional characteristic of some multiferroics, in addition to the Piezo–Ferro couplings, is the so-called "magneto-electric" coupling, which occurs in materials when magnetic electric–elastic or magnetic–electric phases coexist. The conversion of energies held in magnetic and electric fields is encouraged by the magneto-electric effect. The remarkable attribute of magneto-electric multiferroic materials is that they can modify their net magnetization if an external electric field is supplied while also adjusting the electric polarization when an external magnetic field is applied. Additionally, the magneto-electric coupling coefficient quantitatively illustrates the fact that the magneto-electric effect in multiferroics can be triggered either electrically or magnetically. The variation in the sample's magnetization (*M*) brought on by the use of an electric field (*E*) is known as the electrically induced magneto-electric effect.

$$\alpha_E = \frac{\partial M}{\partial E} \tag{11.1}$$

The same holds valid for the magnetically induced magneto-electric effect, which is the modification of the sample's electric polarization (*P*) brought about by the use of an attractive field (*H*):

$$\alpha_H = \frac{\partial P}{\partial H} \tag{11.2}$$

Since *E* = *V*/*t*, where *t* is the thickness and *V* is the voltage of the dielectric structure, the following equation for the magnetically induced magneto-electric effect is acquired:

$$\alpha_H = (\frac{\partial P}{\partial H}) \cong \varepsilon_0 \varepsilon_r (\frac{\partial E}{\partial H}) = \frac{\varepsilon_0 \varepsilon_r}{t} (\frac{\partial V}{\partial H}) = \alpha_H^V \varepsilon_0 \varepsilon_r \tag{11.3}$$

where α_H^V is the magnetically induced coefficient can be defined as follows:

$$\alpha_H^V = \left(\frac{\partial E}{\partial H}\right) = \frac{1}{t}\left(\frac{\partial V}{\partial H}\right) \tag{11.4}$$

The main variable used in analyzing experimental measurements and preparing various multiferroics-dependent applications is the voltage magneto-electric coefficient. The connection between the induced voltage and the magneto-electric coupling coefficient is given by: $\alpha_H = \varepsilon_0 \varepsilon_r \alpha_H^V$. However, the more practical voltage magneto-electric coefficient, α_H^V is expressed as [V/A] in SI units and as [V/cm.Oe] in CGS units, that are additionally utilized for the majority of practical calculations and scientific assessments.

The direct and indirect couplings are the two possible types of magneto-electric coupling. The direct magneto-electric coupling specifies multiferroics in which the application of electric or magnetic field results in changes of magnetization or electric polarization of the system, individually. This class of multiferroic materials are being known as "single-stage" multiferroics (Mayeen et al., 2020).

Elasto-magneto-electric multiferroics, also called composite multiferroics, are where the indirect magneto-electric effect is most prevalent. In these kinds of processes, the application of an electric or magnetic field does not directly or individually affect the electric polarization or magnetization. Instead, they are altered by a magneto-electric connection with strain (Zhu, Zhan and Miao, 2014).

11.2 Classification of Multiferroics

Although the key processes causing the appearance of the magneto-electric coupling effect in multiferroic materials are not directly understood, the piezo-ferro coupling is well recognized. The well-known classical differential Maxwell equations were the first to show how electricity and magnetism are related. The term "time-varying magnetic and electric fields" refers solely to the scientific link between the time derivative of magnetic and electric fields and their spatial gradients (i.e., electromagnetic waves).

Pierre Curie also postulated linear magneto-electric coupling in 1894 by taking into account the crystal symmetry, as has already been mentioned. Because the main ferroic components of multiferroic compounds can be distinguished by their behavior whenever subjected to space and time reversal, the symmetry controls the multiferroic properties. The electric polarization would remain constant while the magnetization's sign was changed via a time-inversion symmetry task. Additionally, a spatial symmetry task would maintain the charge while reversing the electric polarity. Accordingly, magnetic materials break the time-reversal symmetry, while electrically ordered dielectrics break the spatial-reversal symmetry. Henceforth, a multiferroic crystal comprising both magnetic and electric order states would need to break both time- and spatial-reversal symmetry. For this symmetry violation condition, it is indeed necessary to review the microscopic causes of the magnetic and electric processes. It is noteworthy that magnetic phenomena resulted from magnetic moments interacting with one another. The orbital movement of an electron inside the atom produces the magnetic moment. Eqn (13.1.1) defines the magnetic dipole moment μ^m associated with a loop of area A and electric current I:

$$\mu^m = I.\,A = \frac{dQ}{dt}\,.\,A \tag{11.5}$$

11.2.1 Single-phase multiferroic

Magnetic and electric order states coexist at every point in single-phase multiferroic materials, which are described as chemically isotropic and homogeneous compounds. Although it is not a required need, piezo-elastic or elastic characteristics may also be present in single-phase multiferroics. However, strict symmetry criteria are needed for magnetic and electric order states to coexist in the same phase. Crystals must be invariant under space reversal symmetry operation in order to support invariant electric polarization in an electric order phase under the time-inversion symmetry and magnetic order phase with spontaneous magnetization. There are 122 potential crystal structures, but only 31 crystallographic point groups allowed magnetic phase, and 31 point groups concur with magnetic phase, according to Shubnikov point groups. Thirteen crystallographic point groups were detected in both the electric and magnetic phases, according to Shuvalov and Belov, allowing both sets to coexist in the same phase (i.e., multiferroic crystals). The 13 crystallographic point groups permitting multiferroic phase are 2, 1, 3, 4, 20, m, m0, 3m0, m0m20, m0m020, 4m0m0, 6, and 6m0m0.

Since the electrically ordered phase of multiferroic needs dielectric materials, of course, all single-phase multiferroic materials are dielectric oxides. However, it is challenging to prepare the single-phase multiferroic materials artificially, and also they rarely occur in nature. This scarcity of single-phase multiferroic was related to symmetry restrictions. Hill, in her article, argued that transition metal d electrons, which are essential for the magnetic phase, reduce the tendency for off-center ferroelectric distortion (Mayeen et al., 2019).

Single-phase multiferroic materials are once more divided into type-I and type-II multiferroic groups. The physical process of electric polarization is related to these classifications. Electric and magnetic phases coexist continuously within a same compound in type-I multiferroic materials. However, they are independent of one another and have a different microscopic genesis (Mayeen et al., 2018, 2020; Robles et al., 2018).

Ferroelectric order is formed by magnetism in type-II multiferroic materials, which in turn suggest a strong magneto-electric connection. The list of single-phase multiferroic materials contain orthorhombic manganites such as $REMn_2O_5$ and $REMnO_3$, where RE is representing the rare-earth element, $LiCoPO_4$ and $BiFeO_3$, and its solid solutions with $BaTiO_3$ (Dagotto and Tokura, 2008).

Table 11.1 The list of some well-known single-phase and composite multiferroic materials.

Single-phase multiferroic	Composite multiferroic
$BiFeO_3$	NiFe2O4/PZT laminated layers
$GaFeO_3$	Fe_3O_4 @ $PbTiO_3$
$BiMnO_3$	$Ba_{0.85}Ca_{0.15}Zr_{0.1}Ti_{0.9}O3$ @ $CoFe_2O_4$
Cr_2O_3	$CoFe_2O_4$ in Ba TiO_3 matrix
LaMnO3	$NiFe_2O_4$ @ $PbZr_{0.52}Ti_{0.48}O3$

11.2.2 Composite multiferroic

There have been many single-phase multiferroic compounds studied, but none of them have been found to exhibit widespread and mechanically useful magneto-electric coupling at ambient temperatures. This prompted the scientific community to look into networks, adopt novel strategies, and create alternating multiferroic materials. Materials known as multiferroic composites have simultaneous electric, magnetic, and piezo-order states inside of them (Nanomaterials, no date; Oliveira et al., 2016; Woldu et al., 2017; Kopeychenko et al., 2020; Liu et al., 2020). However, unlike single-stage multiferroic, the composite is described by the way that the order state is physically isolated from one another inside the composite material. Making multiferroic composites provides a number of benefits, including improved magneto-electric coupling and design flexibility for materials for precise application. The measured magneto-electric effect of composite multiferroics is a few orders higher in magnitude than that of single multiferroics since the majority of multiferroic composites function above or at room temperature. The need that the material be totally dielectric is no longer necessary because the ferroic phases are completely independent inside the composite. The composites' electric phase is an insulator made of dielectric material. In the magnetic phase, however, it can either be a dielectric insulator or a metallic conductor. This exciting feature promotes the synthesis/manufacturing of multiferroic composites with an extended range of available constituent materials. Compared to single-phase multiferroics, better properties and enhancement of magneto-electric coefficient value are observed in composite multiferroics. Core–shell structures belong to composites with core material and an envelope over it. It possesses perfect mechanical contact at the interface between core and shell materials (Henrichs et al., 2020; Vopson, 2015).

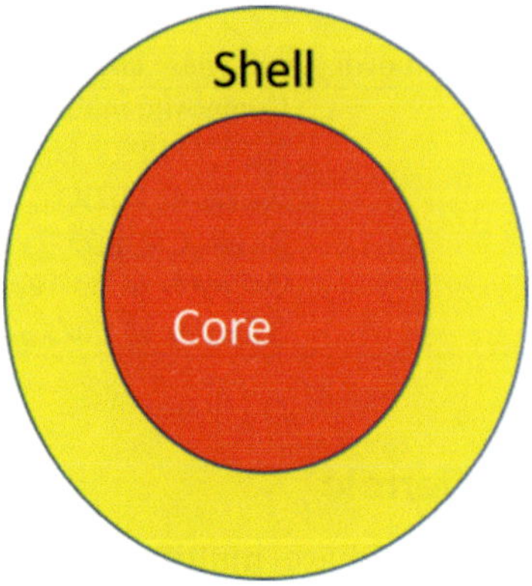

Figure 11.2 Schematic representation of a core–shell nanoparticles.

11.3 Core–Shell Nanoparticles

Nanocomposites fabricated with an inner core and outer shell are widely defined as core–shell nanoparticles (1, 2). As the developments in the core–shell nanoparticles research increases, the definition of the core–shell nanoparticle is kept changing; so it can be broadened to the class of nanomaterials having divergent boundary materials wrapping (either entirely or partially) the inner element as long as they can be distinctly recognized. In general, a core–shell structure is represented as core @ shell. The core–shell nanostructures can be classified into different classes depending on their arrangements and composition of the two elements in the materials. According to the above considerations, there are mainly four types of core–shell nanostructure comprising organic @ inorganic, inorganic @ organic, organic @ organic, and inorganic @ inorganic type. The simplest core shell structure is given in Figure 11.2.

11.3.1 Organic @ Inorganic Core–Shell Nanoparticles

In this type of core–shell nanoparticles, the material is made from polymers like polyurethane, poly(vinyl pyrrolidone), polystyrene, dextrose, and various copolymers. The formation of shell materials can be from metal oxides, silica, metals, and metal chalcogenides so that the resulting core–shell materials will have the twin properties of its individual (organic and inorganic) components. Special attention is given to metal oxide coated organic material because of its numerous benefits such as abrasion resistance, thermal stability, and resistance to oxidation. They also possess properties like flexibility, improved brittleness of shell materials, toughness, and excellent optical properties, which show their polymeric behavior. This type of materials is found to be exploited in the field of catalysis, material science, biotechnology, and magnetic fluids. Another critical example is noble metals (Au and Ag) coated on polymer materials, which enhances the immobilization characteristic of protein as a consequence; they have tremendous application in the fields of biomedical, biotechnology, and immune sensing.

11.3.2 Organic @ Organic Core–Shell Nanoparticles

In this type of core–shell nanoparticles, the inner core and outer shell are made of organic or polymer materials. These nanoparticles have their applications in the fields of biosensing, catalysis, drug delivery, and chemical separation. The main benefit of having a polymer covering over another polymer is that we can alter or modify their physical properties like glass transition temperature or toughness. An example of organic @ organic core–shell nanoparticles is 1,2-distearoyl-sn-glycerol-3-phosphoethanolamine-N-carboxy(poly(ethylene glycol)) coated with biodegradable organic polymer poly(D, L-lactic-co-glycolide), which is used for drug delivery *in vivo* systems.

11.3.3 Inorganic @ Inorganic Core–Shell Nanoparticles

These are the most important and most studied class of core–shell nanoparticles. These are divided into two different classes, such as inorganic @ inorganic silica core–shell nanoparticles, inorganic @ inorganic nonsilica core–shell nanoparticles (Kuila et al., 2018; Robles et al., 2018). The primary difference between these two classes is that in the inorganic @ inorganic silica Core–Shell Nanoparticles, the silica particle will be coated over the core materials with the advantage of its ability to enhance the suspension stability of the core material and reduce the bulk conductivity. Because of its chemical inertness, it can block the core surface from interfering in the redox reaction at the surface of the core material. The optically transparent nature of silica has benefits in the field of SERS applications.

Many metal oxides and metals can also be chosen as shell surface other than silica. We can have gold or silver as core surface as well as core–shell particles with them as the shell surface materials; this has also been explained well by researchers (Islam et al., 2014; Estevez et al., 2017). The benefits of metals like gold/silver coating on any particles are that they will alter the physical properties like biocompatibility, chemical stability, optical properties, and the affinity by functionalization thiol or amino-terminal groups. Apart from these two metals, Ni, Pd, Cu, Co, and Pt are also used as essential shell materials.

11.3.4 Inorganic @ Organic Core–Shell Nanoparticles

The inorganic @ organic core shells is synthesized from metal oxides, metal, metallic compound, or a shell of polymer covering on the silica core. By coating inorganic material with organic material, the advantages are fantastic. One of the benefits is that it will enhance the oxidation stability of the core or else the metal atoms at the surface will be oxidized to form metal oxides. Other than that, this organic coating will provide improved biocompatibility for biological applications. The organic particles covered with polymers have wide application in the fields of paints, additives, cosmetics, pigments, catalysis, and many more. In all these applications, the coating is used for stabilizing the core materials in the suspension media, and this will depend on the nature of force between the particles whether it is attractive or repulsive (Ghosh Chaudhuri and Paria, 2012).

Multiferroic core–shell materials are examples of inorganic @ inorganic (Singh et al., 2014; Siva, Kaviraj and Arockiarajan, 2020; Betal et al., 2015), organic @ inorganic, and inorganic @ organic core–shell. Either core or shell material has to be ferromagnetic or ferroelectric to obtain a multiferroic core–shell system. If the core particle is ferroelectric, then the shell material has to be ferromagnetic and vice versa. Various methods can synthesize these types of materials. In a common core–shell, particle synthesis was a two-stage process; the core particles will be synthesized first followed by the synthesis of shell materials. The two different types of core–shell nanoparticle synthesis include 1) core particles separately prepared and dispersed to the shell material system with surface modification for better coating over the core materials and 2) *in situ* synthesis of the core material; this is followed by shell material coating over it. The first method is the better one because it ensures the purity of the core material.

The possible difficulties in the formation of the core–shell structure are the agglomeration of core particles and the incomplete coating over core material; rather than coating by shell materials, it may form separate particles and reaction rate controllability. Some of the established synthesis methods for core–shell nanoparticles are microemulsion, precipitation, sol–gel autocompletion, polymerization, etc. In the case of multiferroic core–shell structure, it has been synthesized into three types: spherical, cylindrical, and rod type. Here, we have considered only the spherical core–shell nanoparticles (Zeng et al., 2013; Zhu, Zhan and Miao, 2014; Zhang et al., 2016).

11.4 Synthesis Methods for Multiferroic Core–Shell Nanoparticles

Multiferroic core–shell materials are examples of inorganic @ inorganic, organic @ inorganic, inorganic @ organic core–shell. Either core or shell material has to be ferromagnetic or ferroelectric to obtain a multiferroic core–shell system. If the core particle is ferroelectric, then the shell material has to be ferromagnetic and vice versa. Various methods can synthesize these types of materials. The list of synthesis methods is given in Table 11.2. In a common core–shell system, particle synthesis was a two-stage process; the core particles will be synthesized first followed by the synthesis of shell materials. The two different types of core–shell nanoparticle synthesis include: 1) core particles separately prepared and dispersed to the shell material system with surface modification for better coating over the core materials; 2) *in situ* synthesis of the core material. This is followed by shell material coating over it. The first method is the better one because it ensures the purity of the core material.

The main trouble for synthesizing a core–shell nanoparticle is to obtain a proper homogeneous shell coating over the core material and also the control over shell thickness. The possible difficulties in the formation of the core–shell structure are the agglomeration of core particles and incomplete coating over core material; rather than coating by shell materials, it may form separate particles and reaction rate controllability. Some of the established synthesis methods for core–shell nanoparticles are microemulsion, precipitation, sol–gel autocompletion, polymerization, etc. In the case of multiferroic core–shell structure, it has been synthesized into three types: 1) spherical particles, 2) coaxial fibers, and 3) arrays of nanowire, which are vertically aligned on a substrate. The ferroelectric @ magnetic and magnetic @ ferroelectric configurations are explored in spherical particles. Ferroelectrics like barium titanate, lead titanate, strontium titanate, and lead zirconium titanate are mostly used because they exhibit a perovskite structure with spontaneous polarization in the range of 25–50 μC cm^{-2}. Polymers like PVDF are also used as a source for ferroelectricity. Similarly, the most commonly used magnetic counterparts are spinel ferrites like $CoFe_2O_4$, Ni Fe_2O_4, magnetite, and Fe_2O_3.

Table 11.2 List of synthesis methods for multiferroic core–shell nanoparticles.

Core–shell sample	Synthesis technique used	Reference
Fe_2O_3 @ BaTiO3	Sonochemical (shell)	(Buscaglia et al., 2010)
$CoFe_2O_4$ @ BaTiO3	Sol–gel method (shell)	(Corral-Flores, Bueno-Baqués and Ziolo, 2010)
$NiFe_2O_4$ @ BaTiO3	Hydrothermal	(Zhou et al., 2012b)
$CoFe_2O_4$ @ BaTiO3	Coprecipitation (core)	(Corral-Flores, Bueno-Baqués and Ziolo, 2010)
Fe_3O_4 @ $PbTiO_3$	Hydrothermal	(Liu et al., 2010)
NaGdF4:Er3+, Yb3+ @ BiFeO3	Hydrothermal	(Zhang et al., 2017)
$Ba_{0.85}Ca_{0.15}Zr_{0.1}Ti_{0.9}O3$ @ $CoFe_2O_4$	Sol–gel method	(Kumar et al., 2016)
$BaTiO_3$ @ Fe_3O_4	Sonochemical synthesis	(Koo et al., 2009)
$NiFe_2O_4$ @ $PbZr_{0.52}Ti_{0.48}O3$	Sol–gel method	(Liu et al., 2007)
$CoFe_2O_4$ @ Pb(Zr ,Ti)O3	Hydrothermal	(Liu et al., 2010)
$BaTiO_3$ @ γ-Fe_2O_3	Sonochemical synthesis	(Koo et al., 2007)
BaTiO3 @ $ZnFe_2O_4$	Coprecipitation (core and shell)	(Singh et al., 2014)
$BaTiO_3$ @ $BaFe_{12}O_{19}$	Sonochemical synthesis	(Dong et al., 2015)

Homogeneous and continuous ferroelectric shell material over the core magnetic material can be obtained from the hydrothermal method (Zhou et al., 2012a). Zhou et al. reported a combined hydrothermal and calcination process for the synthesis of perovskite lead titanate coated over ferrite nanoparticles ($CoFe_2O_4$ and Fe_2O_3). First, the core material was coated with titanium hydroxide. It was added into the solution containing 0.4 g of potassium hydroxide with 2 mmol of lead nitrate. It was transferred into the autoclave and heated for 6 hours at 160 °C and then washed and dried to get the ferrite @ lead titanate core–shell nanoparticle (Alguero, Gregg and Mitoseriu, 2016).

The precipitation process generally opted for the synthesis of core–shell nanoparticles (Ramam et al., 2014; Etier et al., 2012; Singh et al., 2014; Corral-Flores, Bueno-Baqués and Ziolo, 2010) mainly for inorganic materials like silica. Silica-coated magnetic core with ferroelectric shell and vice versa are also reported. Sushant et al. proposed the synthesis of barium titanate and zinc ferrite core–shell nanoparticle. In this particular reaction, a mixture of zinc nitrate and iron nitrate in a 1:2 ratio was heated at 90 °C and stirred. It was precipitated by adding 0.3 mmol of NaOH. The entire solution was kept for ultrasonication, and during this, 0.1 g of barium titanate was added and maintained for 5 hours. Then it was washed, dried, and sintered at 400 °C to get $BaTiO_3$ @ $ZnFe_2O_4$ nanostructure. Table 11.1 shows some reported synthesis method for multiferroic core–shell nanoparticles.

Sol–gel synthesized multiferroic core–shell materials have received much attention in the scientific world (Cernea et al., 2018, 2020; Corral-Flores et al., 2006; Luz et al., 2017; Liu et al., 2007; Kumar et al., 2016; Corral-Flores, Bueno-Baqués and Ziolo, 2010). Even though the formation of core and shell materials cannot be fully proved, this method provides better connectivity and good dispersion of core materials in the shell precursor solution, and, thereby, it can increase the ME coefficient. In general, the sol–gel synthesis involves the separate formation of the core material and the desired amount of core material will be added to the sol of the shell material. With gel formation and drying, the obtained power will be calcined at a very high temperature in the range of 600–800 °C or higher getting the core shell structure.

The sonochemical synthesis method is widely used for magnetite core and barium titanate shell core–shell structures (Singh et al., 2014; Buscaglia et al., 2010; Koo et al., 2009; Koo et al., 2007). In this case, the shell thickness can be controlled with varying reaction time. Zijing et al. reported the barium titanate coated by bismuth ferrite by using the combined microwave hydrothermal and sonochemical method. Barium ferrite was prepared from nitrates of barium and iron, and pH was controlled by adding an appropriate amount of NaOH. The obtained solution together with barium titanate suspension was taken for ultrasonication and then transferred to vessel for hydrothermal treating at 180 °C. The precipitate thus obtained was washed, dried, and calcined at 850 žC to get the $BaTiO_3$ @ $BaFe_{12}O_{19}$ (Dong et al., 2015).

11.5 Characterization Techniques

The characterization technique used for the multiferroic core–shell nanoparticles is distinctive because of the existence of (magnetic/ferroelectric) shell covering on the (ferroelectric/magnetic) core surface. The most crucial characterization technique used for characterizing this material includes X-ray diffraction spectra for calculating the crystallite size and crystalline structure.

The morphology and thickness of the shell can be understood from TEM and SEM analyses. The presence of elements that constitute the core–shell structure can be identified from EDAX. The decomposition temperature and thermal stability can be obtained from the TGA analysis. For a multiferroic material, both the ferroelectric and magnetic characterizations have to be done. The ferroelectric property can be obtained from the PE loop measurement and magnetic measurement from the VSM/PPMS analysis. The multiferroic property of the samples is tested from the ME coupling measurement using the lock-in amplifier setup.

As usual in the case of single materials, X-ray diffraction spectrum is the preliminary characterization used for identifying the presence of core and shell material intensity peaks. In general, it is used for the identification of unknown elements in the material. Powder X-ray diffraction provides information about favored orientation, the crystalline nature of the material, and the crystalline size, which can be calculated from the Debye Scherrer equation. The crystalline size will vary according to the broadening and change of intensity of the peak. The intensity of the shell materials peaks increases as it is coated over the core material; so the resulting spectrum shows more peaks corresponding to the shell material. SAXS (small-angle X-ray spectrum) and WAXS (wide-angle X-ray spectrum) are also used for the same purpose. WAXS is mainly used for polymer samples with operating 2θ in the range of 5°, but SAXS can be operated for 1° to 10°. Also, it can be used for liquid, solid, or even their combination of samples (Ghosh Chaudhuri and Paria, 2012).

Transmission electron microscopy provides information for a better understanding of the formation of core–shell nanoparticles. The basic knowledge obtained from is the morphology of the sample; it could identify which is core and the TEM image of core shell structure is given in Figure 11.3, gain knowledge about the particle size, the thickness of the shell, whether the shell coating is uniform or not, size of core, and lattice planes. Higher magnification image can also be obtained from HR-TEM; it can give information at the molecular level. The crystallinity of the sample, lattice planes corresponds to core and shell and the d spacing can be inferred from the SAED pattern (Gopi and Sukumaran, 2020).

Scanning electron microscopy (SEM) is the best technique for the direct visualization of nanoparticles. It is generally used to identify the shape and size analysis of the materials. High-resolution image (FSEM) is also possible in SEM with magnification in the range of 10^5–10^6. In the case of core–shell nanoparticles, it is challenging to differentiate the core and shell. Still, it will

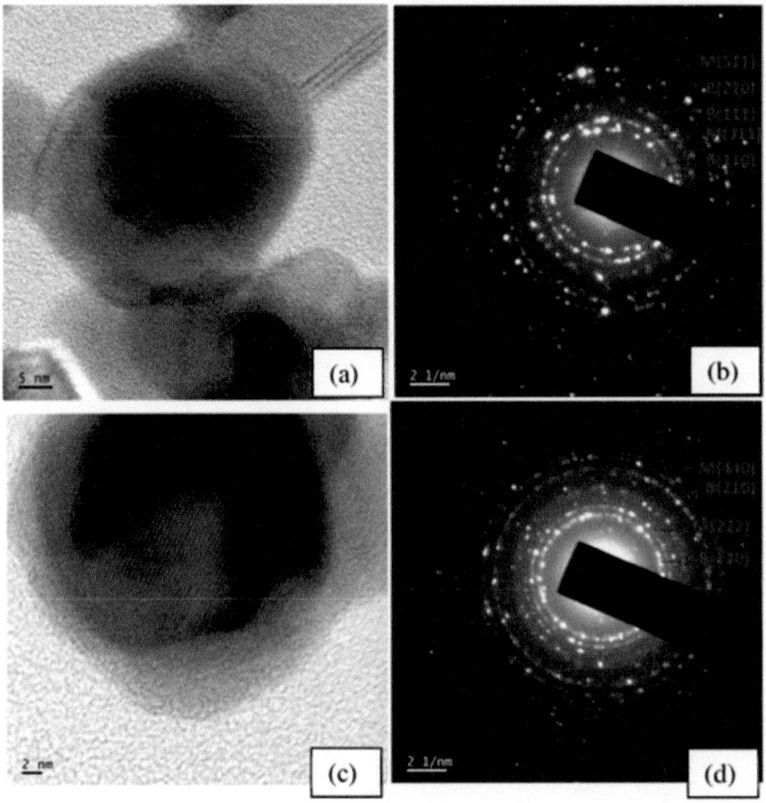

Figure 11.3 (a) TEM and (b) SAED pattern of MFO @ BTO_2 and (c) TEM and (d) SAED pattern of MFO @ BTO_3 (reproduced with permission from ACS).

provide a surface image so that we can get an idea as to whether the surface is rough or smooth. The smooth surface indicates the heterogeneous nucleation of shell over the core material. The rough surface specifies the formation of small particles over the core material (Ghosh Chaudhuri and Paria, 2012). The EDAX spectrum analysis will provide information regarding the purity and stoichiometry of the samples.

Ferroelectricity is the reversal of spontaneous polarization by the application of the electric field. The ferroelectric properties of the multiferroic core–shell nanoparticles obtained from PE loop tracer measurements give the information about the variation of polarization with an applied electric field. The samples will show the hysteresis loop authorizing the ferroelectric nature.

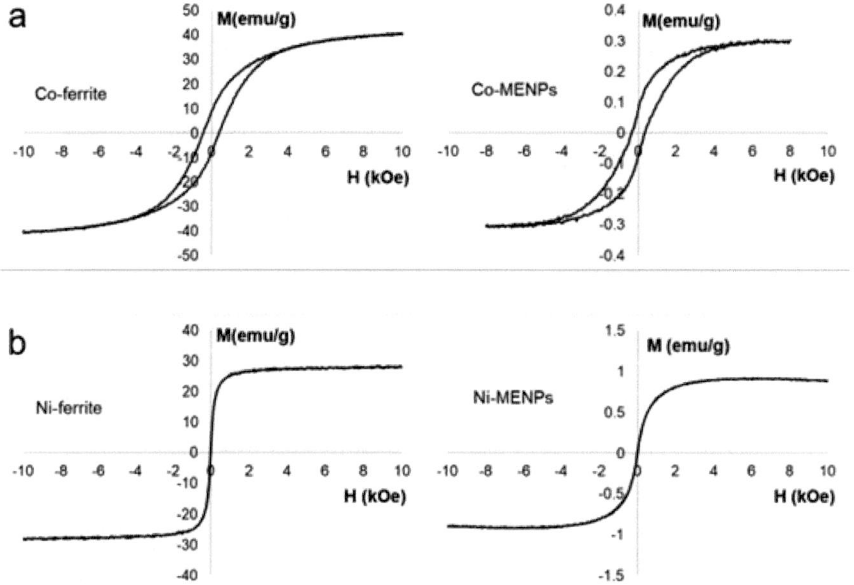

Figure 11.4 DC M-H hysteresis loops measured via AGM of (a) 15 nm co-ferrite and co-ferrite-based 20 nm MENP and (b) 10 nm Ni-ferrite nanoparticles and Ni-ferrite-based 20 nm MENPS (reproduced with permission from ACS).

Some loops may not be saturated because of the leakage current from the magnetic part so that the loop seems round in shape. As the magnetic content increases, the loop shows a lossy behavior. We can get the coercive field, remnant polarization, and maximum polarization from the PE loop (Woldu et al., 2017).

The magnetic property of the multiferroic core–shell nanoparticle can be studied from hysteresis loop measurement. Hysteresis measurements can be done using the VSM/PPMS analysis technique (Islam et al., 2014). The measurement involves the finding of coercivity and remanence of the samples, which is directly linked to the magnetic anisotropy. The temperature dependence of coercivity, thereby blocking temperature and anisotropy constant *K*, can also be found from the magnetic measurements. The magnetic profile of the multiferroic core–shell nanostructure resembles with that of the magnetic material (core or shell) and takes part in the core–shell structure with some modification (Bifeo et al., 2019; Wang et al., 2020). The temperature dependence of magnetization M_s provides information about the surface spin disorders. The magnetic measurement also gives information about whether the materials are the single domain (superparamagnetic) or multiple domains (ferromagnetic) in nature. Most of the magnetic materials used for the multiferroic core–shell nanostructure are ferromagnetic. The magnetic hysteresis is given in Figure 11.4.

Multiferroic core–shell nanoparticles containing magnetostrictive core/shell with ferroelectric shell/core shows the magneto-electric effect, which can be characterized using the lock-in amplifier setup. This coupling facilitates the change of magnetization according to an electric field and modification of electrical polarization with the application of a magnetic field. The coupling can be mathematically defined by calculating the magneto-electric coefficient α. The variation is given in Figure 11.5. The linear magneto-electric coefficient is given by eqn (11.4). The magneto-electric coefficient depends on the magnetic field so that the ME voltage will depend on the applied AC magnetic field and DC magnetic field (Siva, Kaviraj and Arockiarajan, 2020). The recent studies show that the magneto-electric coupling coefficient increases for the multiferroic core–shell structure (Praveena et al., 2020; Abraham et al., 2017; Abraham et al., 2019; Woldu et al., 2017). The

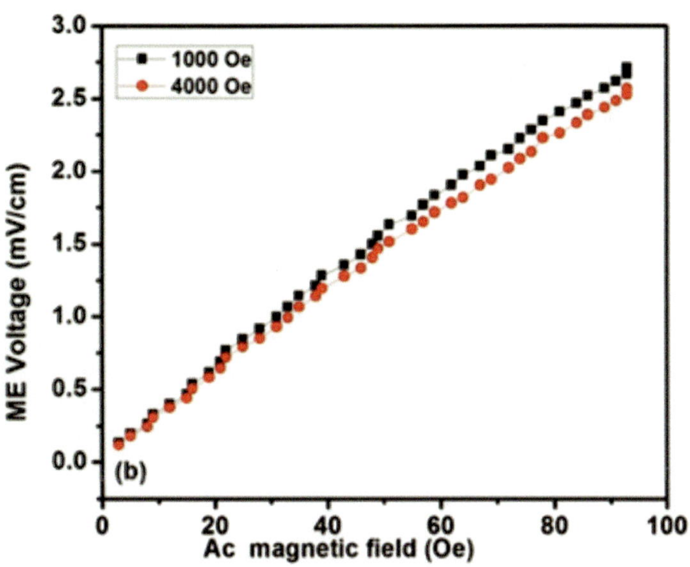

Figure 11.5 ME voltage vs. AC magnetic field with constant DC bias magnetic field for (a) MFO @ BTO1 and (b) MFO @ BTO3 (reproduced with permission from ACS).

increasing value of ME coupling indicates the strongly correlated electron in the materials.

11.6 Applications

The highly correlated multiferroic core–shell nanoparticles are a significant source of technological application because of the coexistence of electric and magnetic materials (Cernea et al., 2020). The requirements of multiferroic materials depend on the area of their applicability. The ordering parameters of ferroelectric and ferromagnetic materials constituting multiferroic core nanoparticles are coupled to get an exciting cross-coupling effect.

11.6.1 Memory storage application

Strongly correlated multiferroic materials are promising contestants for the memory storage applications. Zavaliche et al. have shown studies of the magnetization reversal in bismuth ferrite cobalt ferrite nanostructure produced by the applied electric field. They have also examined the films with $(BiFeO_3)0.65$-$(CoFe_2O_4)0.35$ composition deposited on $SrRuO_3$ bottom electrode upon strontium titanate ($SrTiO_3$) substrate by pulsed laser deposition. So there was an induction of magnetization after applying electric field. Significant changes were observed in the magneto-electric coupling of nanostructure with piezo-electric and magnetic components, and this can be attributed to the change of magnetic state in accordance with the electric field. They have also shown this spatial regulation magnetization induction by the applied electric field for the controlling of data storage (Bedoya-hincapié, Restrepo-parra and López-carreño, 2018).

11.6.2 Energy harvesting application

Multiferroic core–shell materials are exciting materials for energy harvesting applications. The field of energy harvesting industry shows a jump because of the increased maintenance charge due to the very poor lifecycle of batteries. The main idea used for the vibration energy harvesting application is the piezo-electric effect; besides, magnetostrictive vibration-based energy harvesters are also proposed. The change in magnetization produced due to mechanical stress is the fundamental idea of the magnetostrictive energy harvester. The inductive coil picks up the variation in magnetization, and it

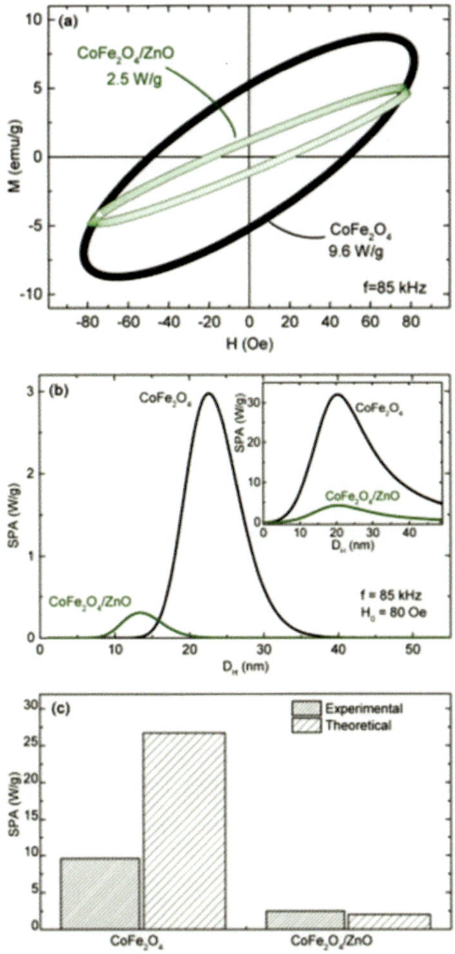

Figure 11.6 (a) Field-dependence of the magnetization during the application of an alternating field with amplitude H0 = 80 Oe and f = 85 kHz. The experimental SPA rates obtained from the area of each loop are indicated. (b) Distribution of SPA rates estimated by considering eqn (13.1.2), the experimental distribution of DH from DLS experiments shown in Figure 11.3, and the experimental conditions. The inset shows the distribution of theoretical SPA as a function of DH. (c) Comparison of the experimental and theoretical SPA rates (reproduced with permission from ACS).

will change the mechanical motion to electromotive force. The magnetostrictive and piezo-electric energy harvesting can be coupled to give effective multiferroic energy harvester (Vopson, 2015).

11.6.3 Biological application

Ion channels are one of the examples for the dealings of electromagnetic fields with biological molecules. The pores formed in the ion channels, which act as membrane proteins, allow the exchange of ion through cellular membranes. The two main characteristics of ion channels are their selectivity and gating. Gating means the process of closing and opening of membrane pores in response to gating variable and selectivity related to the kind of ions they allow to pass through. Voltage-gated channels are the most discussed ion channel, in which the gate will respond to the transmembrane electric field. They are essential for controlling the properties of different kinds of cells and also regulate much more muscle contraction. The defects affected in ion channels may lead to diseases like hypertension, diabetics, neurological disorder, etc. So a proper control over ion channel is required. Kargol et al. proposed that by using multiferroic nanoparticles to produce a change in the electric field following the applied magnetic field, the changes in gating of ion channels could be triggered. This is possible by targeting the nanoparticle to a particular position through the applied magnetic field with the action of antigens on the nanoparticle surface. When an external magnetic field is applied, a mechanical deformation will be produced in the magnetic part in the nanostructure, then that stress transferred to the electric component and thereby create a magneto-electric effect (Bedoya-hincapié, Restrepo-parra and López-carreño, 2018). Lavorato et al. studied the efficiency of the $CoFe_2O_4/ZnO$ core-shell nanoparticles for magnetic fluid hyperthermia (Lavorato et al., 2018). All the above research works show that the materials are highly biocompatible and nontoxic.

11.6.4 Photocatalysis

Photocatalysis plays a focal role in our biological equilibria. It has been the best solution for using our sunlight for removing the pollutants from water and also the process converts water into hydrogen molecule (Liu and Chen, 2017). Jianming Zhang et al. studied the photocatalytic activity of upconverting nanoparticle $NaGdF_4:Er^{3+}$, Yb^{3+} with bismuth ferrite core–shell multiferroic nanoparticle, in which $NaGdF_4:Er^{3+}$, Yb^{3+} act as core and bismuth ferrite acts as the shell. They have checked the photocatalytic activity in near IR and UV region of electromagnetic spectra. Comparing with pure bismuth ferrite powder, the core–shell structure possesses superior photocatalytic activity for the degradation of 4-chlorophenol and methyl orange assays with the NIR and UV irradiation (Zhang et al., 2017). The photocatalytic degradation is given in Figure 11.7.

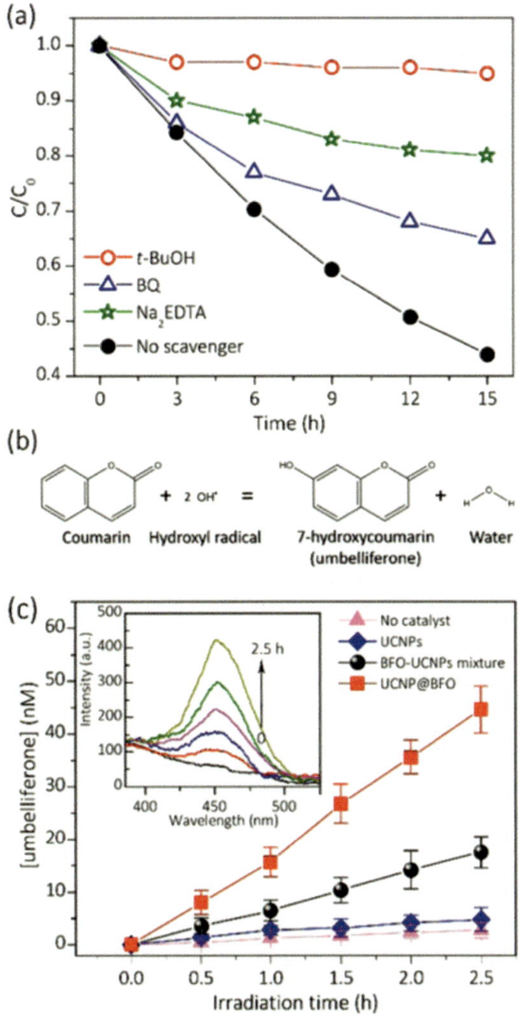

Figure 11.7 Photocatalytic degradation of MO by UCNP @ BFO under 980 nm incident light in the presence of various scavengers (reproduced with permission from ACS).

11.7 Conclusion

The strongly correlated multiferroic core–shell nanoparticle has to become an integral part of our current life. The coupling between the magnetic component and ferroelectric part can have much diverse applications. Composite materials with the core–shell structure are a more encouraging method for obtaining a good multiferroic material. They have found applications in many fields like biomedical application, photocatalysis, memory device applications, energy storage, and many more. The applications of these multiferroic core–shell nanoparticles depend on the useful synthesis and efficiency of the characterization technique. This chapter provides an introduction to multiferroic core–shell nanoparticles, the various synthesis procedures, and characterization techniques used for explaining their properties; there is a lot for exploring. The scientific community is still searching for a better multiferroic core–shell material to meet their device application demands.

References

[1] Abraham, A. R. et al. (2017) 'Realization of Enhanced Magnetoelectric Coupling and Raman Spectroscopic Signatures in 0-0 Type Hybrid Multiferroic Core-Shell Geometric Nanostructures', pp. 0–30. doi: 10.1021/acs.jpcc.6b12461.

[2] Abraham, A. R. et al. (2019) 'Magnetic performance and defect characterization studies of core-shell architectured MgFe2O4@BaTiO3 multiferroic nanostructures', Physical Chemistry Chemical Physics. Royal Society of Chemistry, 21(17), pp. 8709–8720. doi: 10.1039/c8cp04946g.

[3] Alguero, M., Gregg, J. M. and Mitoseriu, L. (2016) Nanoscale Ferroelectrics and Multiferroics: Key Processing and Characterization Issues, and Nanoscale Effects, 2 Volumes. Available at: https://books.google.de/books?id=sAh0CgAAQBAJ.

[4] Bedoya-hincapié, C. M., Restrepo-parra, E. and López-carreño, L. D. (2018) 'Applications of magnetic and multiferroic core / shell nanostructures and their physical properties Aplicaciones de nanoestructuras core / shell magnéticas y multiferroicas y sus propiedades físicas', 85(207), pp. 29–35.

[5] Betal, S. et al. (2015) 'BaTiO3 coated CoFe2O4-Core-Shell Magnetoelectric Nanoparticles (CSMEN) characterization', Integrated Ferroelectrics, 166(1), pp. 225–231. doi: 10.1080/10584587.2015.1092653.

[6] Bifeo, C. O. et al. (2019) 'Journal of Magnetism and Magnetic Materials Comparative study of structural , magnetic and dielectric properties of', Journal of Magnetism and Magnetic Materials. Elsevier B.V., 475(November 2018), pp. 30–37. doi: 10.1016/j.jmmm.2018.11.120.

[7] Buscaglia, M. T. et al. (2010) 'Fe2O3@BaTiO3 core-shell particles as reactive precursors for the preparation of multifunctional composites containing different magnetic phases', Chemistry of Materials, 22(16), pp. 4740–4748. doi: 10.1021/cm1011982.

[8] Cernea, M. et al. (2018) 'materials Synthesis and characterization of novel ferrite – piezoelectric multiferroic core – shell-type structure', Journal of Materials Science. Springer US, 53(13), pp. 9650–9661. doi: 10.1007/s10853-018-2264-x.

[9] Cernea, M. et al. (2020) 'Lead-free BNT–BT0.08/cofe2o4 core–shell nanostructures with potential multifunctional applications', Nanomaterials, 10(4). doi: 10.3390/nano10040672.

[10] Corral-Flores, V. et al. (2006) 'Enhanced magnetoelectric effect in core-shell particulate composites', Journal of Applied Physics, 99(8), pp. 1–4. doi: 10.1063/1.2165147.

[11] Corral-Flores, V., Bueno-Baqués, D. and Ziolo, R. F. (2010) 'Synthesis and characterization of novel CoFe2O4-BaTiO3 multiferroic core-shell-type nanostructures', Acta Materialia, 58(3), pp. 764–769. doi: 10.1016/j.actamat.2009.09.054.

[12] Dagotto, E. and Tokura, Y. (2008) 'Strongly correlated electronic materials: Present and future', MRS Bulletin, 33(11), pp. 1037–1045. doi: 10.1557/mrs2008.223.

[13] Dong, Z. et al. (2015) 'Fabrication, structure and properties of BaTiO< *inf* >3< *inf* >-BaFe1< *inf* >2< *inf* >O< *inf* >19< *inf* > composites with core-shell heterostructure', Journal of the European Ceramic Society. Elsevier Ltd, 35(13), pp. 3513–3520. doi: 10.1016/j.jeurceramsoc.2015.06.016.

[14] Estevez, J. O. et al. (2017) 'Accepted Manuscript'. doi: 10.1016/j.jmmm.2017.06.126.

[15] Etier, M. et al. (2012) 'Cobalt ferrite/barium titanate core/shell nanoparticles', Ferroelectrics, 438(1), pp. 115–122. doi: 10.1080/00150193.2012.743773.

[16] Ghosh Chaudhuri, R. and Paria, S. (2012) 'Core/shell nanoparticles: Classes, properties, synthesis mechanisms, characterization, and applications', Chemical Reviews, 112(4), pp. 2373–2433. doi: 10.1021/cr100449n.

[17] Gopi, P. M. and Sukumaran, K. M. (2020) 'Structural , optical and dielectric properties of magnesium ferrite – Zinc oxide core-shell

nanocomposite Structural , Optical and Dielectric Properties of Magnesium Ferrite – Zinc Oxide core-shell Nanocomposite', 020021(November).

[18] Henrichs, L. F. et al. (2020) 'First-time synthesis of a magnetoelectric core-shell composite: Via conventional solid-state reaction', Nanoscale, 12(29), pp. 15677–15686. doi: 10.1039/d0nr02475a.

[19] Islam, R. A. et al. (2014) 'Magnetoelectric properties of core-shell particulate nanocomposites Magnetoelectric properties of core-shell particulate nanocomposites', 104111(2008), pp. 8–12. doi: 10.1063/1.3013437.

[20] Koo, Y. S. et al. (2007) 'Magnetodielectric coupling in core/shell BaTi O3 γ- Fe2 O3 nanoparticles', Applied Physics Letters, 91(21). doi: 10.1063/1.2817940.

[21] Koo, Y. S. et al. (2009) 'Strain-induced magnetoelectric coupling in BaTiO3 / Fe 3 O4 core/shell nanoparticles', Applied Physics Letters, 94(3), pp. 2007–2010. doi: 10.1063/1.3073751.

[22] Kopeychenko, E. I. et al. (2020) 'Nanocrystalline Heterogeneous Multiferroics Based on Yttrium Ferrite (Core) with Calcium Zirconate (Titanate) Shell', Russian Journal of General Chemistry, 90(6), pp. 1030–1035. doi: 10.1134/S1070363220060158.

[23] Kuila, S. et al. (2018) 'temperature Magnetoresistance in CoFe 2 O 4 / BiFeO 3 core-shell nanoparticles near room temperature', 154104. doi: 10.1063/1.5031170.

[24] Kumar, A. S. et al. (2016) 'Multiferroic and magnetoelectric properties of Ba 0.85 Ca 0.15 Zr 0.1 Ti 0.9 O 3 –CoFe 2 O 4 core–shell nanocomposite', Journal of Magnetism and Magnetic Materials. Elsevier, 418, pp. 294–299. doi: 10.1016/j.jmmm.2016.02.065.

[25] Lavorato, G. et al. (2018) 'Bifunctional CoFe2O4/ZnO Core/Shell Nanoparticles for Magnetic Fluid Hyperthermia with Controlled Optical Response', Journal of Physical Chemistry C, 122(5), pp. 3047–3057. doi: 10.1021/acs.jpcc.7b11115.

[26] Liu, M. et al. (2007) 'Synthesis of ordered arrays of multiferroic NiFe 2 O 4 -Pb(Zr 0.52 Ti 0.48)O 3 core-shell nanowires', Applied Physics Letters, 90(15), pp. 1–4. doi: 10.1063/1.2722043.

[27] Liu, R. et al. (2010) 'Multiferroic ferrite/perovskite oxide core/shell nanostructures', Journal of Materials Chemistry, 20(47), pp. 10665–10670. doi: 10.1039/c0jm02602f.

[28] Liu, Y. et al. (2020) 'Converse magneto - electric effects in a core – shell multiferroic nanofiber by electric field tuning of ferromagnetic resonance', pp. 1–10. doi: 10.1038/s41598-020-77041-x.

[29] Liu, Y. and Chen, X. (2017) Black Titanium Dioxide for Photocatalysis. 1st edn, Semiconductors for Photocatalysis. 1st edn. Elsevier Inc. doi: 10.1016/bs.semsem.2017.04.001.

[30] Luz, G. V. S. et al. (2017) 'Optical and structural characteristics of ZnO nanopowders for different preparation methods', (December). doi: 10.5185/amlett.2017.1581.

[31] Mayeen, A. et al. (2018) 'Dalton Transactions', pp. 2039–2051. doi: 10.1039/c7dt03389c.

[32] Mayeen, A. et al. (2019) 'magnetoelectric applications'. Royal Society of Chemistry, pp. 16961–16973. doi: 10.1039/c9dt02856k.

[33] Mayeen, A. et al. (2020) 'ferroelectric nano fi bers for electrical energy storage', 837. doi: 10.1016/j.jallcom.2020.155492.

[34] Nanomaterials, C. (no date) *No*
主観的健康感を中心とした在宅高齢者における 健康関連指標に関する共分散構造分析Title.

[35] Oliveira, P. N. et al. (2016) 'Synthesis and physical property measurements of CoFe2O4:BaTiO3core-shell composite nanoparticles', Ferroelectrics. Taylor & Francis, 499(1), pp. 76–82. doi: 10.1080/00150193.2016.1172882.

[36] Praveena, M. G. et al. (2020) 'Interface assisted strain-induced magnetoelectric coupling in core-shell nanostructures of CoFe2O4 @ZnO', Journal of Magnetism and Magnetic Materials. Elsevier B.V., 513(June), p. 167252. doi: 10.1016/j.jmmm.2020.167252.

[37] Ramam, K. et al. (2014) 'Core Shell Structured Multiferroic Nanocomposites for Smart Energy Harvestingă: Electric Powering for Portable Electronic Devices', International Review of Applied Engineering Research, 4(1), pp. 57–64.

[38] Robles, J. et al. (2018) 'Exchange-coupled Fe3O4 / CoFe2O4 nanoparticles for advanced magnetic hyperthermia Exchange-coupled Fe 3 O 4 / CoFe 2 O 4 nanoparticles for advanced magnetic hyperthermia', 056719, pp. 2–8.

[39] Singh, S. et al. (2014) 'Magnetodielectric effect in BaTiO 3 /ZnFe 2 O 4 core/shell nanoparticles', Journal of Alloys and Compounds. Elsevier B.V., 587, pp. 437–441. doi: 10.1016/j.jallcom.2013.10.136.

[40] Siva, K. V., Kaviraj, P. and Arockiarajan, A. (2020) 'Improved room temperature magnetoelectric response in CoFe 2 O 4 -

BaTiO 3 core shell and bipolar magnetostrictive properties in CoFe 2 O 4', Materials Letters. Elsevier B.V., 268, p. 127623. doi: 10.1016/j.matlet.2020.127623.

[41] Tokura, Y. (2011) 'Multiferroics Research', pp. 53–54.

[42] Vopson, M. M. (2015) 'Fundamentals of multiferroic materials and their possible applications', Critical Reviews in Solid State and Materials Sciences, 40(4), pp. 223–250. doi: 10.1080/10408436.2014.992584.

[43] Wang, P. et al. (2020) 'Colossal Magnetoelectric Effect in Coreshell Magnetoelectric Nanoparticles'. doi: 10.1021/acs.nanolett.0c01588.

[44] Woldu, T. et al. (2017) 'A comparative study on structural, dielectric and multiferroic properties of CaFe2O4/BaTiO3core-shell and mixed composites', Journal of Alloys and Compounds. Elsevier Ltd, 691, pp. 644–652. doi: 10.1016/j.jallcom.2016.08.277.

[45] Zeng, H. et al. (2013) 'Tailoring magnetic properties of coreshell nanoparticles Tailoring magnetic properties of core / shell nanoparticles', 792(2004), pp. 14–17. doi: 10.1063/1.1776632.

[46] Zhang, J. et al. (2017) 'Efficient Upconverting Multiferroic Core@Shell Photocatalysts: Visible-to-Near-Infrared Photon Harvesting', ACS Applied Materials and Interfaces, 9(9), pp. 8142–8150. doi: 10.1021/acsami.7b00158.

[47] Zhang, X. et al. (2016) 'Tailoring large magnetodielectric response in core/shell CrO2/Cr2O3 nano-rods', Journal of Alloys and Compounds. Elsevier Ltd. doi: 10.1016/j.jallcom.2016.09.152.

[48] Zhou, J. P. et al. (2012a) 'Hydrothermal synthesis and properties of NiFe 2O 4@BaTiO 3 composites with well-matched interface', Science and Technology of Advanced Materials, 13(4). doi: 10.1088/1468-6996/13/4/045001.

[49] Zhou, J. P. et al. (2012b) 'Hydrothermal synthesis and properties of NiFe 2O 4@BaTiO 3 composites with well-matched interface', Science and Technology of Advanced Materials, 13(4). doi: 10.1088/1468-6996/13/4/045001.

[50] Zhu, Q., Zhan, Q. and Miao, H. (2014) 'Multiferroic CoFe2O4 - BiFeO3 core - shell nanofibers and their nanoscale magnetoelectric coupling Multiferroic CoFe 2 O 4 – BiFeO 3 core – shell nano fi bers and their nanoscale magnetoelectric coupling', (February). doi: 10.1557/jmr.2014.36.

Further reading

1. Sang Wook .et.al (2007) 'Multiferroicsă: a magnetic twist for ferroelectricity', nature publishing group pp. 13–20.
2. Gawande, M. B. *et al.* (2015) 'Core– shell nanoparticles: synthesis and applications in catalysis and electrocatalysis', *Chemical Society Reviews*. Royal Society of Chemistry, 44(21), pp. 7540–7590. doi: 10.1039/C5CS00343A.

12

Quantum Magnetism and Frustration

George Mathew

School of Physics, IISER Thiruvananthapuram, India

Abstract

A number of many-body problems that are highly complex in three dimensions can be solved in a low-dimensional system. Simple one-dimensional Hamiltonians that describe such systems have yielded exact, closed-form solutions for their dynamic and static properties. Low-dimensional magnets have gained significant attention for their distinct characteristics that facilitate the analysis of quantum model ground and excited states, the examination of potential new phases of matter, and exploration of the interplay between thermal and quantum fluctuations. The chapter's main topic is theoretical ideas and outcomes concerning one-dimensional magnetism, and their relevance to real materials and experiments.

Keywords: Quantum magnetism, Low dimensional magnets, Frustrated Quantum Systems

12.1 Introduction

What is the underlying principle that keeps a magnet working? Pierre Curie was the one pondering about it. He conducted various experiments over magnetic substances, varying temperatures, and magnetic fields to answer the above question. He was startled by the strange behavior of some non-magnetic materials, which scales up the magnetic behavior on applying magnetic field and reduction in magnetic behavior with temperature. This observation led him to propose a law, which later became the renowned Curie's law, which laid the foundation of the study of magnetism. Curie's

DOI: 10.1201/9781003337102-12

proposition suggested that each atom in a material behaves like a small magnet, and an external magnetic field has the ability to align these atomic magnets, inducing magnetic behavior in an otherwise non-magnetic material. However, this magnetic alignment is disrupted with an increase in temperature. Curie's observations also indicated that this randomization effect decreases with a decrease in temperature, making the material more responsive to the applied magnetic field at lower temperatures. In his research, Curie extended his studies to include compounds that exhibit spontaneous magnetization without an external magnetic field, called ferromagnets. These ferromagnets lose their magnetic behavior above a specific temperature, referred to as the Curie temperature.

During the same period, Pierre Weiss, a French physicist, proposed an explanation for the spontaneous magnetization observed in ferromagnets. Weiss theorized that ferromagnets have their own internal magnetic field that compels atomic magnets to align spontaneously in a particular direction. However, the theory was flawed as it indicated an exceptionally high magnitude of the internal magnetic field. Hans Christian Oersted's research had already demonstrated that an electric current flowing through a wire produces a magnetic field. To maintain the flow of electrons, the current needs to be powered by a battery, and over time, the wire becomes heated due to the continuous flow of current. In contrast, ferromagnets generate a magnetic field without requiring a battery, and they do not heat up over time. The explanation for this phenomenon was provided by quantum mechanics. Within atoms of ferromagnets, electrical currents are generated by the orbital motion of electrons around the nucleus. This current is similar to the current flowing through a superconductor, which requires no external power source to operate and flows without dissipating heat. The principles of quantum mechanics state that electrons are assigned to stable, well-defined orbitals around the nucleus, which explains why there is no dissipation of energy in the form of heat.

In the 1930s, physicists began utilizing quantum mechanics concepts to elucidate the characteristics of magnetic materials. The majority of actual materials exhibit weak diamagnetism, which indicates that they generate an internal magnetic field that opposes the external magnetic field when placed within it. This concept was demonstrated by Andre Geim's experiment of a levitating frog. Quantum mechanics proved to be successful in explaining diamagnetism and also explained a closely related phenomenon called paramagnetism, in which the material develops an internal magnetic field in the same direction as the external magnetic field.

12.1.1 Ferromagnetism

The study of ferromagnetism presents a complex and captivating problem in magnetism. Its analysis delves into the profound concept of quantum mechanics, specifically the idea of symmetry. In quantum mechanics, the state of a quantum object is described by a spatially varying mathematical function called the wave function. Although the wave function itself lacks physical meaning, the square magnitude of the wave function at a specific location does hold significance. It provides the probability of finding the quantum object at that location. Regarding symmetry, imagine two identical particles situated at different points in space. What occurs if we exchange their positions? For a group of particles known as bosons, the final state will be the same as the initial state since they are identical. However, for another group of particles, fermions, such as electrons, an unexpected phenomenon takes place. On swapping two fermions, the wave function changes sign, but the physical reality remains unaltered, given that only the square of the wave function carries physical significance.

Consider two neighboring atoms inside a ferromagnetic material, that is, hydrogen-like atoms with a single electron on the outer-most orbital. Two such atoms are brought into each other's immediate vicinity. We now know that on swapping these two electrons, the wave function of the composite system will change sign. Informations on spatial and magnetic orientations (spin) of electrons are encoded within its wave function and therefore the wave function has two parts: the spatial and the spin part. The spatial part contains information regarding the position of electrons in space and the spin part contains information regarding the orientation of spin. The natural question arises as to which part changes sign on swapping the electrons – spatial or spin?

For ferromagnetic materials, a sign change of the spatial part of the wave function on swapping two electrons is the most energy-efficient process, owing to the fact that, in this configuration, the energy cost of electrostatic repulsion is minimized since the two electrons avoid each other quite effectively. As a result, the sign of spin part remains unchanged on swapping, which leads to a configuration where the magnetic orientation or spins align to a particular direction. This is the mechanism behind the magnetism of ferromagnetic material. This mechanism which involves a strange property describes about what happens to electrons on swapping them; the property that makes the spins to align in a ferromagnetic material and that does not have a classical counterpart is known as the *exchange interaction*. Exchange interactions are so strong in ferromagnets such that magnetism can sustain up to large temperatures, for instance, 770 °C in iron.

Unfortunately, not all ferromagnetic materials, such as iron, exhibit magnetism. This discovery gave rise to the concept of domains and domain walls. Magnetic structures tend to divide into small regions called domains, where all atomic magnets or spins are oriented in the same direction. However, different domains may have distinct orientations, which makes the ferromagnetic material appear non-magnetic. The formation of domain walls is an effort to reduce the magnetic system's energy. If a ferromagnetic material existed as a single domain, it would generate a magnetic field throughout the surrounding space, requiring energy expenditure. Therefore, to minimize energy, ferromagnetic material can separate into several domains. Domain formation results in the development of domain walls between domains, which is energetically expensive due to the twist of spins from one domain to the next across the wall. Whether the magnetic system has a single or multiple domains is determined by the energy balance between the two states. For permanent magnets, a single domain state is preferable since the formation of domain walls is more costly than remaining in a single domain state in terms of energy. These magnets retain their magnetism up to a specific temperature known as the Curie temperature and are referred to as hard magnets. Pure iron is a soft magnet because energy considerations prevent it from forming a single domain. Soft magnets are more attractive for applications that require magnetization to be turned on and off, such as in transformer cores, because magnetization and demagnetization are easier.

12.1.2 Antiferromagnetism

Antiferromagnets are materials in which the exchange interaction favors the neighboring spins to be in antiparallel alignment rather than the parallel alignment in ferromagnets. This configuration of spins is known as Néel state in honor of French physicist Louis Néel who first suggested this idea. Quantum mechanics plays its trick again when it comes to antiferromagnets in the sense that the antiparallel alignment can be in two ways – either "up-down" or "down-up" state. Strangeness of quantum mechanics lies in the fact that it explains reality through the combination of unrealistic possibilities, just like in the famous Schrodinger's cat experiment. The antiferromagnetic ground state or the Neel state can be expressed as follows:

Wave function = $|\uparrow\downarrow\rangle - |\downarrow\uparrow\rangle$.

This wave function can only be represented in a single form, and, hence, this state is known as singlet state. It is not possible to know the state of individual spins. The only thing that can be assumed is that if the first spin is in the "up" state, then the second spin will be in the "down" state and vice versa.

The word spin was introduced back in the 1920s to describe the intrinsic magnetism of an electron. At first, it was thought that the spinning of electron on its own axis results in magnetism, but, later, it was proved that since electron is a point particle, the above concept of spinning makes no sense. During the scientific pursuit to understand how atomic vapors glow when heated or discharged, spin was experimentally confirmed. In the case of light emitting from a sodium vapor lab, it was already discovered that the sodium line splits into two on closer examination under high-resolution spectrometer. Ultimately, Wolfgang Pauli called this the "two-valued quantum degree of freedom" due to the fact that there is something two-valued about the electrons.

The experiments went on further and it was the Dutch physicist Pieter Zeeman who took the experiments to a higher level by applying the magnetic field to the sodium vapor. Due to the fact that different energy levels shift by varying amounts in a magnetic field, the frequencies of the emitted lines change with the application of magnetic field. Known as Zeeman effect, it arises when magnetic fields interact with electrons within the atom. The electron is further restricted by quantum mechanics to only having certain fixed configurations, with certain rotational speeds permissible. This is equivalently expressed as the following fact that angular momentum of electron takes up only integer values. Due to the same energy (degenerate states), these angular momentum states disappear within the same emission line in the absence of an external magnetic field. Application of an external magnetic field lifts the degeneracy, which is known as the Zeeman effect. But, in some cases, additional transitions were noticed, which could not be explained with the orbiting of electron around the nucleus and, hence, named as the *anomalous Zeeman effect*. These additional transitions pointed out the existence of some additional degrees of freedom [1].

12.2 Magnetism in Solids

The successful interpretation of magnetism relies on two primary concepts. The first is that discrete magnetic moments are generated by the ions in solids. The second is that these microscopic moments interact via quantum mechanical forces in addition to the usual dipole–dipole interaction, known as exchange forces. The geometrical arrangement and separation of the magnetic ions determine the magnetic ordering in solids. Magnetism in solids also arises from the electron, which carries charge and possesses an intrinsic magnetic moment due to its spin. A magnetic state results from the competing spontaneous moments' net effect, with spin orientation randomized by thermal energy, while the exchange force tries to align orientations. Magnetism can be classified into cooperative and non-cooperative types based on whether ordering results from the exchange couplings between the moments or an applied external magnetic field.

There are two types of exchange coupling – direct exchange and indirect exchange. When the moments or ions are near enough that their wave functions overlap significantly, exchange coupling operates through direct exchange. This type of coupling is short-range and strong and decreases rapidly with distance. In contrast, moments that are far apart are coupled through indirect exchange. To mediate the exchange coupling, an intermediary is required, which in metals is the itinerant electrons, while in insulators, it can be non-magnetic ions. This type of coupling is known as the RKKY effect in metals and the super-exchange effect in insulators [2].

The magnetic response of materials depends on the orbitals of atoms overlapping to form bands in solids. Materials can be classified as metals or insulators based on the character of their orbitals. In metals, there is a significant amount of orbital overlap, leading to delocalization of electrons and free electron movement within the solid. In contrast, insulators have narrow orbitals, and their magnetic moments are localized. There are also mixed systems where both localized and delocalized electrons contribute to magnetism. These mixed states and itinerant magnetism belong to strongly correlated systems, where electron–electron interactions play a vital role.

12.2.1 The Curie–Weiss law

Weiss proposed that there is a magnetic field produced by molecules, which is directly proportional to the magnetization, in order to explain the temperature dependence of the paramagnetic susceptibility in ferromagnetic insulators at high temperatures. However, this phenomenon is rarely observed because there is no material that remains paramagnetic at very low temperatures.

At some low enough temperature, there is either a ferromagnetic or antiferromagnetic order. Ferromagnetic order sets in at some temperature known as the Curie temperature

$$\chi = \frac{C}{T-\theta}, \; T > T_c.$$

It is seen that in most paramagnetic compounds, the Curie-like behavior is found to hold good only up to certain temperature T_C. Below this temperature, a spontaneous magnetization develops and, in principle, the magnetic susceptibility should diverge.

The paramagnetic response of a system consisting of *N* identical localized magnetic moments can be calculated without considering their interactions.

To determine the magnetization of a system of *N* identical, non-interacting, localized magnetic moments, we can use the Helmholtz free energy.

$$F = -k_B T \ln Z. \tag{12.1}$$

The magnetization can be calculated by taking the partial derivative of the Helmholtz free energy with respect to the magnetic field. For a system of *N* magnetic moments, the magnetization can be expressed as follows:

$$M = -\frac{N}{V}\frac{\partial F}{\partial B}. \tag{12.2}$$

Here, *F* is the Helmholtz free energy, k_B is the Boltzmann constant, *T* is the temperature, *Z* is the canonical partition function for one magnetic moment, and *V* is the volume of the system.

$$Z = \sum_n e^{-\frac{E_n}{k_B T}}. \tag{12.3}$$

Considering a magnetic moment with total momentum *J*, we have 2*J* + 1 possible J_z values and

$$Z = \sum_{-J}^{J} e^{-\frac{1}{k_B T} g\mu_B J_z B}. \tag{12.4}$$

Substituting everything, one finds

$$M = \frac{N}{V} g\mu_B J B_J\left(\frac{\mu_B g J B}{k_B T}\right) \tag{12.5}$$

where $B_J(x) = \frac{2J+1}{2J}\coth\left(\frac{2J+1}{2J}x\right) - \frac{1}{2J}\coth\left(\frac{1}{2J}x\right)$ is the Brillouin function and $x = \frac{\mu_B g J B}{k_B T}$. For $\mu_B B \gg k_B T$, the Brillouin function tends to 1 and the magnetization saturates, that is, all momenta are aligned with the *B*-field. On the other hand, for $\mu_B B \ll k_B T$, one obtains the inverse-temperature dependence of magnetization known as the Curie law, characterized by susceptibility

$$\chi_c = \frac{M}{H} = \frac{N}{V}\frac{\mu_o(g\mu_B)^2}{3}\frac{J(J+1)}{k_B T}. \tag{12.6}$$

These results are suitable for describing insulators that contain rare-earth ions like Yb and Er. However, for transition metal ion-containing insulators, *S* should be used instead of *J* to achieve better agreement. This is because of angular momentum quenching, where *L* equals 0. At high temperatures, materials with a net magnetic moment follow Curie's law. But at lower temperatures, magnetic order develops, and there is deviation from Curie's law. French physicist Pierre-Ernest Weiss proposed the existence of a molecular field within a solid when an external magnetic field is applied. This molecular field is proportional to the material's magnetization and is responsible for magnetic ordering within the material. The total magnetic field acting on a magnetic moment is $\mathbf{H}_{\text{tot}} = \mathbf{H}_{\text{ext}} + \lambda M$, and from Curie susceptibility,

$$\chi_C = \frac{M}{H_{\text{tot}}} = \frac{M}{H_{\text{ext}} + \lambda M} = \frac{C}{T} \tag{12.7}$$

with

$$C = \frac{N}{V}\frac{\mu_o(g\mu_B)^2}{3}\frac{J(J+1)}{k_B} \tag{12.8}$$

$$M = \frac{CH_{\text{ext}}}{T - \lambda C} \tag{12.9}$$

and, therefore, the susceptibility in the external field follows the Curie–Weiss law

$$\chi_M = \frac{C}{T - \lambda C}. \tag{12.10}$$

When the temperature is high, the susceptibility of a material follows the Curie inverse law. However, as the temperature approaches a critical value $T_C = \lambda C$, the susceptibility $\chi_M$diverges, indicating a phase transition to the magnetically ordered phase. In the magnetically ordered phase, the magnetization *M* is nonzero even when the external magnetic field is zero. However, eqn (12.5) shows that without considering the molecular field, *M* equals zero when the external field is zero.

The origin of these molecular fields was unknown at that time. Although the dipole–dipole interaction between magnetic moments could theoretically explain magnetic ordering, this energy scale is too small to account for magnetism observed at room temperature. A quick estimate shows that a critical temperature of 1 K is necessary for ordering to persist solely through dipolar interactions. Therefore, dipolar interactions are responsible for ordering below 1 K. However, some systems exhibit magnetic ordering at room temperature, and the repulsive Coulomb energy between two electrons could be as high as 10^5 K. Our understanding of magnetic ordering will soon be based on a combination of the quantum effect known as the Pauli principle and electrostatic energy.

Even though both 3d and 4f electron system of paramagnetic ions have similar outer electron structure, there are significant differences in their magnetic behavior. For instance, if you measure the magnetic susceptibility and try to relate it to the Brillouin function expression for paramagnetic susceptibility or magnetization, there is very good agreement between the calculated and theoretical values in the case of rare earth ions (4f), whereas this is not the case with 3d ions. In the case of 4f ions, the 4f electrons are surrounded by 5s and 5p electrons, whereas in the case of 3d ions, the 3d electrons itself form the outermost electrons. The 4f electrons are shielded from the neighboring electric charges due to the 5s and 5p electrons, and, hence, the 4f electrons are relatively isolated when compared with the 3d electrons. The ligands (which are charged ions) that surround the paramagnetic ions produce crystalline electric fields at the paramagnetic sites and result in Stark splitting of the ground state of the paramagnetic ion. The magnetic behavior of the paramagnetic ion is almost completely driven by the nature of its ground state. Since the ground state of 3d ions is split by the Stark effect due to the crystalline electric field of the ligands, the nature of the paramagnetic ground state is different, and, hence, the magnetic behavior is different for 3d ions compared to 4f ions. The presence of the ligands lifts the degeneracy, thereby making the ground state non-degenerate. Any non-degenerate quantum mechanical state is represented by a real wave function. If the wave function is real, then the z-component of the angular momentum is represented by $-i\text{h}\frac{\partial}{\partial\phi}$ in a spherically symmetric atom, i.e., $l_z = -i\text{h}\frac{\partial}{\partial\phi}$ is an imaginary operator. So the eigenvalue of l_z vanishes for such a system, and, therefore, the orbital angular momentum for such a system is said to be quenched. This is the phenomenon of quenching of orbital angular momentum by the crystalline electric field of the ligands. Orbital angular momentum quenched means that the crystal electric field somehow keeps the paramagnetic ion locked, making it unable to reorient in the applied magnetic field. So there is no contribution to the total magnetic moment from the orbital state, and, therefore, it is only the spins that contribute to the magnetic moment. In the case of rare-earth compounds, even though the crystal field is not zero, its effect on the ion is not as strong as in the case of 3d ion group because of the shielding of the 4f electrons by the outer 5s and 5p electrons.

12.2.2 Exchange interaction

Our next concern is about the factor responsible for the existence of a molecular magnetic field. The Weiss hypothesis remains a hypothesis; he does not address this question because the molecular magnetic field is basically of quantum mechanical origin. This quantum mechanical molecular field is known as the exchange field, which arises from what is known as exchange interaction.

This theory was proposed by Werner Heisenberg and Dirac in 1926. In order to understand the origin of the exchange mechanism, we have to consider the theory of covalent bonding of a homonuclear diatomic molecule such as hydrogen (proposed by Heilter and London). Consider two hydrogen-like atoms with a single electron on the outer-most orbital and two such atoms are brought into each other's immediate vicinity as shown in Figure 12.1 Electrons are fermions and we need to build an antisymmetric wave function for that composite system.

The two-electron wave function, which has spatial and spin parts, is given by

$$\Psi\,(1,2) = \psi(\boldsymbol{r_1},\boldsymbol{r_2})\chi(1,2) \tag{12.11}$$

where $\psi\,(\boldsymbol{r_1},\boldsymbol{r_2})$ is the spatial part and $\chi(1,2)$ is the spin part of the composite wave function of the two-electron system. Since electrons are fermions,

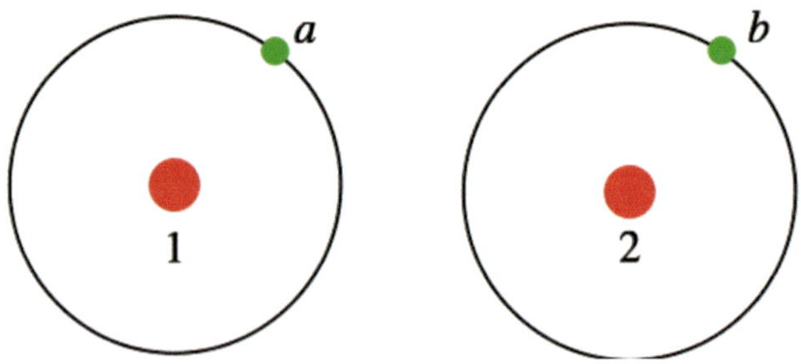

Figure 12.1 Schematic representation of two hydrogen-like atoms with a single electron on the outer-most orbital. The nuclei are represented as **1** and **2** and the respective electrons as ***a*** and ***b***.

$\Psi(1,2) = -\Psi(2,1)$, which means either $\psi(\boldsymbol{r_1}, \boldsymbol{r_2})$ or $\chi(1,2)$ has to be antisymmetric with respect to electron exchange.

The spatial wave function can have the following forms:

$$\text{Symmetric}: \psi_s(\boldsymbol{r_1}, \boldsymbol{r_2}) = \frac{1}{\sqrt{2}}[\varphi_1(\boldsymbol{r_1})\varphi_2(\boldsymbol{r_2}) + \varphi_1(\boldsymbol{r_2})\varphi_2(\boldsymbol{r_1})] \tag{12.12}$$

$$\text{Antisymmetric}: \psi_a(\boldsymbol{r_1}, \boldsymbol{r_2}) = \frac{1}{\sqrt{2}}[\varphi_1(\boldsymbol{r_1})\varphi_2(\boldsymbol{r_2}) - \varphi_1(\boldsymbol{r_2})\varphi_2(\boldsymbol{r_1})] \tag{12.13}$$

where $\varphi_1(\boldsymbol{r_1})$ and $\varphi_2(\boldsymbol{r_2})$ are the single-electron wave functions.

The spin wave function can have the following forms:

$$\chi_s(1,2) = \begin{cases} \chi_\uparrow(1)\chi_\uparrow(2) \\ \chi_\downarrow(1)\chi_\downarrow(2) \\ \frac{1}{\sqrt{2}}[\chi_\uparrow(1)\chi_\downarrow(2) + \chi_\downarrow(1)\chi_\uparrow(2)] \end{cases} \quad \text{Triplet state} \tag{12.14}$$

$$\chi_a(1,2) = \frac{1}{\sqrt{2}}[\chi_\uparrow(1)\chi_\downarrow(2) - \chi_\downarrow(1)\chi_\uparrow(2)] \quad \text{Singlet state.} \tag{12.15}$$

So the total wave function of the composite system can take the following form:

$$\Psi(1,2) = \begin{cases} \psi_s(\boldsymbol{r_1}, \boldsymbol{r_2})\chi_a(1,2) \\ \psi_a(\boldsymbol{r_1}, \boldsymbol{r_2})\chi_s(1,2) \end{cases}. \tag{12.16}$$

The total Hamiltonian for this two-electron system is given by

$$\hat{H} = \widehat{H_1} + \widehat{H_2} + \widehat{H_{\text{ex}}} \tag{12.17}$$

where $\widehat{H_1} + \widehat{H_2} = \frac{p_1^2}{2m_e} + \frac{p_2^2}{2m_e} + V(\boldsymbol{r}_1, \boldsymbol{r}_2)$ are the single-electron Hamiltonians and $\widehat{H_{\text{ex}}}$ is the Hamiltonian for exchange interaction.

$$H = \frac{p_1^2}{2m_e} + \frac{p_2^2}{2m_e} - \frac{e^2}{4\pi\varepsilon_0}\left(\frac{1}{r_{a1}} + \frac{1}{r_{b2}} + \frac{1}{r_{a2}} + \frac{1}{r_{b1}}\right) + \frac{e^2}{4\pi\varepsilon_0}\left(\frac{1}{r_{12}} + \frac{1}{r_{ab}}\right) \tag{12.18}$$

where $\widehat{H_{\text{ex}}}$ is given by

$$\widehat{H_{\text{ex}}} = -\frac{e^2}{4\pi\varepsilon_0}\left(\frac{1}{r_{a2}} + \frac{1}{r_{b1}}\right) + \frac{e^2}{4\pi\varepsilon_0}\left(\frac{1}{r_{12}} + \frac{1}{r_{ab}}\right). \tag{12.19}$$

It would be simpler to model the two-spin system by a spin Hamiltonian, which contains only spin operators and such a model Hamiltonian was proposed by Heisenberg and came to be known as the Heisenberg exchange interaction Hamiltonian. It is given by

$$H_{\text{exchange}} = J\boldsymbol{S_1}.\boldsymbol{S_2} \tag{12.20}$$

where $S_1$and S_2 are the spin operators for the electrons 1 and 2, and J is the exchange integral or overlap integral, which basically informs us about the relative proximity of the electrons to each other.

$$J = E_T - E_S \tag{12.21}$$

In this two-electron model, the difference in energy between triplet E_T and singlet state E_S defines the exchange energy J.

For $J > 0$, the antiparallel alignment of the two-electron spins is preferred and for $J < 0$, the parallel alignment is preferred. Thus, it is the sign of the exchange integral that determines whether a given material is ferromagnetic or antiferromagnetic in nature. This is the basic clue to the understanding of why exchange interaction leads to ferromagnetic or antiferromagnetic ordering.

12.3 Quantum Magnetism and Frustration

Magnetism is fundamentally a quantum effect. If so, what is so peculiar about the name "quantum" magnetism? The term "quantum" in quantum magnetism indicates the low limit (low value like $s = 1/2$) of the spin quantum number for various models [3].

In the field of physics, many-body problems that are difficult to solve in three dimensions become easier to solve in lower dimensions, particularly in one dimension. The physics of lower dimensions, particularly quantum magnetism, offers a better explanation for the behavior of magnets in these dimensions. Theoretical and experimental studies have yielded a better understanding of most aspects of lower-dimensional magnets. Additionally, examples of lower-dimensional magnets provide the best explanation for cooperative behavior in lower dimensions.

The study of low-dimensional magnetism has a history that dates back 95 years. It began in 1925 when a student was tasked with developing a model to explain Weiss' molecular field theory in magnets by their research supervisor. This model is now known as the Ising model [4], named after Ernst Ising. Hans Bethe later created the Bethe Ansatz [5] in 1931 to determine the exact quantum mechanical ground state of the one-dimensional antiferromagnetic Heisenberg model. Despite low-dimensional magnetism's unremarkable start, it has since become one of the most active areas of research in solid-state physics. This is due to two significant developments: one experimental, where several materials exhibit one-dimensional properties, and one theoretical, where it has been discovered that fluctuations are impacted by spatial dimensionality. In general, fluctuations become more noticeable as spatial dimensionality decreases.

Around 1970, it became clear that materials exhibiting low-dimensional magnetic properties exist in nature or can be synthesized in labs. Compared to 2D electron gases and 2D lattices, low-dimensional magnets have an advantage in that while magnetic interactions are limited to lower dimensions, the material as a whole remains bulk, providing sufficient intensity for experiments investigating thermodynamic and dynamic properties. Both experimental and theoretical research works have contributed significantly to the field of one-dimensional magnets. The spin-on nature of the excitation spectrum of the spin-1/2 antiferromagnetic chain was proposed by Faddev and Takhtajan [6] in 1981, while Haldane [7] discovered the principal difference between the chains of integer and half-integer spins two years later. The development of spallation sources in the mid-1980s led to an increase in interest in experimental investigations as it opened a new energy range for neutron scattering experiments. Furthermore, materials called spin ladders, which are intermediate between one and two dimensions, have been discovered. In low-dimensional magnetism, there are still many unanswered phenomena, making it a research area that will remain active for some time

The exchange interaction in spin systems characterizes quantum magnets, in which spins interact with each other. The Heisenberg model introduced earlier (eqn (12.20)) can be generalized to N multi-electron atoms, as shown in eqn (12.22). The exchange interaction strength J determines the interaction between the spin operators $\boldsymbol{S}_i$, located at different lattice sites in a d-dimensional system. The spin can have different values, such as 1/2, 1, 3/2, 2, etc. The dimensionality of the lattice is defined by $d = 1$ for a linear chain, $d = 2$ for a square lattice, and $d = 3$ for a cubic lattice. Real magnetic solids are three-dimensional but can be considered low-dimensional if the exchange interactions have different strengths in different directions. For example, if the intra-chain exchange interactions are stronger than the inter-chain exchange interactions, a material consisting of a chain of spins can be regarded as a one-dimensional system. Figure 12.2 illustrates a linear chain with magnetic centers, where each magnetic center interacts with only two nearest neighbors. The simplest case involves a single magnetic species with either ferromagnetic or antiferromagnetic exchange interactions [8].

Figure 12.2 Schematic representation of a $d = 1$ linear chain with magnetic centers (spins) located at $i - 1$, I, and $i + 1$ and interacting with an exchange constant J.

$$H = \sum_{i=1}^{N-1} J \boldsymbol{S_i}.\boldsymbol{S_{i+1}}. \tag{12.22}$$

The strength of the exchange interaction depends upon the separation between the two spins. The larger the distance, the smaller the interaction strength. If J has the same magnitude for all the nearest-neighbor interactions, then the Hamiltonian in eqn (12.22) takes the following form:

$$H = J \sum_{i=1}^{N-1} \boldsymbol{S_i}.\boldsymbol{S_{i+1}}. \tag{12.23}$$

There are magnetic systems in which the exchange interactions of different strengths exist between the successive pairs of spins. Also the interaction Hamiltonian can extend further to accommodate neighbor interactions. The well-established Majumdar–Ghosh chain [9] includes the nearest as well as the next-nearest-neighbor interactions. The Hamiltonian of the Majumdar–Ghosh chain is given by

$$H_{\mathrm{MG}} = J \sum_{i=1}^{N} \boldsymbol{S}_i.\boldsymbol{S}_{i+1} + \frac{J}{2} \sum_{i=1}^{N} \boldsymbol{S}_i.\boldsymbol{S}_{i+2}. \tag{12.24}$$

Similarly, long-range interactions are included in the Haldane–Shastry model [4] whose Hamiltonian can be represented as

$$H = J \sum_{\langle ij \rangle} \frac{1}{|i-j|^2} \boldsymbol{S}_i.\boldsymbol{S}_j. \tag{12.25}$$

Anisotropy comes into play in real materials and a fully anisotropic Heisenberg Hamiltonian with nearest-neighbor interaction (in 1D) can be represented by

$$H_{XYZ} = \sum_{i=1}^{N} \left[J_x S_i^x S_{i+1}^x + J_y S_i^y S_{i+1}^y + J_z S_i^z S_{i+1}^z \right]. \tag{12.26}$$

The special cases of this Hamiltonian are worth mentioning here. $J_x = J_y = 0$ corresponds to the Ising model, $J_z = 0$ corresponds to the XY model, $J_x = J_y = J_z$ corresponds to the XXX or isotropic Heisenberg model, and $J_x = J_y \neq J_z$ corresponds to the XXZ or anisotropic Heisenberg models.

The favorable alignment of the spins is determined by the sign of the exchange interaction. Let us treat spins as classical vectors to see how exchange interaction (eqn (12.22)) leads to magnetic order. The interaction energy of nearest-neighbor spin pair is given by $JS^2\cos\theta$, where θ is the angle between spin orientations. When J is negative, the lowest energy corresponds to $\theta = 0$, i.e., the interacting spins are parallel, with ferromagnetic state being the ground state, with the ground state energy $E_g = -J\frac{NzS^2}{2}$, where z is the coordination number of the lattice. On the other hand, if J is positive, the ground state corresponds to $\theta = \pi$, with antiferromagnetic state being the ground state, with the ground state energy $E_g = -J\frac{NzS^2}{2}$.

The nature of magnetism is not classical but quantum mechanical. Therefore, current trends focus on models that consider quantum effects, although classical models had a significant impact on the early development of 1D magnetism. This trend is reflected in the materials investigated, as most experimental studies were conducted on compounds containing either Cu^{2+} or Ni^{2+} ions, which correspond to spin-1/2 and spin-1, respectively. Among the spin-1/2 chain-like materials, $CuCl_2.2NC_5H_5$ [9] was the first compound studied experimentally. $KCuF_3$ and Sr_2CuO_3 were identified as the best realizations of the spin-1/2 antiferromagnetic Heisenberg model. $CuGeO_3$ was another quasi-1D spin-1/2 antiferromagnet, widely investigated as the first inorganic spin-Peierls material [10]. $SrCu_2O_3$ is the prototype of a spin-1/2 ladder material. Anisotropy is usually negligible in spin-1/2 chain materials, whereas $S = 1$ chain materials are typically anisotropic in spin space due to spin–orbit effects. A list of compounds serving as 1D magnets can be found in previous reviews [11, 12].

The focus of this chapter is primarily on models of localized spins, denoted as $\mathbf{S}_n$, which possess an exchange interaction energy between spin pairs, $J_{i,j}(\boldsymbol{S_i}.\boldsymbol{S_j})$, commonly known as the Heisenberg model. While real systems may exhibit a transition from 1D to 2D depending on the strength of inter-chain coupling, our focus remains on the weak coupling limit and excludes phase transitions into phases beyond a strictly one-dimensional character. Sticking on to this point, the most prominent single model, the $S = 1/2$ ($S^\alpha = 1/2\ \sigma^\alpha$) XXZ model in 1D, is

$$H = J\sum_n \left\{ \frac{1}{2}\left(S_n^+ S_{n+1}^- + S_n^- S_{n+1}^+\right) + \Delta S_n^z S_{n+1}^z \right\} \tag{12.27}$$

where the scalar product has been decomposed into longitudinal and transverse terms

$$\boldsymbol{S}_1.\boldsymbol{S}_2 = S_1^z S_2^z + \frac{1}{2}\left(S_1^+ S_2^- + S_1^- S_2^+\right). \tag{12.28}$$

Spin correlation functions or dynamical structure factors are the connecting factor between theory and experiment. For a spin chain, the spin correlation function can be defined as follows:

$$S^{\alpha,\alpha}(q,\omega) = \sum_n \int dt e^{i(qn-\omega t)} \langle S_n^\alpha(t) S_0^\alpha(t=0)\rangle \tag{12.29}$$

$$S^{\alpha,\alpha}(q) = \sum_n e^{iqn} \langle S_n^\alpha S_0^\alpha \rangle = \frac{1}{2\pi}\int d\omega S^{\alpha,\alpha}(q,\omega). \tag{12.30}$$

$S(q,\omega)$ can be determined from the scattering cross-section of scattering experiments and the total intensity can be obtained by integrating $S(q,\omega)$ over frequency and wave vector

$$\frac{1}{4\pi^2}\int d\omega S^{\alpha,\alpha}(q,\omega) = \frac{1}{2\pi}\int dq S^{\alpha,\alpha}(q) = \left\langle (S_0^\alpha)^2 \right\rangle, \tag{12.31}$$

which is simply equal to 1/3 $S(S + 1)$ in the isotropic case.

12.3.1 One-dimensional magnetism

The chemical community has played a vital role in designing and synthesizing low-dimensional materials, especially chain compounds, which serve as excellent test cases for theoretical models that are difficult to apply in higher dimensions [13]. In the 1970s, various families of magnetic compounds with linear chain structures were discovered, and their properties were studied using existing models [14]. The discovery of bimetallic chain compounds with alternating spin carriers [15–17] in the 1980s was a significant breakthrough as it introduced a new concept in magnetism known as 1D ferrimagnetism.

12.3.2 S = 1/2 Heisenberg chain

The XXZ model, defined by eqn (12.27), describes the S = 1/2 Heisenberg chain and is significant in both theoretical and experimental aspects. It serves as a prototype for low-dimensional quantum magnetism and is applicable to real materials. The Bethe Ansatz technique provides exact solutions for 1D magnetism, particularly for the XXZ model. A more general form of eqn (12.27) incorporating the external magnetic field applied is as follows:

$$H = J\sum_n \left[(1+\gamma)S_n^x S_{n+1}^x + (1-\gamma)S_n^y S_{n+1}^y + \Delta S_n^z S_{n+1}^z\right] - g\mu_B H \sum_n S_n. \tag{12.32}$$

12.3.2.1 Ferromagnetic phase

The XXZ chain, defined in eqn (12.27), is an important model for both theoretical and experimental studies, serving as a prototype for low-dimensional quantum magnetism and describing real materials. When $\Delta < -1$, the XXZ chain is in the ferromagnetic Ising phase, where the ground state consists of all spins aligned in either the z or $-z$ direction, breaking the symmetry. When an external magnetic field is applied in the z-direction, the Zeeman term $H_z = -g\mu_B H \sum_n S_n^z$ must be added to the Hamiltonian as in eqn (12.32). This results in an additional energy contribution of $-g\mu_B H S_{\text{tot}}^z$, without affecting the wave functions, since H_{XXZ} commutes with the total spin component S_{tot}^z.

In the ferromagnetic phase, the low-lying excited states are magnons, with the total spin quantum number $S_{\text{tot}}^z = \frac{1}{2}N - 1$, and their dispersion law is given by

$$\varepsilon(q) = 2JS\left(1 - \cos q - (\Delta + 1)\right) + 2g\mu_B HS. \tag{12.33}$$

These states form the exact eigenstates of the XXZ Hamiltonian.

12.3.2.2 Neel phase

When $\Delta > +1$, XXZ chain is in the antiferromagnetic Ising or Neel phase. The ground states have $S_{\text{tot}}^z = 0$ and the finite sublattice magnetization is given by

$$N^z = \sum_n (-1)^n S_n^z. \tag{12.34}$$

The primary contrast between ferromagnets and antiferromagnets is that the latter's sublattice magnetization cannot fully order due to quantum fluctuations that do not commute with the XXZ Hamiltonian. When close to the Ising limit $\Delta \to \infty$, the antiferromagnetic Ising phase's elementary excitations can be described by starting with one of the two Neel states. Turning one spin causes two bonds to break, resulting in a state with an energy of Δ that is degenerate with all spins that result from turning any subsequent number of spins. These states are called two-domain wall states because each of the two broken bonds connects two different Neel states. There are a total of $N(N-1)$ of these states, $N^2/4$ with $S^z_{\text{tot}} = 1$, and $S^z_{\text{tot}} = -1$ (number of turned spins odd) and $N^2/2 - N$ states with $S^z_{\text{tot}} = 0$ (number of turned spins even). Although these states are no longer eigenstates when $\varDelta^{-1}$ is finite, they can be studied using the perturbation theory for $\varDelta^{-1} << 1$, leading to the excitation spectrum in the first order of $1/\varDelta$ [18].

$$\omega(q,k) = \Delta + 2\cos q \cos 2\Phi = \varepsilon\left(\frac{q}{2} + \Phi\right) + \varepsilon\left(\frac{q}{2} - \Phi\right) \tag{12.35}$$

with

$$\varepsilon(k) = \frac{1}{2}\Delta + \cos 2k \tag{12.36}$$

where q is the total momentum and Φ is the wave vector related to the superposition of domain walls with different distances.

Eqn (12.35) formulates the excitation spectrum, which consists of two components. The dispersion relation given by eqn (12.36) describes the propagation of domain walls independently with momenta $k1$ and $k2$. Villain [19] was the first to describe these propagating domain walls, which marked the initial emergence of magnetic solitons in the quantum realm. The argument 2k in the dispersion is due to the fact that a domain wall can hop two sites owing to the transverse interaction. The elementary excitations in the antiferromagnetic Ising phase form a continuum, with the two domain walls' relative momentum constituting an internal degree of freedom.

12.3.2.3 XY phase

In the absence of an external field, the XXZ chain exhibits the XY phase when $-1 < \varDelta < 1$, which is characterized by an easy plane type of uniaxial symmetry and a gapless excitation continuum. A thorough analysis of this phase requires the use of Bethe Ansatz and bosonization techniques.

For general $\varDelta$, the XXZ chain can be represented as an interacting one-dimensional fermion system. However, in the case of $\varDelta = 0$ (the XX model), the fermion chain becomes non-interacting and is therefore entirely amenable to exact analysis. This chain of free fermions can be described by the dispersion law

$$\varepsilon(k) = Jcosk - g\mu_B H \tag{12.37}$$

where H is the external magnetic field.

Experimentally relevant quantities can be easily derived from the exact expression of free energy in terms of the basic fermion dispersion, eqn (12.37),

$$F = -Nk_BT\left[\ln 2 + \frac{2}{\pi}\int_0^{\frac{\pi}{2}} dk\ln\cosh\left(\frac{\varepsilon(k)}{2k_BT}\right)\right]. \tag{12.38}$$

Such an experimentally relevant quantity is the specific heat whose low-temperature behavior is linear in T as shown in the following:

$$C(T) \approx \frac{\pi T}{6v_F} \tag{12.39}$$

where $v_F = (\partial\varepsilon/\partial k)_{k=k_f} = J$ is the Fermi velocity.

Thus, we can deduce that moving a fermion just across the Fermi point costs arbitrarily low energy, leading to an excitation spectrum, which is gapless.

12.4 Frustrated Quantum Systems

Conceptually simple, frustration has far-reaching consequences in the magnetic ground state properties of the system. In magnetic systems, atomic moments are the one responsible for the magnetic interactions and through bond formation. Frustration comes into play when all bonds cannot be satisfied simultaneously. Figure 12.3 shows the best illustration of this fact in a triangular geometry. If the interaction is ferromagnetic, all bonds are satisfied at once, whereas in the case of antiferromagnetic interactions, which are in favor of antiparallel alignment, it is not possible. The topic of discussion here is the fluctuations exclusively at zero temperature, which are the quantum fluctuations and how they destroy the classical order, establishing new types of order [20].

Consider a spin-1/2 chain with N-spins coupled by Heisenberg interaction

$$H = \sum_{i<j}^{N} J\boldsymbol{S}_i.\boldsymbol{S}_j$$

where $J < 0$ and $J > 0$ correspond to ferromagnetic and antiferromagnetic interactions, respectively. The energy of the ordered spin configuration will be $-1/4$ J for $N = 2$. This is the energy gained upon breaking the spin rotational symmetry by considering that the alignment will be in the z-direction and can

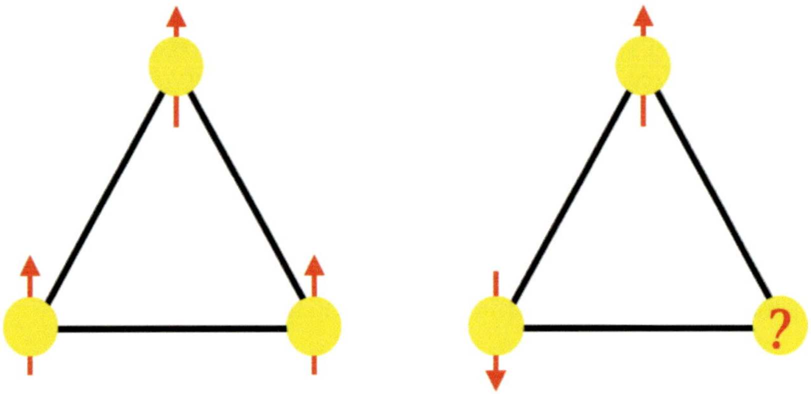

Figure 12.3 Schematic illustration showing a triangular lattice exhibiting geometrical frustration in which nearest-neighbor interactions prefer anti-aligned spins. However, on a triangle, it is impossible for all three spins to be antiparallel at the same time. As a result, the spins either fluctuate or order depending on the circumstantial conditions.

be represented as $-JS_i^zS_j^z$. But when the spin is considered as a quantum mechanical entity, it can have an alternate state, the singlet state, which has no classical counterpart

$$|s> = \frac{1}{\sqrt{2}}(|\uparrow\downarrow> - |\downarrow\uparrow>)$$

whose energy is −3/4 J. This state does not break spin rotational symmetry and forms the ultimate expression for quantum fluctuations.

Frustration can result in a variety of unusual material properties. It arises from the presence of competing forces that cannot be simultaneously satisfied. The study of frustration was primarily triggered by antiferromagnets, in which it typically has a straightforward geometric origin, as illustrated in Figure 12.3.

12.4.1 Broken symmetries

In frustrated systems, broken symmetries can reveal the ground state and the resulting magnetic or non-magnetic order [21]. For example, conventional magnetic order will break spin rotational symmetry, while a state where all spins form a singlet pair with their nearest neighbor will break the translational symmetry of the lattice. There are many possible combinations of spin and lattice symmetries, which will be discussed below.

12.4.1.1 Broken spin, unbroken lattice

Conventional ferromagnetic ordering involves the spontaneous breaking of continuous spin rotational symmetry, resulting in a preferred direction on the surface of the sphere. This category also includes non-conventional forms of order, such as spin nematic, in which spin rotation symmetry is broken without the formation of a magnetic moment. Instead, there may be quadrupolar or even higher multi-polar fields.

12.4.1.2 Unbroken spin, broken lattice

The valence-bond crystal (VBC) is the canonical ground state of a system in which the kinetic energy of singlet formation on only a fraction of bonds can exceed the magnetic order potential energy, thereby preserving the SU(2) spin rotational symmetry at the expense of breaking the translational symmetry spontaneously. Extensive research has been conducted on VBC phases for various models, and their analysis is now highly established and refined.

12.4.1.3 Broken spin and lattice

The breaking of spin rotation and lattice translation in this exotic ground state are fundamental phenomena and not just trivial consequences of each other.

12.4.1.4 Unbroken spin and lattice: the spin liquid

The concept of a magnetic system having a ground state that preserves symmetry was long considered unproven. The idea of a "spin-liquid" phase was proposed by Fazekas and Anderson in the context of the $S = 1/2$ triangular lattice [22]. One-dimensional systems provide a good example of the basic characteristics of spin liquids. Chain models, which maintain SU(2) spin symmetry and one-site translation, have spin correlations that decay rapidly with distance and can exhibit fractional magnetic excitations. However, the one-dimensional case is unique because the Mermin–Wagner theorem [23] states that continuous symmetries cannot be spontaneously broken in 1D due to strong quantum fluctuations, making long-range order impossible. Therefore, symmetries can be used to categorize magnetic ground states.

12.4.2 Frustrated systems

Frustrated magnetism has become a significant area of research due to the ability to produce real materials with frustration in the laboratory. The primary magnetic components in these compounds are metallic species with unpaired electrons in partially filled outer d or f shells. Itinerant magnetism in inter-metallics was among the first studies in this field, but the focus gradually shifted to local-moment magnetism in insulators following the discovery of high-temperature superconductivity.

The classification of frustrated materials can be based on their composition, either inorganic or organic. The adjustment of magnetic interactions in inorganic materials can be a complex procedure, and achieving high purity at high temperatures is challenging. Furthermore, the energy range of magnetic interactions is typically hundreds of Kelvin or Tesla, which surpasses the available continuous magnetic fields in laboratory instruments. This presents an obstacle for studying the impact of external magnetic fields on spin dynamics. However, synthesizing compounds with low interaction energies provides a chance to study these effects [3, 24, 25].

Challenges associated with inorganic magnetic materials can be overcome by using organometallic compounds. The organic synthesis route offers several advantages, including the ability to perform reactions at lower temperatures and ambient pressure, and a more efficient removal of impurities during crystal formation. Organometallic compounds typically have large organic side groups that separate metal ions, leading to weaker magnetic interactions. These systems allow for greater control over magnetic properties due to the wide range of organic ligands available. However, the small crystal size of these systems can complicate certain experiments, and the cost of deuteration of organic groups, which is necessary for neutron scattering experiments, can be prohibitive.

12.4.2.1 Spin liquids

The concept of a spin-liquid state emerged [22] from the application of the resonating-valence-bond (RVB) state ideas and examining the valence bond basis. The reasoning behind it is simple: a valence bond configuration of a frustrated system could be a low-energy competitive basis state; numerous such states are degenerate at the lowest order; fluctuations between these states can raise the kinetic energy; such fluctuations lift the degeneracy of the manifold, and this energetic separation determines the true nature of the ground state. The discussion of the spin liquid can continue by examining the spectrum of this manifold of low-lying states, and the direct consideration of RVB states can be postponed until later in this section.

Let us establish a relationship between spin correlations and energy gaps in a magnetic system. The spin–spin correlation function, denoted by $f(r) =< \boldsymbol{S}_o \boldsymbol{S}_r >$, measures the state of a spin located at r with respect to a spin located at o. Generally, for long distances, the correlation function depends only on the magnitude of r. If the system has long-range magnetic order, $f(r) \rightarrow m^2$ as $r \rightarrow \infty$, where m is the ordered moment, and the ground state is composed of many different spin components with the same energy. However, if the magnetic order is short-ranged, the correlation function decays exponentially, $f(r) = A\exp\left(-\frac{r}{\xi}\right)$,, where ξ is the characteristic spin correlation length, which is inversely proportional to the energy gap Δ between the singlet ground state and the lowest triplet state of the system. Algebraic order is a situation between these two extremes, where the correlation function follows a power-law decay to zero, $f(r) = Br^{-\alpha}$, for some positive power α. In one-dimensional systems, this power-law decay is known as "quasi-long-range order" and also requires degeneracy of triplets with singlets in the ground-state manifold.

The spins in a spin liquid can exhibit both classical and quantum fluctuations. When the spins are large, classical fluctuations caused by thermal energy dominate. In this scenario, the spins randomly reorient themselves over time, resulting in various microstates. As the thermal energy decreases, the fluctuations cease, leading to either freezing or ordering of spins. However, for small spins, quantum fluctuations, driven by the uncertainty principle, persist down to 0 K, and the spins are phase coherent unlike thermal fluctuations. If these quantum fluctuations are strong enough, the spins enter a superposition state, where they point in many random directions. This state is called a quantum spin liquid (QSL). In QSL, even widely separated spins are highly entangled, leading to exotic phenomena such as fractional quantum numbers. For materials with spin $S = 1/2$ having Heisenberg symmetry, the spins continue to fluctuate and avoid order even at 0 K due to strong quantum effects [26–29].

12.5 Conclusion and Future Perspectives

The field of quantum magnetism and frustration has been discussed in this article, with a focus on low-dimensional magnetism such as one-dimensional magnetic systems and spin liquids. These systems serve as an ideal testing ground for theoretical and experimental approaches, and significant progress has been made in understanding their properties. However, this chapter only scratches the surface of the many aspects of low-dimensional magnetism that have yet to be explored. The main aim of this chapter is to provide an overview of the fundamental concepts in this field.

Significant progress has been made in the field of low-dimensional magnetism through both theoretical and experimental efforts. Neutron scattering experiments have successfully predicted and measured various excitation spectra in model systems, demonstrating the fruitful interplay between experiment and theory that has played an important role in advancing this field. Fluctuations are central to the behavior of low-dimensional magnets due to their low dimensionality. The study of these systems benefits from the availability of exactly solvable idealized models that are physically realizable, allowing for a close comparison between theory and experiment.

References

[1] Blundell, Stephen J. *Magnetism: a very short introduction*. OUP Oxford, 2012.

[2] Hurd, C. M. "Varieties of magnetic order in solids." *Contemporary Physics* 23.5 (1982): 469-493.

[3] Gosuly, S.P. *Neutron scattering studies of Low-dimensional quantum spin systems*, Diss. UCL (University College London), 2017

[4] F.D.M Haldane, Phys. Rev. Lett. 60, 635 (1988); B.S. Shastry, Phys. Rev. Lett. 60, 639 (1988)

[5] W. Lenz: Z. Physik 21, 613 (1920); E. Ising: Z. Physik 31, 253 (1925).

[6] H. Bethe: Z. Physik 71, 205 (1931)

[7] L.D. Faddeev, L.A. Takhtajan: Phys. Lett. 85A, 375 (1981)

[8] R. Georges et al, One-dimensional Magnetism: An Overview of the Models. Magnetism: Molecules to Materials I: Models and experiments, 1-47

[9] C.K. Majumdar and D.K. Ghosh, J. Math. Phys. 10, 1388 (1969)

[10] M. Hase, I. Terasaki, K. Uchinokura: Phys. Rev. Lett. 70, 3651 (1993)

[11] M. Steiner, J. Villain, C.G. Windsor: Adv. Phys. 25, 87 (1976)

[12] H. J. Mikeska and M. Steiner: Adv. Phys. 40, 191 (1991)

[13] L.J. De Jongh, A.R. Miedema, Adv. Phys. 1974, 23,1

[14] E.W. Hatfield, W.E. Estes, W.E. Marsh, M.W. Pickens, L.W. Ter Haar, R.R. Weller in Extended Linear Chain compounds, Vol. 3, Plenum Press, New York, 1983, p. 43

[15] E. Coronado, M. Drillon, R. Georges in Research Frontiers in Magnetochemistry, World Scientific, Singapore, 1993, p. 27

[16] E. Coronado in Magnetic Molecular Materials, Vol. E 191, Kluwer Academic Publishers, Dordrecht, 1991, p.267

[17] O.Kahn, Molecular Magnetism, VCH Publishers, New York, 1993, p. 379

[18] N. Ishimura, H. Shiba: Progr. Theor. Phys. 63, 743 (1980)

[19] J. Villain: Physica B 79, 1 (1975)

[20] Normand, Bruce. "Frontiers in frustrated magnetism." *Contemporary physics* 50.4 (2009): 533-552.

[21] Read, N., and Subir Sachdev. "Valence-bond and spin-Peierls ground states of low-dimensional quantum antiferromagnets." *Physical review letters* 62.14 (1989): 1694.

[22] Fazekas, P. W. A. P., and Philip W. Anderson. "On the ground state properties of the anisotropic triangular antiferromagnet." *Philosophical Magazine* 30.2 (1974): 423-440.

[23] Mermin, N. David, and Herbert Wagner. "Absence of ferromagnetism or antiferromagnetism in one-or two-dimensional isotropic Heisenberg models." *Physical Review Letters* 17.22 (1966): 1133.

[24] P.R. Hammar et. al., Magnetic studies of the two-dimensional, S= 1/2 Heisenberg antiferromagnets (5CAP) 2 CuCl 4 and (5MAP) 2 CuCl 4. *Journal of applied physics* 81.8 (1997): 4615-4617.

[25] I.U. Heilmann, G. Shirane, Y. Endoh, R.J. Birgeneau, S.L. Holt: Phys. Rev. B 18, 3530 (1978)

[26] Balents, Leon. "Spin liquids in frustrated magnets." *Nature* 464.7286 (2010): 199-208.

[27] Rüegg, Ch, et al. "Thermodynamics of the spin Luttinger liquid in a model ladder material." *Physical review letters* 101.24 (2008): 247202.

[28] Hermele, Michael, T. Senthil, and Matthew PA Fisher. "Algebraic spin liquid as the mother of many competing orders." *Physical Review B* 72.10 (2005): 104404

[29] Hermele, Michael, et al. "Stability of U (1) spin liquids in two dimensions." *Physical Review B* 70.21 (2004): 214437.

13

Magnetoelectric Multiferroics: A Promising Highly Correlated System

Bosely Anne Bose[1,2], Bhadrapriya B. C.[1,2], and Nandakumar Kalarikkal[1,2]

[1]School of Pure and Applied Physics, Mahatma Gandhi University, India
[2]International and Interuniversity Centre for Nanoscience and Nanotechnology, Mahatma Gandhi University, India

Abstract

Correlated systems refer to those systems in which the properties exhibited by them are somewhat related. Multiferroics are considered to be one among the highly correlated systems since one or more primary ferroic properties coexist in these materials. They are broadly classified into single-phase and composite (diphasic or polyphasic) multiferroics. Magnetoelectric multiferroics are one among the highly explored multiferroic materials in which ferromagnetic and ferroelectric orders occur simultaneously with or without ferroelasticity. These materials are highly beneficial in a technological point of view and are used for developing multifunctional devices, sensors, energy scavengers, memory devices, etc. In this chapter, we are discussing the basic theory and applications of magnetoelectric multiferroic systems.

Keywords: Multiferroics, Magnetoelectrics, Correlated system

13.1 Introduction

Strongly correlated systems are those for which the inter-particle interactions are not weak and these interactions play a major role in determining their properties. Conventional superconductors, high-temperature superconductors, magnetic materials, low-dimensional systems, etc., are some examples of highly correlated systems.

DOI: 10.1201/9781003337102-13

In magnetic systems, the Coulomb interaction between electrons can give rise to different spin ordering patterns. Two common examples of these patterns are ferromagnetism and antiferromagnetism. Since high magnetic fields greatly alter the electronic states, in magnetic and superconducting systems, high magnetic fields have become an extremely useful means to explore novel phenomena and to investigate new materials.

13.1.1 Electrons in solids

The first quantum mechanical explanation for electron behavior in solids was given by the Sommerfeld theory, which treats electrons as free and does not include periodic potential of the ions and the electron–electron interaction. Then comes the band theory, which has taken into account the ion field. According to the band theory of solids, just as electrons in atoms occupy certain allowed levels, the electrons in solids occupy bands of allowed states. In a metal, there is no gap between the occupied and unoccupied states (valence and conduction bands), whereas in an insulator or semiconductor, there is a gap [1].

In condensed matter systems, conduction electrons govern most of the physical and chemical properties. Most of the properties can be understood by finding the eigenstates of these electrons. The evaluation of Schrodinger equation for a many-electron system, i.e.,

$$\sum_{i=1}^{N} -\frac{\hbar^2}{2m}\nabla_i^2\varphi + V\varphi = E\varphi$$

is a hectic task, and, hence, it immediately calls for further approximations. The Born–Oppenheimer approximation is often applied in the study of molecular systems. It assumes that the motion of atomic nuclei is much slower compared to the motion of electrons due to their large mass difference. As a result, the dynamics of electrons and atomic nuclei can be treated independently. By making this approximation, the molecular wave function can be expressed as a product of two components: one for the electronic wave function that depends on electron positions and one for the nuclear wave function that depends on the positions of atomic nuclei [2]. Another important approximation is the one electron approximation, also known as the independent electron approximation, according to which: each electron in the system occupies a one-electron state independently of the others. The main features of this approximation are:

a) Each electron occupies a one-electron state: This means that the behavior of each electron is determined by its own wave function, neglecting the interactions with other electrons.
b) Pauli's exclusion principle: The one-electron states obey the Pauli exclusion principle, which states that no two electrons can occupy the same quantum state simultaneously. This principle ensures that each electron has a unique set of quantum numbers.
c) No overlap between one-electron states: The one-electron states are derived by diagonalizing the Hamiltonian, resulting in distinct energy levels. This implies that there is no overlap or mixing between the different one-electron states.
d) No scattering between states: Since there is no overlap between the one-electron states, there is no scattering or interaction between electrons occupying different states.
e) Time independence: In the one-electron approximation, the system is assumed to be independent of changes in time. This means that the one-electron states and energies do not depend explicitly on time.

Indeed, external perturbations can have several effects on a system described by a Hamiltonian:

- External perturbations can modify the Hamiltonian of a system to short time.
- Perturbations can introduce particles that are not eigenstates of the Hamiltonian.
- External perturbations can couple to the particles in the system and induce scattering processes.

As a consequence of scattering, the electron wave function is described as a sum (superposition) of various one-electron states. This superposition is denoted by the following expression:

$$\varphi = \sum_n C_n f_n$$

where C_n represents the probability amplitude associated with the *n*th one-electron state (f_n).

- This mixed state will relax into a pure state f_n with a probability $|c_n|^2$.
- The probability of the system relaxing into a particular pure state f_n is given by the squared magnitude of the probability amplitude $|c_n|^2$.

The one-electron approximation helps simplify the treatment of a many-electron system by reducing the N-electron Schrödinger equation to multiple coupled one-electron Schrödinger equations [3].

13.1.2 Electron-electron interactions

The conventional band theory was challenged by a class of materials called Mott insulators. These materials, which should be conductive by band theory predictions, are in fact insulators especially at low temperatures. The interactions between electron interactions were not considered in the conventional band theory. Sir Nevill Francis Mott was the first to point out the flaw in the central approximation in band theory, hence the name Mott insulators. He directed attention to the importance of inter-electron forces. NiO, VO_2, V_2O_3, etc., are examples for Mott insulators [4, 5]. Mott insulators are a typical example for strongly correlated systems.

Heavy fermion materials are a class of compounds that exhibit strong electron−electron correlations and have unusually large effective masses for conduction electrons. These compounds contain rare-earth or actinide atoms with partially filled 4f or 5f electron shells. Unlike the valence electrons involved in chemical bonding, the f electrons are typically localized in the inner shells and do not participate directly in bonding interactions. At temperatures much below both the Debye temperature and the Fermi temperature, the heat capacity of metals may be written as the sum of electron and phonon contributions:

$$C = \gamma T + AT^3$$

where γ is called the Sommerfeld parameter, which is a material-dependent constant and is proportional to the effective electron mass $\mathbf{m^*}$. In common metals, the ratio **m*/me** is typically on the order of 10, indicating that the effective mass is slightly larger than the bare electron mass. However, in inter-metallic compounds with rare-earth or actinide atoms, this ratio becomes enormous, implying a substantial increase in the effective mass [6]. The f electron shells, which are partially filled in these compounds, are inner shells and generally do not participate in chemical bonding. At room temperature and higher, the f-shell electrons tend to remain localized on their respective atomic sites. Consequently, the system behaves as a weakly interacting collection of f-electron moments and conducting electrons with ordinary masses. However, at low temperatures, peculiar phenomena emerge. The f-electron moments become strongly coupled to the conduction electrons and to each other. They form an antiferromagnetic singlet electron pair with the conduction electrons. This coupling leads to a significant enhancement in the effective mass of the conducting electrons, typically ranging from 10 to 1000 times the bare electron mass [7]. Examples of heavy fermion materials include $CeAl_3$, $CeCuSi_2$, $CeCu_6$, UBe_{13}, UPt_3, UCd_{11}, $U2Zn_{17}$, $NpBe_{13}$, and many others.

Multiferroics are another well-known example of highly correlated systems. The discovery of multiferroic materials challenged the long-accepted belief that ferroelectricity and magnetism tend to be mutually exclusive and interact only weakly with each other when they coexist [8]. Multiferroics exhibit a combination of ferroelectricity, magnetic order, and ferroelasticity, which are properties related to electric polarization, magnetization, and elastic deformation, respectively. One of the remarkable features of multiferroics is the coupling between these properties, known as the magnetoelectric effect. This effect refers to the induction of magnetization by an electric field or the induction of polarization by a magnetic field. The magnetoelectric effect can be described using both linear and non-linear terms in the free energy expansion. In general, coupling parameters will be expressed as tensors since the magnetic field $\boldsymbol{H}$ and electric field $\boldsymbol{E}$ are vectors quantities. It leads to an expression for the magnetoelectric interaction energy $\boldsymbol{F}_{ME}$ [9]. When an electric field is applied, magnetization is induced or when an applied magnetic field induces a polarization, then such an effect is called magnetoelectric effect. The coupling between magnetic and electric order parameters can produce a magnetoelectric effect, and such coupling is referred to as a linear magnetoelectric effect

$$F_{ME} = \boldsymbol{\alpha}_{ij} E_i H_j \ + 1/2 b_{ij} k E_i H_j H_k \ + 1/2 \gamma E_i E_j H_k \\ + 1/6 \boldsymbol{\delta}_{ijkl} E_i E_j H_k H_l.$$

The above equation can be rearranged by polarization P and magnetization M or staggered magnetization L for the magnetoelectric coupling of the multiferroics. The (δijkiEiEjHkHl) term produces an effect called magnetodielectric effect, which refers to the change in the dielectric response of a material that varies with the applied field quadratically [10]. Also, this term appears in any free energy expansion since both E and H are scalar and bilinear. In most of the cases, other terms are not allowed. The term refers to an odd number of magnetic or electric fields that violate at least one of these two symmetries. Magnetic fields violate temporal reversal symmetry, whilst electric fields violate symmetry of spatial inversion. The other magnetoelectric coupling terms, however, may not be zero in crystals with specific transformational properties.

13.1.3 Theoretical approach to highly correlated systems

In order to calculate various microscopic properties of solids, independent electron theory and density functional theory are used. However, for strongly correlated systems, these methods seem to be inaccurate [11]. Systems of correlated electrons are described using the Hamiltonian model. The simplest model of correlated electron is the Hubbard Hamiltonian that describes the electron with spin directions moving between localized states at lattice sites. But the calculation of simple property such as phase diagram of a strongly correlated system is challenging in the case of Hamiltonian model. Later, the Hubbard model was mapped on to a self-consistent quantum impurity model, which forms the basis of the dynamical mean field theory (DMFT) of correlated electrons [12].

13.2 Multiferroics as a Strongly Correlated System

H. Schmid introduced the term multiferroics in 1994 [13]. In 1894, Pierre Curie predicted the magnetoelectric effect in non-moving crystals based on symmetry considerations [14]; but Dzyaloshinskii theoretically confirmed this effect in 1959 [15] and Astrov experimentally demonstrated the effect [16]. Magnetoelectric multiferroic materials belong to a newer class of multiferroic materials in which ferro/antiferromagnetic and ferroelectric properties occur simultaneously. In the thermodynamical point of view, the magnetoelectric coupling occurs in materials where two or more phases (magnetic, electric, and elastic phases) coexist. The strong magnetoelectric (ME) coupling with unique functionalities and superior performance characteristics attract the researchers toward magnetoelectric multiferroic materials and devices. The strong ME coupling offers great opportunities in applications of spintronics devices, sensors, energy harvesters, etc. [17].

13.2.1 General introduction to multiferroics

However, magnetism and electricity were combined into one common discipline, but electric and magnetic ordering in solids are considered separately, i.e., the charge of ions and electrons constitute electrical properties, whereas the electron spin governs the magnetic properties. By the emergence of a new type of coupling, linear magnetoelectric effect leads to the discovery of new compounds and new class of symmetry groups. In solids, magnetoelectric phenomenon arises from either the development of spontaneous magnetization in the presence of an electric field or the generation of electric

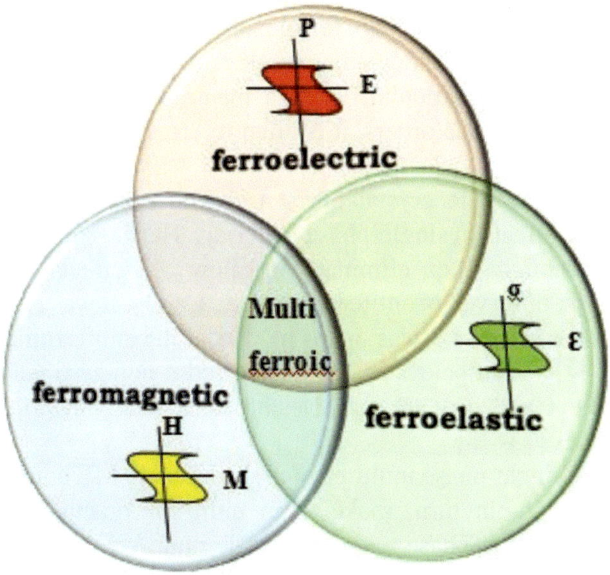

Figure 13.1 Illustration of different ferroic orders of a multiferroic system.

polarization in the presence of a magnetic field. Also, there may exist two types of ordering: the spontaneous ordering of orbital and spin magnetic moments called ferromagnetism as well as the spontaneous ordering of electric dipole moments called ferroelectricity. Apart from these two, there is a third type of ordering called ferroelasticity that arises due to spontaneous deformation. If any of these three types of ordering coexist in one material, they are called multiferroics (Figure 13.1) [18]. After boracites, the first known multiferroics, various other multiferroics were either found naturally or developed artificially.

Other than ferromagnetism, ferroelectricity, and ferroelasticity, there is a fourth type of ferroic order called ferrotoroidicity. $LiCoPO_4$ and $LiFeSi_2O_6$ are examples of such types of ferroic order, in which there is a spontaneous ordered arrangement of magnetic vortices.

13.2.2 Classification of multiferroics

Multiferroic materials are mainly classified into single-phase and composite multiferroics. Magnetic and electric orders coexist in single-phase multiferroics. But in composite multiferroics, magnetic and electric order phases within the composite are physically separated.

13.2.2.1 Single-phase multiferroics

The combination of at least two ferroic orders, such as ferromagnetism, ferroelectricity, and ferroelasticity, in a single compound is known as a single-phase multiferroic. In particular, when a material simultaneously exhibits both magnetic and electric orders, it is often referred to as a magnetoelectric multiferroic or a magnetoelectric.

The symmetry of the crystal plays a crucial role in the existence of magnetoelectric effects in single-phase materials. Hence, small modifications in the structure might alter, eliminate, or allow ME effects [22]. The ME effects were first observed in antiferromagnetic Cr_2O_3. It was experimentally observed that the value of α was 4.1 ps m^{-1} near the antiferromagnetic Neel temperature ($\sim$307 K), below which it acquired a non-zero value [23, 24]. $TbFeO_3$ and $Fe_xGa_{2-x}O_3$ are examples of single-phase materials that show weak ME effects [25–28]

Most of the single-phase multiferroics show feeble ME coupling at room temperature, which, in turn, make them unfit for practical applications. The only known single-phase magnetoelectric multiferroic with a large ME coupling coefficient (α) at room temperature is $BiFeO_3$ [29]. Other examples of single-phase multiferroics are CuO, Sm_2CuO_4, Gd_2CuO_4, $KNiPO_4$, lead−iron perovskites, aurivilus-phase ferrites, cobaltates, rare-earth manganites ($BiMnO_3$ and $YMnO_3$), hexaferrites, etc.

Single-phase multiferroics are indeed classified into type 1 and type 2 based on their different physical mechanisms of electric polarization. Type 1 multiferroics have distinct ferroelectric and magnetic ordering temperatures that are widely separated. This characteristic allows for the manipulation of spin-charge properties, providing opportunities for tuning and controlling their behavior. However, type 1 multiferroics are relatively rare in nature. This is because of the fact that conventional ferroelectrics typically develop a spontaneous polarization through the off-center displacement of transition metal ions with a d^0 electronic configuration. On the other hand, magnetism arises from ions with partially filled *d* orbitals. Consequently, it is unlikely for the same ions to exhibit both ferroelectric and magnetic ordering.

The "lone pair model," proposed by Hill and Seshadri, explains the multiferroic behavior observed in certain perovskite compounds [30]. One example is BiMnO3, which is an insulating ferromagnet. In $BiMnO_3$, the perovskite lattice is distorted due to the presence of lone pair electrons on the Bi^{3+} site. This structural deformation, caused by the lone pairs, leads to ferromagnetic ordering of the Mn^{3+} ions. Different mechanisms have been proposed for different types of multiferroics. Hexagonal $YMnO_3$ is another type 1 multiferroic compound where ferroelectricity arises from the tilting of the MnO_5 bipyramids [31]. BiFeO3 is a well-known example of a type 1 multiferroic, exhibiting both magnetic and ferroelectric order at room temperature. It displays ferroelectric order below 1100 K and undergoes antiferromagnetic ordering around 640 K. The hexagonal manganites $REMnO_3$ (where RE = Ho, Er, Tm, Yb, Lu, and Sc) belong to the class of type 1 multiferroics. They exhibit high bulk ferroelectric transition temperatures (>900 K) and undergo antiferromagnetic ordering at temperatures between 70 and 130 K [32].

In contrast, type 2 multiferroics have lower magnetic and ferroelectric transition temperatures, which limit their potential applications in device fabrication. Orthorhombic $REMnO_3$ is an example of a type 2 multiferroic, specifically for RE = Tb and Dy. In these compounds, the ferroelectric order emerges at an incommensurate to commensurate magnetic transition [33]. Another class of type 2 multiferroics are the $REMn_2O_5$ compounds, where RE represents rare-earth elements, Bi, and Y. These compounds exhibit multiferroic behavior but have relatively low magnetic and ferroelectric transition temperatures compared to type 1 multiferroics.

Occurrence of single-phase multiferroic materials in nature is rare, and also the artificial preparation of such materials is difficult. The known single-phase multiferroics are limited by several other aspects such as antiferromagnetic or weak ferromagnetic arrangements, weak coupling of order parameters, and low temperature properties.

13.2.2.2 Composite multiferroics

In these materials, the ferroic orders are physically separated. Composite multiferroics possess either solicited or new characteristics that are absent in single-phase materials. Mechanical force-mediated coupling in a composite of ferromagnetic and ferroelectric materials makes it multiferroic [34]. A composite material can be made magnetoelectric by combining the magnetostrictive effect with the piezoelectric effect. Introducing indirect coupling (strain-mediated coupling) [35] between two materials such as a ferromagnet and a ferroelectric can enhance magnetoelectric effects. Strain-mediated coupling needs close proximity of a piezomagnetic (or magnetostrictive) material and a piezoelectric (or electrostrictive) material. This can be achieved in the form of composites [35, 36], laminates [37–39], or epitaxial multilayers (MLs) [40]. In these composites, the ME coupling coefficient is dependent upon the elastic coupling occurring at the interface between piezoelectric and magnetostrictive phases. The coupling constant depends on the frequency of the applied AC magnetic field [41].

In the early 1970s, several attempts were made to synthesize bulk composites of piezoelectric barium titanate and magnetostrictive ferrites [42]. However, due to the presence of micro-cracks, defects, impurities, and leakage currents in the sintered composites, the bulk composites showed typically weak ME coupling, limiting their potential applications in the development of new devices [42–45].

The theoretical model proposed by Harshe, Dougherty, and Newnham for magnetoelectric (ME) heterostructures involves alternating layers of magnetostrictive and piezoelectric phases. The researchers were able to fabricate such structures and study their properties [46]. The use of layered structures, as opposed to bulk composites, offers advantages such as reduced leakage currents. In these structures, the piezoelectric layer can be electrically poled, leading to an enhanced magnetoelectric coupling [47]. In recent years, several research groups have reported strong magnetoelectric coupling in layered composites [48–59]. Improvements in magnetoelectric coupling have been achieved by incorporating various materials into laminar composites. For example, the use of Tb1−xDyxFe2 (Terfenol-D), FexGa1−x (Galfenol), Metglas, and ferrites in these structures has resulted in significant enhancements in the magnetoelectric coupling behavior. In addition to the heterostructures mentioned above, a notable example of a self-assembled nanostructured multiferroic involves CoFe2O4 nanopillars arranged in a hetero-epitaxial manner within a BaTiO3 matrix. This arrangement demonstrated exceptional magnetoelectric coupling properties [60].

13.2.2.3 Multiferroic perovskites

A perovskite is a simple primitive cubic lattice with the chemical formula ABO_3 in which A and B corresponds to two cations of different sizes and valencies (+2 and +4, respectively) and O is the oxygen atom (anion) that binds to both A and B. An ideal perovskite can crystallize in Pm3m space group. The soft mode theory explains the origin of ferroelectricity in multiferroics. According to the soft mode theory, the low-frequency polar phonon mode becomes softer and gradually tends to zero frequency at critical temperature (T_C). This results in the transformation of material from an ideal cubic structure to a ferroelectric structure with lower symmetry. This structural instability breaks the centrosymmetry and thereby results in the deformation of perovskite (Jahn−Teller effect) [61]. Long-range Coulomb interactions are lower than the A−O and B−O distance in both B-site-driven and A-site-driven ferroelectrics [62]. Quantum-mechanical exchange interactions govern the magnetism of multiferroics. Quantum-mechanical exchange interactions include double exchange interaction, which induces ferromagnetism and super-exchange interaction that either induces ferromagnetism or antiferromagnetism. Due to long-range Coulombic interactions, free electrons from perovskite materials with ferromagnetic ordering increase the conductivity and destroy the ferroelectricity. Thus, double exchange interaction in multiferroic perovskite is difficult to exist [63, 64]. For super-exchange interaction, a combination of ferroelectricity and magnetism is observed since the compounds are insulators. In case of magnetic perovskite oxides, partially filled *d* electrons of B site cation splits into two energy states of octahedral oxygen, i.e., three low energetic t_{2g} orbitals (d_{xy}, d_{yz}, and d_{zx}) and two highly energetic e_g orbitals (d_{x2-y2} and d_{3r2-z2}). It is clear that the magnetic order of materials is determined by the configuration of *d* electrons and ferromagnetism is produced only if the neighboring orbitals of cations with same spin are orthogonal [65].

Bismuth ferrite ($BiFeO_3$) [66], which is multiferroic perovskite, exhibits ferroelectric behavior that arises from the 6s electron pair of bismuth, and due to the partially filled *d* orbital of iron atoms, it shows a magnetic behavior. $LaMn_3Cr_4O_{12}$ [67] belongs to a class of magnetoelectric cubic perovskite that exhibits multiferroicity in crystallographic cubic symmetry, i.e., the symmetry group and magnetoelectric effect are linked with each other. Lu_2MnCoO_6 [68] is a double perovskite that shows both ferroelectricity (below 35 K) and magnetism (42 K) in which below 35 K, the net electric polarization and net magnetism are strongly coupled to each other.

13.2.3 Mechanisms involved in multiferroics

On the basis of mechanisms inducing the multiferroicity, materials can be classified into four types: lone pair mechanism, geometric effects, charge order, and spin-driven mechanisms. Ferroelectricity is due to electronic lone pairs, geometric effects, charge order, or magnetism. In these types, the magnetic and ferric orders occur independently and the materials are denoted as type 1 multiferroics. In spin-driven mechanism, the magnetic and ferroelectric transitions emerge jointly and multiferroic materials are termed as type 2 [69].

13.2.3.1 Lone pair mechanism

The host ion creates an anisotropic distribution of unbound valence electrons around it, which results in a spatial asymmetry. The presence of a lone pair of valence electrons results in a spatial asymmetry. This mechanism is the reason for room temperature ferroelectricity in $BiFeO_3$. The lone pair electrons present in the system have high polarizability and are responsible for the ferroelectric contribution. Below the Curie temperature T_C = 1103 K [70], a spontaneous polarization of ~100 μC cm^{-2} [71] is generated by the lone pairs (a lone pair of valence electrons in the 6s orbital of $BiFeO_3$). $BiFeO_3$ is the only room-temperature single-phase multiferroic material among the lone pair systems. Below the Neel temperature T_N = 643 K, the long-range periodic antiferromagnetic material structure is observed. $BiMnO_3$ and $PbVO_3$ also show this mechanism.

13.2.3.2 Geometric ferroelectricity

Geometric ferroelectricity is based on structural instabilities in materials due to space-filling effects and geometric constraints, which leads to the formation of a polar state. In materials like h-$RMnO_3$ (where R represents various rare-earth elements), ferroelectric order is observed at temperatures equal to or greater than 1200 K (T_C) with a polarization value of P_s = 5.6 μCcm^{-2}. This is followed by magnetic ordering at temperatures equal to or less than 120 K (T_N) driven by unit cell tripling. Similar to the above case, magnetoelectric coupling is observed in h-$LuFeO_3$ thin films despite of its larger magnetic moment and room temperature magnetic order. $BaNiF_4$, despite the small polarization value (~0.01 μC cm^{-2}) due to an asymmetry between Ba^{2+} and F^- sites, allows for coupling with a weak ferromagnetic moment. This enables the reversal of the electric polarization along with the ferromagnetic moment. $Ca_3Mn_2O_7$ exhibits ferroelectric polarization driven by the cooperation between two non-polar lattice modes in which this polarization interacts with the canted magnetic moment.

13.2.3.3 Charge ordering

Periodic super structure is formed by the non-uniformly distributed valence electrons around the host ions in the crystal lattice. Charge ordering is often observed in transition metal compounds, especially in those formally containing transition metal ions with a different valence. Inequivalency in sites and bonds after charge ordering can lead to ferroelectricity [72]. $LuFe_2O_4$ is an example where charge ordering plays a role in inducing ferroelectric behavior. It has been suggested that the Fe atoms in $LuFe_2O_4$ form a superlattice with an alternating sequence of Fe^{2+} and Fe^{3+} ions. Thus, an electric polarization, and hence ferroelectricity, is generated by this kind of charge ordering. The site-centered charge order appears on top due to the inequivalency in bonds because of the structure of the material [73]. It was the prime candidate for the charge-order-driven multiferroicity. But the occurrence of ferroelectricity is still unknown. Also there exists a situation in which dimerization occurs on the top of the ions with different charge. Ca_3CoMnO_6 [74] and $TbMn_2O_5$ [75] belong to this class. Today, charge-order-driven multiferroicity emerged as an interesting concept for the researchers.

13.2.3.4 Spin-driven mechanism

The ferroelectricity in multiferroics is because of a particular type of magnetism and exists only in a magnetically ordered state. Magnetic order can break inversion symmetry. A polar state is generated by the transfer of the non-centrosymmetry from the magnetic to the electric lattice due to the interaction of spins and charges. Electric polarization induces a magnetic ordering due to a shift from displacive ferroelectrics to magnetically induced so-called improper ferroelectric.

Three kinds of multiferroicity have been established [76–78], out of which the most intensely discussed mechanism is the inverse Dzyaloshinskii–Moriya (DM) interaction [79]. In the DM interaction, an antisymmetric magnetic interaction is promoted in a crystallographic environment in the case of non-centrosymmety, and in an inverse DM interaction, a non-centrosymmetric displacement of charges is driven by an acentric spin structure [80, 81].

Spin−orbit coupling plays a key role in both the DM and inverse DM interactions. The optimization of the spin configuration as a result of polarization from inverse DM interaction is determined by antisymmetric exchange, expressed in terms of antisymmetric product $\mathbf{S}i\ {*}\mathbf{S}j$ of neighboring spins Si and Sj. This interaction leads to a one-to-one correlation between (antiferro-)magnetic order and electric polarization. The multiferroicity based on the inverse DM interaction was initially discovered in materials such as Cr_2BeO_4, o-$TbMnO_3$, and $CaMn_7O_{12}$. Among these, $CaMn_7O_{12}$ exhibits the highest polarization achieved thus far with this mechanism, reaching a value of 0.3 μC cm^{-2} [82].

The optimization of the symmetric spin product Si *Sj can lead to an acentric displacement of charges, which is distinct from the Heisenberg-like exchange striction observed in the DM interaction [83]. In materials like $TbMn_2O_5$, ferroelectricity is generated by this type of charge displacement. |$\mathbf{S}i$ $\mathbf{S}j$| $>$ |$\mathbf{S}i$. $\mathbf{S}j$| represent that the relativistic antisymmetric mechanism is dominant over the non-relativistic symmetric mechanism. It is exhibited by systems like o-$TbMnO_3$. Under pressure, when a cycloidal magnetic order (parameterized by SiůS*j*) transforms into a collinear antiferromagnetic order, the polarization in these materials can increase with the order of magnitude [84]. The violation of magnetic order by inversion symmetry can occur in various ways, and the most effective arrangement for promoting ferroelectricity is through the existence of spin distributions that facilitate this coupling.

13.2.4 Magnetoelectric multiferroics: recent advances

The coupling between the ferroic orders in multiferroics facilitates the manipulation of one ferroic property with the change of the other. Different order parameter couplings include linear coupling (magnetoelectric coupling, piezoelectricity, and piezomagnetism) and non-linear coupling (electrostriction and magnetostriction). Different order parameters in a ferroic system can be transformed by applying requisite field, which is often associated with hysteresis [85].

The correlation between different phases (e.g., magnetic and electric) can be made useful in developing multifunctional materials. Moreover, multiferroicity opens up an avenue to explore the novel physics of strongly correlated electronic systems.

The magnetoelectric effect is defined as the dielectric polarization of a material in the presence of a magnetic field or as an induced magnetization in the presence of an external electric field. The coupling between magnetic and electric phases allows the modification of electric polarization and net magnetization, respectively, by the application of an external magnetic field and an external electric field. Mathematically, this effect is described in terms of the magnetoelectric coupling coefficient α [86]. When an electric field (*E*) is applied, there occurs a change in magnetic induction (*B*) results in an electrically induced magnetoelectric coupling coefficient

$$\alpha_{ij}E = \partial B_i/\partial E_j.$$

When a magnetic field (*H*) is applied, there occurs a change in electric polarization (*P*) results in a magnetically induced magnetoelectric coupling coefficient

$$\alpha_{ij}H = \partial P_i/\partial H.$$

Magnetoelectric coupling coefficient is represented in the form of a second rank tensor of magnetoelectric susceptibility α_{ij} with nine components

$$\alpha_{ij} = \begin{matrix} \alpha_{11} & \alpha_{12} & \alpha_{13} \\ \alpha_{21} & \alpha_{22} & \alpha_{23} \\ \alpha_{31} & \alpha_{32} & \alpha_{33}. \end{matrix}$$

Since the materials' negative permittivity and permeability do not exist, all the components of the magnetoelectric coupling coefficient tensor must be positive. But in vacuum, the permittivity and permeability are positive values. The voltage magnetoelectric coefficient is the main parameter used in analyzing experimental measurements and designing various applications based on multiferroics.

Multiferroic materials with magnetoelectric coupling find many real-world applications such as multiple state memory devices, magnetic field sensors, resonators, etc. [87–89]. Boracites ($Ni_3BB_7O_{13}I$, $Cr_3B_7BO_{13}Cl$, etc.) were probably the first known magnetoelectric multiferroics [90]. Boracites have the general formula $M_3B_7O_{13}X$, where M is a bivalent cation such as Mg^{2+}, Cr^{2+}, Mn^{2+}, Fe^{2+}, Co^{2+}, Ni^{2+}, Cu^{2+}, Zn^{2+}, etc., and X a monovalent anion like OH^-, F^-, Cl^-, Br^-, I^-, or NO^{3-}. Later on, several other multiferroics were either found in nature or synthesized artificially [91].

Recent research has been focused on studying the interaction and contraindication between transition metals with empty *d* orbitals (ferroelectricity) and those with partially filled *d* orbitals (magnetism) [92]. The ferroelectricity is mainly composed of two conventional mechanisms: d^0-ness and lone pair stereochemical activity. The d^0-ness effect is shown by Ti^{4+} ion in $BaTiO_3$. On the other hand, the lone pair effect is demonstrated by compounds like $PbTiO_3$ by Pb^{2+} ion. A similar effect is observed in compounds like GeTe, where the Ge^{2+} ion exhibits the lone pair effect. Similarly, the magnetism also shows two established routes. The first route involves materials with partially filled *f*-levels on rare-earth ions and the second route involves materials with partially filled *d*-levels in transition metals.

In the field of composite multiferroics, researchers have introduced new materials that combine non-magnetic ferroelectric materials, such as $BaTiO_3$, with non-ferroelectric magnets like $CoFe_2O_4$. Such composite materials find applications in memory devices, ME sensors, drug delivery, etc. [93]. One area of focus is the layer-by-layer growth of multiferroic superlattices. For example, the combination of ferroelectric and antiferromagnetic materials, such as $LuFeO_3$ and $LuFe_2O_4$, respectively, in a superlattice structure offers tremendous opportunities due to the availability of crystal structures [94]. The superlattice of $SrTiO_3/PbTiO_3$ with electric field controllable chiral polarization vertices is introduced in multiferroic materials. These structures exhibit magnetization coupled to vortices and have broad applications [95]. Nanocomposites with vertical alignment are also attracting attention. These materials have three-dimensional hetero-epitaxy, which provides enhanced coupling and coercive field. Researchers are exploring various chemistries in this area [96]. Synthesis of nanopillars of lateral dimensions less than 20 nm and the influence of dislocations and antiphase boundaries in sites of magnetic disorder are of great interest. As the next step of research in the field of multiferroics, nano-patterned hybrid composites are of great interest. These composites consist of ferromagnetic metals combined with ferroelectric polymers. The goal is to achieve switching of electric polarization or magnetization by either a magnetic field or an electric field at room temperature. This research has the potential to enable new functionalities and applications in the field of multiferroics [97, 98].

13.2.5 Applications of magnetoelectric multiferroics

A variety of applications have been proposed in the field of magnetic/multiferroic heterostructures that bring novel devices in electric field-tunable radiofrequency (RF)/microwave signal processing, memory devices, magnetic field sensor, magnetoelectric random access memory (MERAM) magnetic sensors, voltage-tunable magnetoresistance, high-frequency inductors, and also in high-frequency signal processing devices [99]. The multiferroics along with the tunability of ferromagnetic resonance in ferromagnets provide the opportunity to realize voltage-tunable RF or microwave devices, such as filters, phase shifters, and antennas. This can be observed in $FeGaB/PbZnNbO_3–PbTiO_3$, as the strain induced by electric field changes magnetic anisotropy results in the switching of ferroelectric polarization when there is a correlation of change in resonance frequencies [100]. A significant development in ultralow power logic-memory devices has been achieved by combining two recent key discoveries: the inverse Rashba–Edelstein (spin Hall) effect (IREE) and multiferroic electric-field control of magnetism. It has made notable contributions in this area, leading to the realization of a magnetoelectric, spin–orbit (MESO) coupled logic device. The IREE component in MESO converts the spin to charge, whereas charge to spin is converted by the multiferroics. Enhancing the output of inverse Rashba effect by two to three orders of magnitudes is one of the challenges faced by the spintronics. Magnetoelectric thin films have gained significant research attention due to their promising applications in electric-write/magnetic-read memories and spintronics.

In magnetic memory devices (MRAM), when the magnetic field is applied, the magnetic states switch between +M and –M and data are written. Thus, the variation of magnetoresistance in the magnetic field helps to read the data. Due to high coercivity that results in large magnetic field for switching magnetic states, the materials in MRAMs (hard ferromagnets) consume large energy. In contradiction, energy efficient and faster writing by polarization switching ferroelectric memories (FeRAMs) is introduced. Since there is a size limitation, they exhibit slow reading speeds because the read operation is destructive and also due to the subsequent reset [101]. Therefore, a memory device with ferroelectric-write and magnetic-read operations (best functions of both FeRAMs and MRAMs) enhance the writing speed as well as the decrease in consumption of energy by device miniaturization. Thus, electrically written magnetically read memory devices originate. By integrating the best functions of both FeRAMs and MRAMs, such devices aim to store binary data effectively and will provide an opportunity toward the fabrication of the next-generation memory devices. By exploiting the coupling between electrical ($\pm P$) and magnetic-order ($\pm M$) parameters in these materials, various functions can be explored, such as the magnetic or electrical control of polarization or magnetization, non-destructive magnetic reading operations, and the design of multistate memories with electrical writing capabilities. High-density and efficient non-volatile memory elements and selection of material with respect to device fabrication are relevant in design and construction of memory devices [102].

Multiferroic materials that combine the features of both ferrite and ferroelectric are promising candidates in the field of tunable microwave and millimeter-wave devices. These devices leverage the unique characteristics of multiferroics, either through the converse magnetoelectric effect or the excitations of hybrid spin-electromagnetic waves in ferrite-ferroelectric layered structures, i.e., by applying an electric field to the ferroelectric layer, electrical tuning can be achieved, whereas the application of a magnetic field enables magnetic tuning.

Single-phase magnetoelectric materials found applications in memory devices and spintronics. However, these materials exhibit magnetoelectric coupling high above room temperature due to low Curie temperature in these materials. The emergence of stress-mediated electromagnetically coupled multiphase magnetoelectric composites of both ferromagnetic and ferroelectric components find a way around this difficulty. Metglas alloys and piezo-fibers with one-dimensional electrodes with high magnetoelectric and large sensitivity to magnetic field that correspond to back–forth mechanism of magnetoelectric lamina are the milestones in the advancement of magnetic sensor applications. The output ME voltage signal in response to an applied magnetic field governs the sensing potential of the magnetoelectric sensor. In the absence of an incident field, intrinsic or extrinsic noises are generated.

The magnetoelectric multiferroics also find applications in drug delivery mechanisms. On application of an electric field, the properties of cell membrane can be modified since it is electrically polarizable, and, therefore, it can exploit the porosity of the membrane (electroporosity) [103]. A lower electric field is used to open the pores in healthy cells compared to cancer cells. Based on this property, the drugs are transferred to cancer cells via the membrane. The magnetoelectric nanoparticles made up of nanocomposites of $CoFe_2O_4/BaTiO_3$ along with a drug in core–shell arrangement helps to reduce the control of release of drugs. For this, electrical properties of particles are controlled by a magnetic field kept at a certain distance. Thus, localized electric field produced by the MENs helps to open the pores of cancer cell membrane and put them in while drug is released inside the target [104].

13.3 Conclusion

Multiferroic materials invite attention for fundamental research and technological applications because of the combination of magnetic, polar, and piezo-elastic properties. Multiferroics mechanisms involved in multiferroic property, classification of multiferroics, magnetoelectric coupling and magnetoelectric coupling coefficient of different types of multiferroics are discussed in this chapter. While progress has been made in the field of multiferroics, there are still significant challenges to overcome. Improving the properties of single-phase multiferroics, such as increasing the magnitude of the ferroelectric and magnetic order parameters, remains a focus of research. Additionally, developing commercially viable multiferroic materials and processing techniques is necessary for practical applications. Ongoing research efforts are aimed at improving the properties of multiferroic materials, discovering new materials with enhanced characteristics, and developing scalable fabrication methods. These advancements will pave the way for the practical implementation of multiferroics in a wide range of technologies and applications.

References

[1] Charles Kittel, "Introduction to Solid State Physics", 7th Ed - 2015

[2] J. M. Ziman, "Principles of the Theory of Solids"

[3] M Ali Omar, "Elementary solid state physics"

[4] N.F.Mott, "Metal-Insulator Transition", Rev.Mod.Phy. 40,677(1968)

[5] Martens, K., I. P. Radu, S. Mertens, X. Shi, L. Nyns, S. Cosemans, P. Favia, et al. "The VO_2 Interface, the Metal-insulator Transition Tunnel Junction, and the Metal-insulator Transition Switch On-Off Resistance." Journal of Applied Physics 112, 12 (2012)

[6] Charles M. Steiner, "Heavy Fermion Superconductivity", Phys. 569, 2011

[7] Mengshu Liu, "Properties of Heavy Fermion Systems", 2010

[8] G Lawes and G Srinivasan, 'Introduction to magnetoelectric coupling and multiferroic films' J. Phys. D: Appl. Phys. 44 (2011)

[9] M. M. Vopson, ID, Y. K. Fetisov, G. Caruntu and G. Srinivasan,"Measurement Techniques of the Magneto-Electric Coupling in Multiferroics", Materials 2017, 10, 963

[10] Lawes G, T. Kimura, C.M. Varma,Mas Subramanian, "Magnetodielectric effects at magnetic ordering transitions" Prog. Solid State Chem. 2009, 37,40-54

[11] R. O. Jones and O. Gunnarsson, 'The density functional formalism, its applications and prospects', Rev. Mod. Phys. 61, 689 1989

[12] Gabriel Kotliar and Dieter Vollhardt, "Strongly Correlated Materials Insight From Dynamic Mean Field Theory", Physics Today, March 2004

[13] H. Schmid, "Multi-ferroicmagnetoelectrics, Ferroelectrics" 162 (1994) 317–338

[14] P. Martins, S. Lanceros-Mendez, "Polymer-based magnetoelectric materials", Adv. Funct. Mater. 23 (2013) 3371–3385

[15] I.E. Dzyaloshinskii, "On the magneto-electrical effect in antiferromagnets", Sov. Phys. JETPUSSR 10 (1960) 628–629

[16] D.N. Astrov, "The magnetoelectric effect in antiferromagnetics", Sov. Phys. JETP 11 (1960) 708–709

[17] L.E. Fuentes-Cobas, D. Chateigner, 'Advances in Magnetoelectric Materials and Their Application', Handbook of Magnetic Materials, 2015

[18] Daniel Khomskii, 'Classifying multiferroics: Mechanisms and effects', Physics 2, 20 (2009)

[19] Anne S. Zimmermann, Dennis Meier & Manfred Fiebig, "Ferroic nature of magnetic toroidal order" Nature Communications 5, 4796 (2014)
[20] Cheong, S.-W. &Mostovoy,"M. Multiferroics: a magnetic twist for ferroelectricity". Nat. Mater. 6, 13–20 (2007).
[21] Van Aken, B. B., Rivera, J.-P., Schmid, H. &Fiebig, M., "Observation of Ferrictorrodic Order", Nature 449, 702–705 (2007)
[22] A. Smolenskii And I. E. Chupis, "Ferroelectromagnets", Sov. Phys. Usp. 25, 475 (1982).
[23] AstrovDN.."Magnetoelectric effect in chromium oxide". Soviet Phys. JETP 13:729–33(1961)
[24] Folen, V. J., Rado, G. T. &Stalder, E. W. 'Anisotropy of the magnetoelectric effect in Cr_2O_3', Phys. Rev. Lett. 6, 607—608 (1961)
[25] Schmid H.Bellingham,WA: SPIE WS, Weiglhofer, A Lakhtakia"Magnetoelectric effects in insulating magnetic materials: Introduction to Complex Mediums for Optics and Electromagnetics, ed., pp.167–95(2003).
[26] Spaldin NA, FiebigM.."The renaissance of magnetoelectric multiferroics". Science 309:391–92 (2005)
[27] Fiebig M. 'Revival of the magnetoelectric effect'. J. Phys. D 38:R123–52 Annu. Rev. Mater. Res. 2010.40:153-178. (2005)
[28] Eerenstein W, Mathur ND, ScottJF.2006. 'Multiferroic and magnetoelectric materials' .Nature442?759– 65).
[29] Wang J, Neatson JB, Zheng H, Nagarajan V, Ogale SB, Epitaxial $BiFeO_3$ 'multiferroic thin film heterostructures'. Science 299:1719–22. (2003)
[30] Seshadri R and Hill N A 'Visualizing the role of Bi 6s 'lone pairs' in the off-center distortion in ferromagnetic $BiMnO_3$', Chem. Mater. 13, 2892, (2001)
[31] Van AkenB 'The origin of ferroelectricity in magnetoelectric $YMnO_3$' Nature Mater. 3 164–70) (2004)
[32] KatsufujiT 'Dielectric and magnetic anomalies and spin frustration in hexagonal $RMnO_3$' (R =Y, Yb, and Lu) Phys. Rev. B 64 104419(2001)
[33] Kimura T 'Magnetoelectric phase diagrams of orthorhombic $RMnO_3$ (R =Gd, Tb, and Dy)' Phys. Rev. B 71 224425],(2005)
[34] Van Den BoomgaardJ 'An in situ grown eutectic magnetoelectric composite material: I. Composition and unidirectional solidification' J. Mater. Sci. 9 1705–9](1974)

[35] Van Run, A. M. J. G., Terrell, D. R. &Scholing, J. H. 'An in situ grown eutectic magnetoelectric composite material'. J. Mater. Sci. 9, 1710—1714 (1974)
[36] Nan, C.-W. 'A three-phase magnetoelectric composite of piezoelectric ceramics, rare-earth iron alloys, and polymer'. Appl. Phys. Lett. 81, 3831—3833 (2002).)
[37] Ryu, J., Va t'squez Carazo, A., Uchino, K. & Kim, H.-E. "Magnetoelectric properties in piezoelectric and magnetostrictive laminate composites". Jpn. J. Appl. Phys. 40, 4948—4951 (2001).
[38] Cai, N., Nan, C.-W., Zhai, J. & Lin, Y. "Large high-frequency magnetoelectric response in laminated composites of piezoelectric ceramics, rare-earth iron alloys and polymer". Appl. Phys. Lett. 84, 3516—3519 (2004).
[39] Srinivasan, G." Magnetoelectric bilayer and multilayer structures of magnetostrictive and piezoelectric oxides". Phys. Rev. B 65, 134402 (2002)
[40] Lee, M. K. "Strain modification of epitaxial perovskite oxide thin films using structural transitions of ferroelectric BaTiO3 substrate". Appl. Phys. Lett. 77, 3547—3549 (2000)
[41] Bichurin, M. I. "Resonance magnetoelectric effects in layered magnetostrictive-piezoelectric composites". Phys. Rev. B 68, 132408 (2003).
[42] Agrawal S, " Magnetoelectric properties of microwave sintered particulate composites", Mater. Lett. 63 2198–200 (2009)
[43] Petrov V M,"Magnetoelectric effects in porous ferromagnetic–piezoelectric bulk composites: experiment and theory", Phys. Rev. B 75 174422, 2007
[44] Ryu J, "Piezoelectric and magnetoelectric properties of lead zirconatetitanate/Ni-ferrite particulate composites" J. Electroceram. 7 17–24(2001).
[45] Van Den Boomgaard J 'An in situ grown eutectic magnetoelectric composite material: I. Composition and unidirectional solidification' J. Mater. Sci. 9 1705–9(1974)
[46] Harshe G, Dougherty J P and Newnham R E " Theoretical modelling of multilayer magnetoelectric composites" Int. J. Appl.Electromagn. Mater. 4 145–59,(1993)
[47] Bichurin M I, Petrov V M and Srinivasan G "Theory of low-frequency magnetoelectric coupling in magnetostrictive-piezoelectric bilayers" Phys. Rev. B 68, 54402, (2003)

[48] Ryu, J., Va t'squez Carazo, A., Uchino, K. & Kim, H.-E. "Magnetoelectric properties in piezoelectric and magnetostrictive laminate composites."Jpn. J. Appl. Phys. 40, 4948—4951 (2001).
[49] Cai N et al Dielectric, ferroelectric, magnetic, and magnetoelectric properties of multiferroic laminated composites Phys. Rev. B 68 224103,(2003)
[50] Chang C-M and Carman G P, "Experimental evidence of end effects in magneto-electric laminate composites" J. Appl. Phys. 102 12490118J. Phys. D: Appl. Phys. 44 (2011) 243001 Topical Review, (2007)
[51] Chen W C, "Coupled magnetic and ferroelectric domains in multiferroic $Ni_3V_2O_8$", Phys.Rev. Lett. 103 087201, (2009)
[52] Das J, "Enhancement in the field sensitivity of magnetoelectric laminate heterostructures", Appl. Phys. Lett. 95 092501, (2009)
[53] Dong S, Li J F and ViehlandD Giant, "Magnetoelectric effect in laminate composites" Phil. Mag. Lett. 86 102901, (2003)
[54] Dong S, "Extremely low frequency response of magnetoelectric multilayer composites Appl. Phys. Lett. 86 1–3,(2005)
[55] Islam R A, "Giant magnetoelectric effect in sintered multilayered composite structures" J. Appl. Phys. 104 044103, (2008)
[56] Liu M, "Strong magnetoelectric coupling in ferrite/ferroelectric Multiferroic heterostructures derived by low temperature spin-spray deposition", J. Phys. D: Appl. Phys. 42 045007, (2009)
[57] Mathe V L, Srinivasan G and BalbashovA M, "Magnetoelectric effects in bilayers of lead zirconatetitanate and single crystal hexaferrites" Appl. Phys. Lett. 92 122505,(2008)
[58] Mori K and Wuttig M, "Magnetoelectric coupling in terfenol-D/polyvinylidenedifluoride composites" Appl. Phys. Lett. 81 100–1, (2002)
[59] Srinivasan G, "Magnetoelectric bilayer and multilayer structures of magnetostrictive and piezoelectric oxides" Phys. Rev. B 64 214408, (2001)
[60] Zheng H, "Multiferroic $BaTiO_3$–$CoFe_2O_4$ nanostructures", Science 303 661–3, (2004)
[61] P. S. Halasyamani, K. R.Poeppelmeier, Noncentrosymmetric Oxides [J]. Chemistry of Materials, 10 (10), 2753–2769. (1998)
[62] R. E. Cohen, "Origin of ferroelectricity in perovskite oxides" [J]. Nature, 358(6382), 136–138 (1992).

[63] C. Zener, "Interaction between the d-Shells in the Transition Metals. II. Ferromagnetic Compounds of Manganese with Perovskite Structure" [J]. Phys. Rev., 82 (3), 403–405. (1951)

[64] P. W. Anderson," New Approach to the Theory of Super exchange Interactions", [J]. Phys. Rev., 115 (1), 2 (1959).

[65] Y. Tokura, N. Nagaosa, "Orbital Physics in Transition-Metal Oxides", [J]. Science, 288(5465),462–468. (2000)

[66] Petit, S., "Viewpoint: Subtle Atomic Shifts", Physics 2013, 6, 93

[67] Wang, X.; Chai, Y.; Zhou, L.; Cao, H.; Cruz, C.-D.; Yang, J.; Dai, J.; Yin, Y.; Yuan, Z.; Zhang, S. "Observation of Magnetoelectric Multiferroicity in a Cubic Perovskite System: LaMn3Cr4O12" . Phys. Rev. Lett. 2015, 115, 087601)

[68] Yaťn Ÿez-Vilar, S.; Mun, E. D.; Zapf, V. S.; Ueland, B. G.; Gardner, J. S.; Thompson, J. D.; Singleton, J.; Sat'nchez-Andut'jar, M.; Mira, J.; Biskup, N. "Multiferroic Behavior in the Double-perovskite Lu2MnCoO6". Phys. Rev. B 2011, 84.

[69] Manfred Fiebig, Thomas Lottermoser, Dennis Meier and Morgan Trassin, 'The evolution of multiferroics', nature reviews materials 1, 16046, (2016)

[70] Wang, J. 'Epitaxial $BiFeO_3$ multiferroic thin film heterostructures', Science 299, 1719–1722 (2003).

[71] Fedulov, S. A. 'Determination of Curie temperature for $BiFeO_3$ ferroelectric'. Dokl. Akad. Nauk SSSR 139, 1345 (in russian) (1961); English translation available in Sov. Phys. -Dokl. 6, 729 (1962)

[72] D. V. Efremov, J. van den Brink, and D. I. Khomskii,"Bond- versus site-centred ordering and possible ferroelectricity in manganites", Nature Mater. 3, 853 (2004).

[73] N. Ikeda, "Ferroelectricity from iron valence ordering in the charge-frustrated system LuFe2O4". J. Phys. Soc. Japan. 69, 1526 (2000); Nature 436, 1136 (2005).

[74] Y. J. Choi, "Ferroelectricity in an Ising Chain Magnet" Phys. Rev. Lett. 100, 047601 (2008)

[75] N. Hur, "Multiferroics: a magnetic twist for ferroelectricity", Nature 429, 392 (2004).

[76] Tokura, Y., Seki, S. &Nagaosa, N., "Multiferroics of spin origin". Rep. Progress Phys. 77, 076501 (2014).

[77] Picozzi, S. &Stroppa, A, " Advances in ab initio theoryof multiferroics". Eur. Phys. J. B 85, 240 (2012).

[78] Kimura, T. "Spiral magnets as magnetoelectrics". Annu. Rev. Mater. Res. 37, 387–413 (2007).

[79] Dzyaloshinskii, I. E. 'A thermodynamic theory of "weak" ferromagnetism of antiferromagnetics', J. Phys. Chem.Solids 4, 241–255 (1958)

[80] Katsura, H.,Nagaosa, N. &Balatsky, A. V. "Spin current and magnetoelectric effect in noncollinear magnets", Phys. Rev. Lett. 95, 057205 (2005).

[81] Mostovoy, M, "Ferroelectricity in spiral magnets". Phys.Rev. Lett. 96, 067601 (2006).

[82] Johnson, R. D. 'Giant improper ferroelectricity in the ferroaxial magnet $CaMn_7O_{12'}$, Phys. Rev. Lett. 108, 067201 (2012)

[83] Sergienko, I. A. &Dagotto, E. 'Role of the Dzyaloshinskii–Moriya interaction in Multiferroic perovskites'. Phys. Rev. B 73, 094434 (2006)

[84] Aoyama, T,"Giant spin-driven ferroelectric polarization in TbMnO3 under high pressure". Nat. Commun. 5, 4927 (2014).

[85] LiLianaMitoseriu, 'Magnetoelectric phenomena in single-phase and composite systems', Bol. soc. esp. Ceram. V., 44 [3] 177-184 (2005)

[86] Fiebig M 'Revival of the magnetoelectric effect'. J. Phys. D 38:R123–52 Annu. Rev. Mater. Res. 2010.40:153-178(2005)

[87] C. W. Nan, M. I. Bichurin, S. Dong, D. Viehland and G. Srinivasan, 'Multiferroic magnetoelectric composites: historical perspective, status, and future directions', J Appl. Phys., 103 (3), 031101 (2008).

[88] J. Zhai, S. Dong, Z. Xing, J. Li and D. Viehland, 'Geomagnetic sensor based on giant magnetoelectric effect', Appl. Phys. Lett., 91, 123513 (2007).

[89] Y. K. Fetisov and G. Srinivasan, 'Electric field tuning characteristics of a ferrite-piezoelectric microwave resonator', Appl. Phys. Lett., 88 (14), 143503 (2006)

[90] E. Asher, H. Rieder, H. Schmid, and H. Stossel,"Roomtemperature ferroelectric and magnetic investigations and detailed phase analysis of Aurivillius phase $Bi_5Ti_3Fe_{0.7}Co_{0.3}O_{15}$ thin films", J. Appl. Phys. 37, 1404 (1966)

[91] G. A. Smolenskii and I. E. Chupis, "Structural and Electronic Properties of Bixo3 (X = Mn, Fe, Cr)" Sov. Phys. Usp. 25, 475 (1982).

[92] Hill N. A. "Why are there so few magnetic ferroelectrics?"J. Phys. Chem. B 104, 6694–6709 (2000).

[93] Vaz, C. "Magnetoelectric coupling effects in multiferroic complex oxide composite structures". Adv. Mater. 22, 2900–2918 (2010).
[94] Mundy, J, "Atomically engineered ferroic layers yield a room temperature magnetoelectric multiferroic". Nature 537, 523–527 (2016).
[95] Yadav, A. K." Observation of polar vortices in oxide superlattices" Nature 530, 198–201 (2016)
[96] Zheng, H. " Multiferroic BaTiO3-CoFe2O4 Nanostructures" Science 303, 661–663 (2004).
[97] Cai, R "Multiferroic nanopatterned hybrid material with roomtemperaturemagnetic switching of the electric polarization." Adv. Mater. 29,1604604 (2017).
[98] Poddar, S. "Room-temperature magnetic switching of theelectric polarization in ferroelectric nanopillars", ACS Nano 12,576–584 (2018)
[99] HaribabuPalneedi, VenkateswarluAnnapureddy, Shashank Priya and Jungho Ryu Review Status, "Perspectives of Multiferroic Magnetoelectric Composite Materials and Applications" Actuators, 5 (1), [9] (2016).
[100] N. A. Spaldin and R. Ramesh, "Advances in magnetoelectric multiferroics", Nature Materials VOL 18, 203–212, (2019)
[101] Jiang, J. Bai ZL, Chen ZH, He L, Zhang DW, Zhang QH, Shi JA, Park MH, Scott JF, Hwang CS, Jiang, "ATemporary formation of highly conducting domain walls for non-destructive read-out of ferroelectric domain-wall resistance switching memories", Nat. Mater. 17, 49–56 (2018).
[102] Mirza Bichurin, Vladimir Petrov, Shashank Priya and Amar Bhalla, "Multiferroic Magnetoelectric Composites and Their Applications", Advances in Condensed Matter Physics, (2012)
[103] Cahill, K. Molecular electroporation and the transduction of oligoarginines. Phys. Biol. 7, 016001 (2010).
[104] Guduru, R.; Liang, P.; Runowicz, C.; Nair, M.; Atluri, V.; Khizroev, S. "Magneto-ElectricNanoparticles to Enable Field-Controlled High-Specificity Drug Delivery to Eradicate Ovarian Cancer Cells." Sci. Rep., 3, 2953, (2013)

14

Strongly Correlated System of Materials: Modeling Challenges and Exciting Phenomena

Ravi Trivedi[1,2] and Brahmananda Chakroborty[3,4]

[1]Department of Physics, Karpagam Academy of Higher Education, Coimbatore 641021, Tamil Nadu, India
[2]Centre for Computational Physics, Karpagam Academy of Higher Education, Coimbatore 641021, Tamil Nadu, India
[3]High Pressure & Synchroton Radiation Physics Division, Bhabha Atomic Research Centre, Trombay, Mumbai-400085, India
[4]Homi Bhabha National Institute, Mumbai

Abstract

Many materials that have strong electron-electron interactions are of great interest in research. This interest is particularly due to their potential applications in spintronics, as seen in materials like dilute magnetic semiconductors (DMS) and colossal magnetoresistive (CMR) manganites. In strongly correlated (SC) materials, the behavior of each electron is heavily influenced by the behavior of other electrons in the system, which interacts with each other through the Coulomb force. This is a complex problem involving many interacting particles, which makes it very different from the behavior of a uniform electron gas. Among strongly correlated (SC) materials, transition metal oxides (TMOs) are particularly interesting due to their tightly bound d electrons, which are important in Mott insulators. Magnons, which are quantized low-energy magnetic excitations, play a crucial role in the behavior of high-temperature superconductors. The complex phase diagram of manganites is not fully understood because of the interactions and competition between spin, charge, and orbital degrees of freedom. Modern electronic

DOI: 10.1201/9781003337102-14

structure theory is key to determining the chemical and physical properties of many materials in theoretical chemistry and materials science. However, our understanding of the geometry and behavior of electronic systems, particularly in strongly correlated systems, where the interaction between d and f orbital electrons is strong, is still incomplete. For systems where electron-electron repulsion is relatively weak compared to kinetic energy, we have a good understanding of electronic behavior based on well-developed methods like the local density approximation, Hartree–Fock approximation, and Thomas–Fermi screening in metals. But when the interactions become stronger, as in d- and f-electron systems, the effectiveness of these theoretical methods decreases significantly. In this chapter, we provide an overview of realistic modeling approaches for strongly correlated systems, explaining the key concepts and highlighting some of the fascinating phenomena observed in these systems, supported by examples.

Keywords: Strongly Correlated materials, DMS, RDMFT, DMFT, Kondo effect, LSDA+U method

14.1 What are the Strongly Correlated Systems and its Importance?

Strongly correlated materials and strong correlation impacts have, as of now, pulled a lot of consideration in both key and applied explorations in light of their universal presence in numerous significant fields [1, 2]. Many wilderness subjects in dense matter physical science, such as high-T_c superconductivity seen in layere materials like CuO_2 and iron pnictides, enormous magneto-obstruction impacts found in MnO_2, and hefty fermion marvels seen in lanthanide or actinide inter-metallic materials, to give some examples, are totally identified with solid relationship impacts of d- and f-electrons. The wrong treatment of static/solid connection is one of the most difficult issues in DFT [3–5]. Density functional theory is quite possibly the most mainstream and effective quantum mechanical way to deal with the matter. In quantum mechanics, we discover that all data we can have about a given framework is contained in the system wave function. In this theory, we will solely be concerned with the electronic structure of different materials. The atomic degrees of freedom show up just as potential; so the wave functional just relies on the electronic directions. In 1964, Hohenberg and Kohn found that the minimum energy state is the absolute energy of an arrangement of interacting electrons, which is a remarkable utilitarian of the electron density.

By definition, a function returns a number when given a number. In the ground-state DFT, we can describe the N electron system by the Hamiltonian

$$\begin{aligned} \hat{H} &= \hat{T} + \hat{V} + \hat{V}_{ee} \\ &= -\sum_{i=1}^{N} \frac{\nabla_i^2}{2} + \sum_{i=1}^{N} v(r_i) + \frac{1}{2} \sum_{i=1}^{N} \sum_{\substack{j=1 \\ i \neq j}}^{N} \frac{1}{|r_i - r_j|} \end{aligned}$$

with the kinetic, potential, and interaction energy operators. The idea of SC (or strongly correlated systems) can have totally different implications in various fields of hypothetical examination. In dense matter material science, the standard beginning stage for first-standard estimations is LDA. The most straightforward estimation is to accept that the density can be dealt with locally as a uniform electron gas; the trade relationship energy at each point in the framework is equivalent to that of a uniform electron gas of a similar density. This estimate was initially presented by Kohn and Sham and holds for a gradually changing density. Utilizing this guess, the trade connection energy for a density is given by $\rho(r)$. The critical thought of KS DFT is a definite planning [6] between the physical, interfacing, many-electron framework and a model arrangement of non-collaborating fermions with a similar density. Just one term, the purported exchange connection (X_c) energy useful (containing all the confounded many-body effects), should be approximated so countless approximations have been created over the most recent 20 years, both by scientific experts and physicists, frequently focusing on different frameworks, different properties, and different wonders. If in any system where the interaction of electron–electron dominates over kinetic energy and giving very effective spatial correlations, we mean its strong electronic correlation. To explore the physics in this type of system, we always need many Slater determinants by using a very small occupation number. Researchers have synthesized and discovered a large number of new materials with a lot of properties in which high-temperature superconductors are most promising [7, 8] discovered by Berdnoz and Müller. On the other hand, we also have new materials with large change in magnetoresistivity having industrial applications [9]. A very simple characteristic of these materials is that they have large susceptibility due to a minute change in external perturbation, which corresponds to a physical quantity. Now this characteristic belongs to a phase transition like metal-to-insulator transition in these materials. This class of materials is known as strongly correlated, which might be defined as those systems for which traditional speculations are dependent on the one-electron scenario; for example, the LDA approximation is not able to explain due to insufficiency in depicting electronic construction. The valence conditions of these materials are portrayed by a bunch of partially filled tight groups, generally starting from 3d orbitals of progress metals or 4f orbitals of lanthanides implanted in moderately wide groups. A significant outcome of these thin groups is that the Coulomb repulsion among the electrons involving these groups gets tantamount to or bigger than the active energy. At the end of the day, the proportion between the Coulomb aversion and the data transfer capacity turns out to be significantly bigger than 1. Hence, an appropriate treatment of the Coulomb association is important to depict the electronic design. Materials with solid electronic relationships are materials in which the development of one electron relies upon the place and developments of any remaining electrons because of the long-range Coulomb interactions. In profoundly corresponded electron frameworks, basic approximations separate and completely new marvels and functionalities can show up.

14.2 Difficulties in Modeling Strongly Correlated Systems

Assessment of the intermittent table uncovers two sorts of valence electrons or restricted electrons; the previous structure expansive groups though the last limited groups. It might be recognized with electrons beginning from the s nuclear orbitals, which is found in direct metals like aluminum and sodium or in semiconductors like Si/GaAs. The subsequent sort might be related with electrons beginning from 3d (TM) or 4f orbitals (lanthanides). Materials having s or p valence electron characters, the one-molecule LDA functions admirably, albeit with a few inadequacies; for example, very few band holes are notable [10]. In materials where d or f electrons play role as a valence electrons, the LDA approximation is not good. This disappointment can somewhat be followed back to the solid orbital polarization related with electron restriction as found in 3d and 4f components [11]. In pure transition metals like iron and nickel, the 3d electrons are fairly mobile and can be accurately described using the Local Density Approximation (LDA). However, in insulating transition metal oxide compounds, the 3d electrons become localized due to strong on-site interactions [12].

14.3 Different Approaches to Describe Strongly Correlated Systems

To tackle this issue, we ought to unavoidably go beyond the local-density estimation, which utilizes a scientific articulation acquired by the electron gas in which the density of states of the electron gas is supplanted by the nearby electron density of the whole system. LDA is a long way from being awesome and there are numerous instances of supposed unequivocally corresponded system where the LDA bombs in depicting the energized and the minimum energy state properties [13]. In SC materials, the state of every electron firmly relies on the state of different electrons of the framework, corresponding to one another through the Coulomb repulsion. Hence, it is a many-particle issue, and the circumstance is totally different from the conduct of a homogeneous electron gas. Now, we do not have a decent broad methodology for seeing "strongly correlated systems." We can explain several systems; for example, certain quantum Hall frameworks or some magnetic materials. Be that as it may, we see inadequately the vast majority of the others. For instance, the idea of high-temperature superconductors stays baffling, despite numerous long periods of concentrated work. In particular, we do not have a brought-together view on the frameworks for which associations are significant. Ab-initio description of strongly correlated systems can be understood by utilizing four general methodologies, in particular density functionals, LDA+U, dynamical mean field theory (DMFT), reduced density matrix functional theory (RDMFT), self-interaction corrected (SIC), etc.

14.3.1 Reduced density matrix functional theory (RDMFT)

RDMFT is an elective method to overcome the many-body system problems. It depends on Gilbert's hypothesis [14], which ensures that the assumption estimation of any system in the lowest energy state is an interesting utilitarian of the one-body reduced density matrix (1-RDM). RDMF is applied to open-shell frameworks. It presents a spin-confined detailing by suitably communicating inexact relationship energy functionals as far as spin subordinate occupation numbers and spin-free normal orbitals. The RDMF theory has showed up as an elective way to deal with many-body frameworks. It has illustrated the extraordinary potential for developing DFT outputs for any finite system [15] or extended framework [16, 17]. The 1-RDM uses γ, as the variable [18]. This quantity is determined using the following equation:

$$\chi_N\left(\mathbf{r},\mathbf{r}'\right) = \mathbf{N}\int d^3\mathbf{r}_2\ldots d^3\mathbf{r}_\mathrm{N}\psi^*\left(\mathbf{r}'\mathbf{r}_2\ldots\mathbf{r}_\mathrm{N}\right)\Psi\left(\mathbf{r}'\mathbf{r}_2\ldots\mathbf{r}_\mathrm{D}\right)$$

where Ψ indicates the wave function for minimum energy structure. By comparing DFT, the merit of this system is that the exact much-body kinetic energy can be expressed in terms of γ. The prediction of an insulating state for SC materials without any magnetic moment is one of the main failures of the LDA approximation. Using the LDA approximation exchange-correlation functional, we get an incorrect analysis of metallic ground state, i.e., it is not able to predict the metallic nature for the SC materials for Mott insulators, where transition metal oxide (TMO) can be taken as the main example. People got a very small energy gap with spin-polarized calculations in NiO or MnO as compared to the experimental bandgap. However, all TMOs show a metallic character with spin-polarized calculations. It is worth mentioning to note that, experimentally, all these types of materials are insulating in nature at a Curie temperature above the Neel temperature. By using the RDMFT mechanism, researchers accurately predicted energy gaps, demonstrating insulating behavior for Mott insulators with no magnetic order [19]. It can be said that RDMFT techniques explore the fundamental physics and explain the minimum energy state rather than other simulation methods. But the overall success of RDMFT is hindered by the absence of a method for the measuring of spectral information. This happens because many Slater determinates are used to predict the ground state in RDMFT instead of single-particle Kohn-Sham-like equations. While we can calculate the energy gap by using the chemical potential, it gives a discontinuity at integer particle numbers [20]. There is no other spectral information on photo emission spectroscopy through computational techniques, which can be compared with experiments. However, recently, using this approach, the photo emission spectra of a set of TMOs are calculated and all results are matching with experimental data [21].

14.3.2 Dynamic mean field theory (DMFT)

Magnetic atoms having a strong interaction in d or f orbitals play an important role in the electronic structure and transport properties of these devices. The Kondo effect [22] is a very interesting effect due to the interaction in dynamical correlations. For example, due to Kondo resonances at the Fermi level, people observed Fano line shapes through scanning tunneling microscope (STM) experiments with magnetic atoms and molecular complexes [23–27]. At low temperatures, the effective coupling becomes AF (antiferromagnetic) between the conduction electrons and magnetic moments, which results in the formation of a singlet state with the total spin of the system being zero. Hence, the nature of the system becomes quite different instead of free magnetic moment due to the formation of this so-called Kondo singlet state.

Now due to this Kondo effect, the magnetic susceptibility does not go according to the Curie–Weiss law as a function of temperature. So the Kondo effect is very much interesting in the many-body case [28]. It comes into the picture whenever non-interacting conduction electrons are coupled with the local magnetic moment. We can observe this situation, i.e., the Kondo effect, when such a transition metal impurity (TM) is introduced in a metal host, where d- or f-shell of the TM gives an effective local magnetic moment. Similarly, if a molecule with an unpaired electron also joins with metallic leads, this interaction may happen. For example, if we analyze atomic states of the Co 3d shell, which is participating in the minimum energy state of the system, the main contribution (45%) of Co atom with eight electrons and a total spin of $S = 1$ (d8, $S = 1$), as displayed in Figure 14.1 (a). The fluctuation between the (d8, $S = 1$) and the (d7 $S = 3/2$) states are important firstly for the three spectral characteristics very near to E having a sharp Kondo-like peak right at E_F, as given in Figure 14.1 (b). Similarly, an orbital Kondo effect in

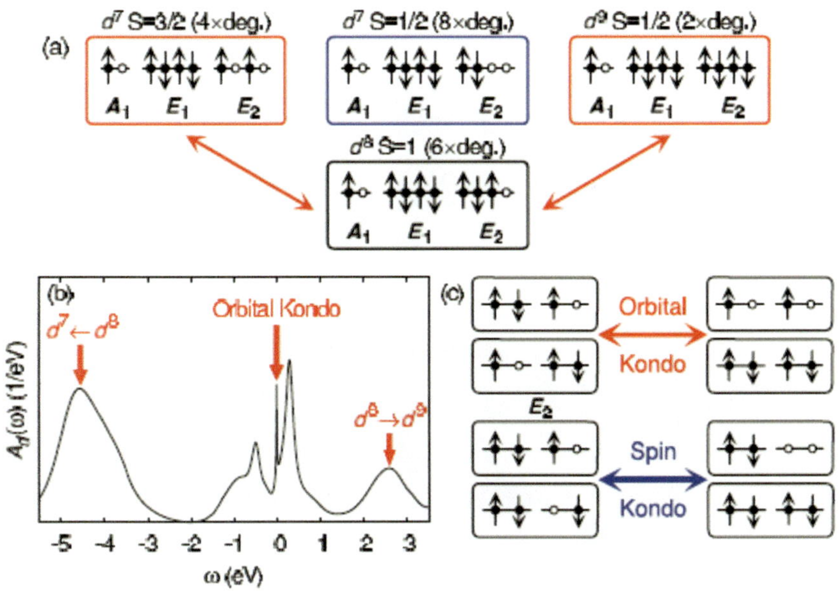

Figure 14.1 (a) Atomic states fluctuations. (b) Spectral function around E_F for $d = 3{:}6$. (c) Orbital and spin flip fluctuations lead to an orbital Kondo effect and to an under-screened spin Kondo effect, respectively. Arrows indicate spectral features arising from the fluctuations between atomic states [28].

the doubly degenerate *E* or *F* levels of the Co 3d orbital is shown in the upper panel of Figure 14.1 (c).

Figure 14.2 displays the Hubbard model at zero temperature and half filling, having the same number of electrons as lattice sites. Here, *U* and *W* are the interaction energy and bandwidth of non-interacting electrons, respectively. We can see two cases here: one case where electrons are completely independent, forming a half ellipse with the Fermi level *E* presented at the middle of the band, and the second case where the number of electrons can be written as quasiparticles whose DOS still accumulates the free electrons. Model Fermi liquid is for the narrowing of the peak. (c) In SC metals, three peaks can be seen in the spectrum with the Hubbard bands, due to local "atomic" excitations widened by the hopping of electrons and the quasiparticle peak close to the E_F. (d) Whenever the electronic interaction is strong, the Mott metal–insulator transition occurs due to the quasiparticle peak. The total numbers of electrons in that system are divided into two sets: weakly correlated electrons and strongly correlated high-frequency Hubbard bands.

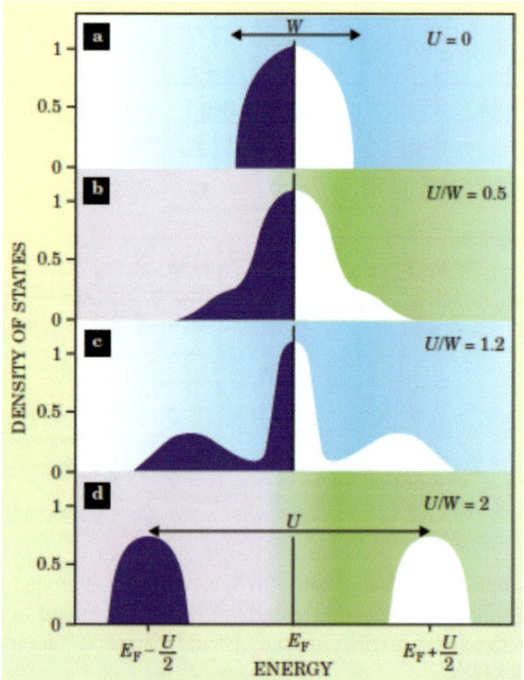

Figure 14.2 Density of states (DOS) as a function of local Coulomb interactions [29].

14.3.3 Self-interaction corrected (SIC) density functionals

A great deal of the hopeful applicants are SC material frameworks, which, from multiple points of view, are as of yet a test to be appropriately depicted hypothetically in regard to their minimum energy characteristics. One significant illustration of these nanomaterials is progress TMO, specifically Mn, Fe, Co, and Ni oxide. They are charge-moving entities, notable for solid relationship impacts related to the TM 3d electrons. Attraction of progress metal (TM) oxides is typically portrayed regarding the Heisenberg model, with direction-free connections between the twists. Nonetheless, the materialness of a particular model is not completely explained for TM oxides since spin-polarized oxygen is typically overlooked. In a traditional model dependent on the Anderson standard, oxygen effects are considered.

To understand this system, R. Logemann et al. [30] considered the total spin Hamiltonian within oxygen sites and also gave a very impressive method where the electron spin polarization renormalizes the exchange connections between TM locales. Shockingly, the exchange connections in nickel oxide rely on the state of magnetic if polarization on oxygen is ignored, taking about non-Heisenberg conduct. The exact inverse, Mn oxide carries on as a magnet signature when oxygen spin polarization is dismissed, yet gives solid non-Heisenberg impacts when spin polarization on oxygen is taken into account. In hematite, the two methods bring about non-Heisenberg conduct. Within the

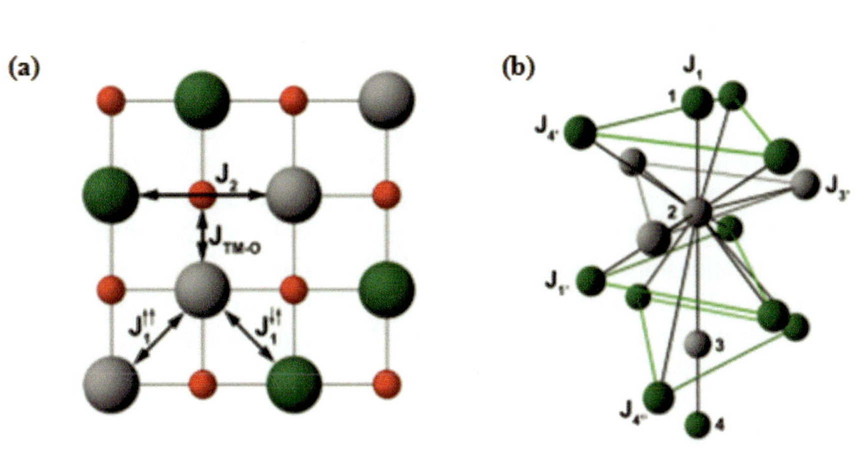

Figure 14.3 (a) A 100-plane of the rock salt structure of NiO and MnO. The exchange interactions are indicated by black arrows. (b) Fe atoms within a single unit cell of hematite [30].

sight of spin polarization on oxygen, likewise a TM–O interaction happens. In the event that no O–O interactions are available, the O destinations follow the magnetic field made by the TM molecules and give the comparing compelling exchange connections. Logemann et al. [30] studied NiO and MnO using three different magnetic states: (i) ferromagnetic (FM); (ii) AFI: an antiferromagnetic (AFM) order in the [001] direction; and (iii) AFII: an AFM order in the [111] direction, which is the magnetic ground state. In these states, the exchange interactions between TM (transition metal) and oxygen atoms, denoted as J1 and J2, can be identified. Rak and Brenner [31] also investigated the magnetic moments in this oxide using density functional theory combined with Monte Carlo (MC) simulations. To describe the magnetic moments in these materials, it's necessary to evaluate the spin interactions among all the magnetic ions present in the structure. Like other transition metal oxides with a rock salt structure, such as cobalt oxide and nickel oxide, the primary interaction in this oxide is the antiferromagnetic (AFM) super-exchange mechanism. However, in the MgO complex, a random distribution of magnetic cations makes the super-exchange mechanism challenging to understand based on the Goodenough-Kanamori rules [32, 33], which typically predict ferromagnetic interactions with neighboring ions.

14.3.4 LSDA+U method

Throughout the long term, numerous thoughts were created to treat emphatically associated frameworks. Among them, generally unmistakable inside the casing work of DFT are self-cooperation revised neighborhood turn thickness estimate (SIC-LSDA) [34] mixture utilitarian [35] and local spin density approximation with effective *U* (LSDA+U). Apart from these strategies, LSDA+U is especially helpful as it takes less time to request and henceforth can, without much of a stretch, be utilized for bigger frameworks of cutting-edge revenue. Moreover, with LSDA+U, it is feasible to deliberately build the strength of the on-location shock *U* to acquire knowledge into the impact of Coulomb relationships. The total energy in the LSDA+U strategy comprises the LSDA total energy in addition to a "Hubbard-like" term [36]. A significant issue inside the LSDA+U strategy is that the electronic collaborations are now mostly remembered for the LSDA energy; hence, a basic expansion of the "Hubbard-like" term to the LSDA energy would prompt twofold tallying (DC) blunders. Subsequently, an extra "DC term" is deducted from the LSDA+U absolute energy to stay away from this change. By deducting the mean field part from Hubbard, an ideal DC term will leave just an orbital autonomous LSDA potential [37]. The LSDA+U strategy has ended up being ready to duplicate the right energy structure for the Mott protectors [38]. We know that strongly correlated materials have an incompletely filled d or f orbital and d-d or f-f band hole. The hole is brought about by an enormous on-location Coulomb aversion, which parts the d or f groups in a lower (involved) and upper (empty) Hubbard band. This technique is the LSDA+U technique. Not just ground-state properties like the hole, magnetic moment or charge dissemination is influenced by the U [39]. In Figure 14.4, the magnon range for three transition metal oxides appears. The magnon energies are stifled with expanding U, which is because of improved charge limitation around the change metal molecules [40]. Essenberger et al. introduced LSDA+U estimations of the magnon frequencies of change TMOs utilizing the frozen magnon approach.

By changing the values of U, the effect of onsite coulomb repulsion on different properties like energy gap and magnetic moments has been investigated and found similar types of behavior with gap and magnetic moments, as shown in the figures. Charge delocalization and localization are the reason behind this. The gap for NiO, CoO, FeO, and MnO with variations of U and different approaches is illustrated in Figure 14.4.

Gadolinium is a decent test to check how actually stable the approximations are, which were utilized in the inference of the LDA+U recipe introduced previously. The Gd electronic configuration has 7f electrons; so the upper sub-shell is totally filled and the down-spin sub-shell is vacant. That implies that the ground state is very much depicted by a solitary Slater-determinant wave work and the LDA+U hypothesis as a single electron theory is legitimate here. The benefit of the LDA+U technique is the capacity to treat all the delocalized electrons and restricted electrons in 4f orbitals in a similar simulation technique. For such a strategy, it is essential to be certain that the general energy places of these two sorts of band are repeated accurately. Then again, CeSb pnictides have an extraordinary behavior. Notwithstanding the irregularities inalienable to all monopnictides, CeSb has a huge magnetic anisotropy along with a little crystal field parting, an incredibly confounded magnetic stage outline, and the biggest realized Kerr point [41]. As the local spin density approach always ignores the ground-state properties, it is very difficult to get exact ideas about the magnetic moments [42]. Be that as it may, the avoidance of 4f orbitals from the premise set (regarding them as totally limited pseudo-center states) additionally did not prompt a good depiction of electronic properties of CeSb. Then again, an experimental p–f model with a restricted 4f level of 2–3 eV beneath E hybridizing with the conduction band electrons had the option to depict CeSb subjectively [42].

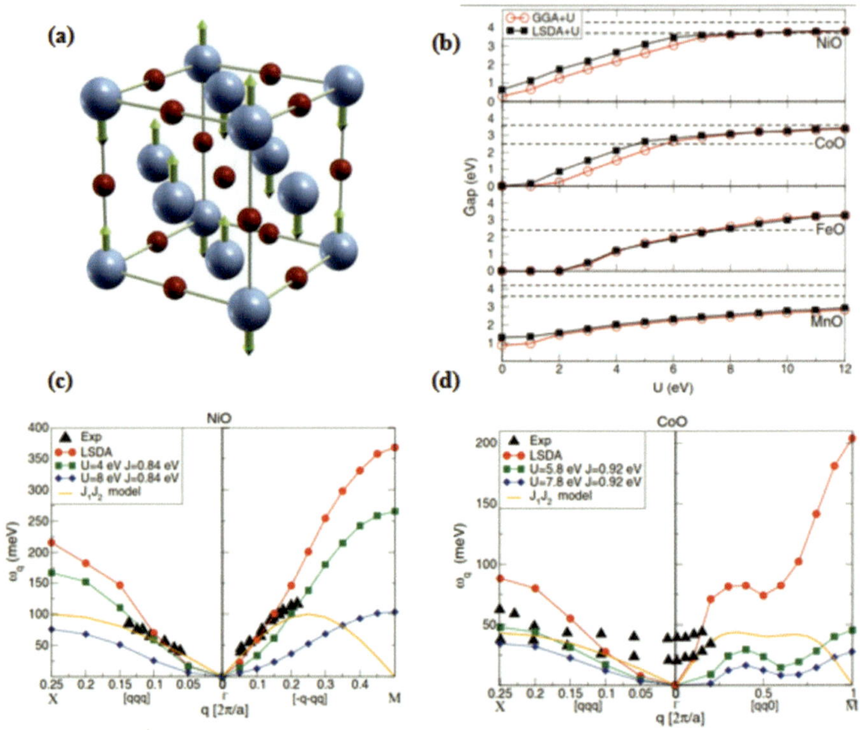

Figure 14.4 (a) AFM structure of the TMO with oxygen and TM atoms. (b) Energy gap with LSDA+U, and open GGA+U (reported experimental values are given by a dashed line). (c) Spectrum of magnon for NiO and CoO, with two separate values of U with green squares and blue diamonds, experiment with black triangles, and the J1, J2 results with orange line [40].

14.4 Exciting Phenomena of a Strongly Correlated System

In recent years, researchers have done tremendous work theoretically as well as experimentally and reported that a lot of transition metal oxides and some other materials have dominant states where physical interactions among charges, spin, etc., are active, and these phenomena create very interesting effects on materials.

14.4.1 Colossal magnetoresistance

In the many ongoing years, compounds having d- and f-electrons have been examined theoretically as well as experimentally in the area of condense matter physical science. It was found that materials having d- and f-electron gatherings are in similar classification of SC materials. The main highlight of both d- and f-electron systems is absolutely the presence of the dynamic orbital level of opportunity; however, in the f-orbital system, because of the solid twist circle cooperation in uncommon earth and actinide atoms, the physical science is, by all accounts, not quite the same as that of the d-electron material system. Numerous materials portrayed by solid electron–electron connections are of mechanical interest. On account of weakened magnetic semiconductors and the huge magnetoresistive manganites, among others, it comes due to possible applications in spintronics devices. Thus, it is critical to comprehend the interchange among charge and twist levels of opportunity of the many interfacing electrons in this class of materials. It has been generally perceived that the orbital level of opportunity assumes a vital part in the comprehension of novel attraction seen on the moving metal oxides [43–48]. An abnormal material of such a twist charge–orbital complex is manganese oxide, displaying surprising giant magneto-opposition (CMR) marvels. In the past preposterous decade, the investigation of manganites has been quite possibly the main space of examination in dense matter physical science. So, the CMR impact is supposed to happen when the magnetic signature is changing from paramagnetic to ferromagnetic, after giving a little magnetic field. In light of the idea of a two-stage rivalry, the presence of the FM metallic stage in manganites has been typically legitimized by the alleged twofold trade (DE) system [48], in view of a solid Hund's standard coupling between portable e_g electrons and limited *t* spins.

Coupling between the e_{2g} electrons and Jahn–Teller (JT) mutilations in the MnO octahedral [45, 46] generates the insulating nature in manganites and prompts different sorts of electronic states and additionally orbital magnitude. The area of manganites has been uncovered because of rivalry and interchange among twist, charge, and orbital levels of opportunity. There is undoped perovskite manganite, RMnO, with a very different earth atom, R, which is the main compound of CMR manganites. For R = La, it has been obviously seen that the A-type antiferromagnetic (AF) stage seems [47, 48] to be the C-type requesting of (3_{x3}) and (3_{y2-r2}) orbitals. By substituting La by basic earth particles like Sr and Ca, openings are viably doped into the *e*-electron band, and because of the DE component, the FM metallic stage shows up with its associative CMR impact. In the muddled stage outline for manganites, such countless magnetic stages show up. A central question to seeing such wealth is the opposition among vagrant and restricted propensities contained in manganites. As referenced in the model

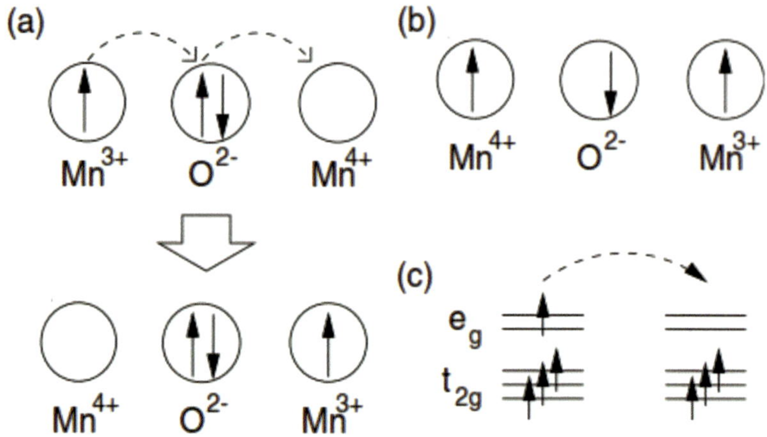

Figure 14.5 (a) Schematic diagram of a double exchange interaction involving Mn and O ions. (b) Intermediate state. (c) Improved mobility of e–e with localized spins [49].

development, electrons can gain kinetic energy when the electrons align ferromagnetically, leading to a metallic ferromagnetic (FM) phase in manganites. On the other hand, to minimize magnetic energy between localized t_{2g} spins, an antiferromagnetic (AF) phase with insulating behavior emerges. The electron movement is illustrated in Figure 14.5 (a). This process is schematically represented as $Mn^{3+} \rightarrow Mn^{4+}O^{1}\uparrow,^{3}\downarrow\ Mn^{3+\ 2}\uparrow$ [49], where 1, 2, and 3 denote the electrons associated with either the oxygen between manganese atoms or the electron level of the Mn atoms. Figure 14.5 (b) shows a second-order measure in which the two states portrayed above go from one to the next, utilizing a halfway state $Mn^{3+1}\uparrow O^{3}\downarrow^{3+1}\uparrow$Mn. Most of the hypothetical works performed with regard to manganites basically fail to remember the presence of oxygen and utilize a manganese-just Hamiltonian. It is fascinating to see that FM states show up in this setting even without the oxygen. It is evident that the electrons require a polarized background to enhance their kinetic energy, as illustrated in Figure 14.5 (c). Studies on the structural properties of $BiMnO_3$ [50–52] have not provided detailed insights into how these properties influence its magnetic behavior. To manipulate the properties of $BiMnO_3$, it is essential to use thin films, where the additional strain from lattice mismatch significantly affects the magnetic behavior. This influence is similar to the colossal magnetoresistance observed in manganese oxides, making it a valuable tool for controlling the electronic and magnetic properties near the phase transition [50–52].

14.4.2 Mott insulator

The Mott (metal-to-insulator) transition is a very basic problem in the strongly correlated system because it is necessary to have a coherent framework to explain the wave and particle nature of electron. In reality, frameworks close to the Mott transition show odd properties, for example, metallic conductivities less than the base anticipated inside an energy band, eccentric optical conductivities, and otherworldly capacities outside of what the band hypothesis can depict. An essential inquiry is the manner by which the nearby density of states or, for the most part, the otherworldly capacity changes as the proportion of the connection strength U to data transmission W builds (see Figure 14.6). The limits are sufficiently clear. At the point when the electrons are not bound with a single atom or a covalent bond, its intensity looks like the nearby thickness of conditions of band hypothesis; when the electron is restricted, the density of states tops at the ionization energy and the electron proclivity of the molecule. Those nuclear-like states create the Hubbard groups.

To see the value in the DMFT method of the Mott transition, consider the subjective highlights of the stage chart shown in Figure 14.6. This is

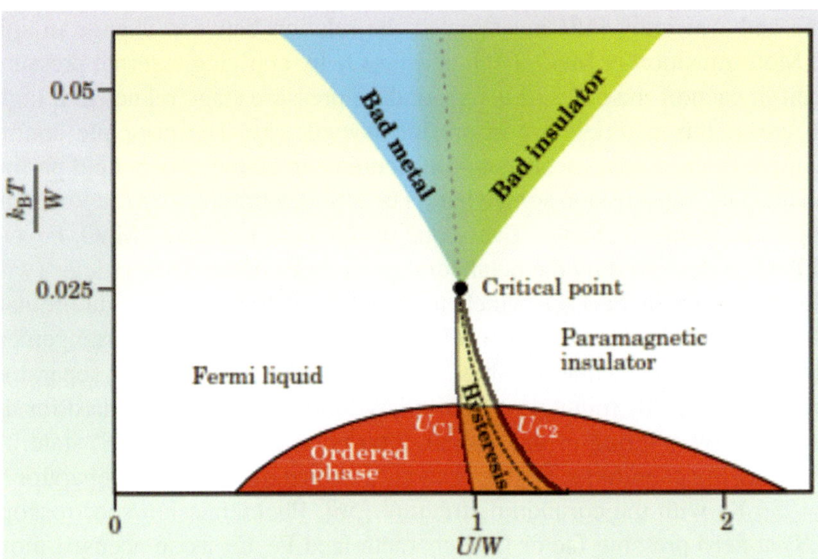

Figure 14.6 Schematic presentation of a metal-to-insulator transition with Coulomb repulsion U is plotted for a correlated electron material that is described using a Hubbard model [29].

basically showing a metal-to-insulator transition with Coulomb repulsion in the Hubbard model in various temperature systems. The vanishing of metallic nature and the end of the great energy Mott–Hubbard hole are unmistakable wonders that happen in various locales of the stage graph at low temperatures as lines at energies UC1 and UC2 that outline a hysteretic area, and at high temperatures as two particular hybrid lines that progressively separate metallic from insulating stages.

14.4.3 Metal insulator transition

A Mott transition occurs in a strongly correlated (SC) electron system due to the difference between kinetic energy and potential energy, reflecting the dual wave-like and particle-like nature of electrons in a solid. In this context, the Coulomb interaction energy (U) and the kinetic energy (t), which describes electron hopping between sites in a solid, are key factors for understanding and predicting phase diagrams in real materials. Boer and Verwey [53] studied the effect of electron−electron interaction on the material (with completely filled 3d shells) properties. To comprehend the materials with solid electronic relationships, DMFT [54] is utilized, which permitted the planning of the geometrical form onto a quantum contamination while safeguarding quite possibly the main dynamical highlights, the on-location connection. Kune and Anisimov [55] investigated the relation between change in spin and Mott transition in MnO, FeO, and Fe_2O_3 by applying external pressure. Albeit in each of the cases, the surrounding pressure stage relates to a high-spin cover, the materials act in an unexpected way. The principle control boundary in these cases seems, by all accounts, to be the crystal field parting (and the data transmission somewhat). The HS−LS progress can be joined by a separator-to-metal change. Every one of the three materials, MnO, Fe_2O_3, and FeO, is described by the octahedral geometries of the TM particles (with little mutilation in Fe_2O_3), which leads to the pressing factor subordinate e_g−t_{2g} parting. MnO and Fe_2O_3 offer similar neighborhood d^5 arrangement of the TM particles, while FeO has a conventional d^6 design. Mott separators display huge nearby minutes, which satisfy the main Hund's standard for the relating halfway filling of the d shell, a purported high twist (HS) state. At encompassing conditions, Fe_2O_3 is an antiferromagnetic (AFM) separator (T N = 956 K) with the corundum structure [56]. Photoemission spectroscopy (PES) at zero pressing factor [57] characterized Fe_2O_3 as an accused move cover of the charge hole of 2.0–2.7 eV, gathered from the electrical conductivity information [58]. Utilizing LDA+DMFT computations [59], the authors have reported a first-order HS–LS transition by decreasing the spin moment from 4.5μ B to 1μ B. In Figure 14.7 (a), the molecular spectra show a gap in the high-spin (HS) phase and a metallic state in the low-spin (LS) phase. Expanding the pressing factor inside the HS stage has the following impacts: (i) crystal field splitting expansion between the e_g and t_{2g} stages (well noticeable in both upper and lower Hubbard groups) and (ii) band widening. Both these impacts lead to consistent pressing of the hole down to nothing. When the hole vanishes, electronic remaking happens and the framework changes into a loser energy state of LS. The volume breakdown can be perceived as a result of exhausting of the counter holding e_g groups.

The interest in $NiS_{2x}Se_x$ strong arrangements emerges from the Mott transition, which they show as a component of Se focus *x* or applied pressure. NiX_2 (X = S, Se) can be seen as NiO with the O atom supplanted by a X_2 dimer, which obliges two openings in its p_σ against holding orbital prompting a X_{2-2} valence state [59]. The similarity is upheld by various photoemission examinations [60]. The LDA+DMFT spectra [61, 62] catch very well the

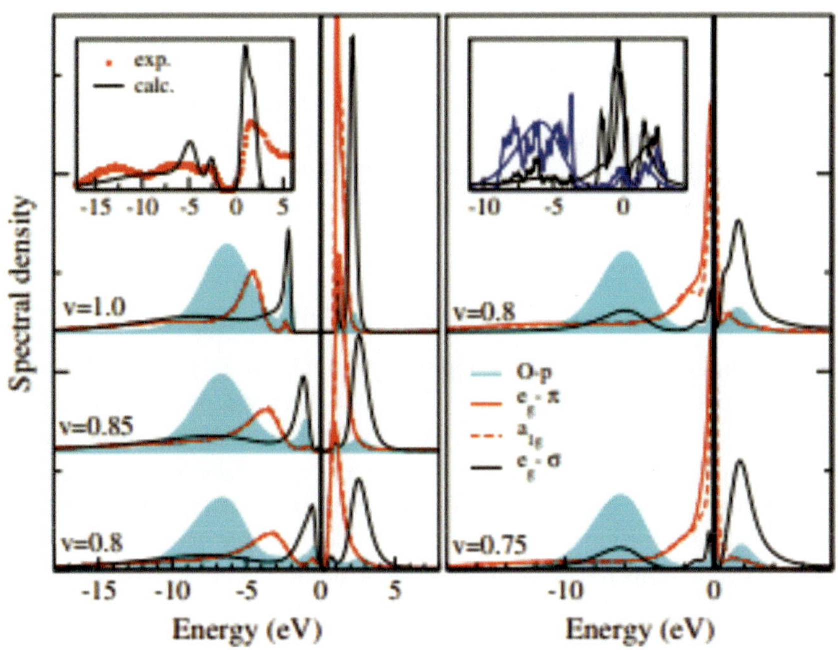

Figure 14.7 (a) The single particle spectra of Fe_2O_3 with HS solution in the left and LS solution in the right panel at various specific volumes [59].

test highlights. The primary understanding of the one-molecule spectra of NiO depended on little group definite diagonalization examinations [63]. The focal inquiry concerning $NiS_{2x}Se_x$ is: why is NiS_2 a cover while $NiSe_2$ is metallic? A few recommendations show up in the writing: (i) bigger d-band width in $NiSe_2$, (ii) cover of the Se-p with the upper Hubbard band in $NiSe_2$, (iii) and more modest collaboration strength U because of an enormous polarizability of the Se-Se dimer. Our estimations [63] recommend that none of these situations really happens. Concerning (i), other than the way that the d-transfer speed is definitely not an especially significant amount in CT frameworks, the d tops in the LDA just as the LDA+DMFT spectra are smaller in NiSe2. The situation (ii) is barred as the upper Hubbard band, which compares to the 3.5 eV top in NiS_2, is not the least electron-expansion excitation of $NiS_{2x}Se_x$. Lastly, proposition (iii) is precluded since the presence or nonattendance of the hole in the spectra of NiS_2 and $NiSe_2$ does not rely on the U worth inside the contemplated scope of 4.5–7 eV (the determined U worth is = 5 eV). Our understanding of the metal-protector change depends on the perception that the least electron-expansion energy is controlled by the situation of the S-S (Se-Se) dimer hostile to holding state; for example, it mirrors the holding strength (holding hostile to holding parting) in the dimer. A more fragile bonding in Se−Se dimer prompts the cover of the dimer hostile to holding band with the valence band and hence causes the vanishing of the hole. The pressing-factor-initiated metal-protector progress can be perceived as follows. While the S-S holding against holding parting stays steady, since the dimer carries on as an inflexible item, a decrease of dimer distances prompts a general widening of the S-p groups just as expanded p-d hybridization.

14.4.4 High-temperature superconductivity in cuprates

High-T_c superconductivity in copper-oxygen layers is illustrated in Figure 14.8. In key structures such as La_2CuO_4, the valence state of copper (Cu) is +2, resulting in a d^9 electronic configuration. In this structure, Cu is surrounded by six oxygen atoms arranged in an octahedral configuration; however, the apical oxygens above and below the Cu are not depicted in Figure 14.8. Distortion from an ideal octahedron due to the shift of the apical oxygens splits the eg orbitals, with the highest occupied d orbital being $x2-y2$. The lobes of this orbital align directly with the p orbitals of the adjacent oxygen atoms, forming strong covalent bonds with significant hopping integrals. In this way, the electronic condition of the cuprates can be portrayed

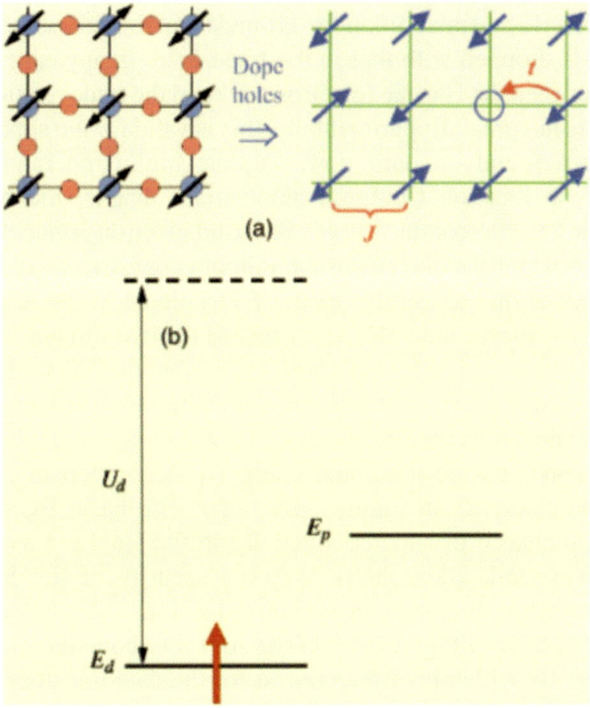

Figure 14.8 (a) One band model of a two-dimensional copper oxygen layer. (b) Copper d and oxygen p orbital. A single hole with $S = 1/2$ occupies the copper d orbital in the insulator [67].

by the alleged three-band model, where, in every unit cell, there are the Cu d$x2-y2$ orbital and two oxygen p orbitals [64, 65]. Creating proficient first-standard approximation that can treat strongly correlated systems precisely is, as of now, quite possibly the most dynamic field in hypothetical science and dense matter physical science.

The local spin density approximation (LSDA) in the DFT mathematical approach has been assuming significant parts in the consolidated matter physical science as an incredible and viable technique for diminishing the many-electron issue to a viable one-electron issue [67]. Fundamental electronic properties cannot be portrayed as expected by the LSDA for certain classes of materials, where the electron relationship is solid. For instance, the LSDA cannot recreate the antiferromagnetic ground province of NiS_5 $>$ and some high-T_c related materials like $LazCu0_4$ and $YBazCu_30_6$ [68]. One of the beginnings of this burden might be followed back to the issue of self-cooperation. As is notable, the self-connection is missing in the Hartree–Fock (HF) estimation on the grounds that the self-association in the Hartree field is dropped with that in the trade field. In any case, the relating scratch-off between the Hartree (electrostatic) and the trade relationship terms in the LSDA is not great. Experimentally, this is valid for the superconductors that have been found up until now. Cuprate high-temperature is incited by the solid on location Coulomb association, that is, the beginning of high-temperature superconductivity is the solid electron connection. Takashi Yanagisawa [69] clarified the electron minimum energy state models for high T_c cuprates based on the improvement of variational Monte Carlo strategy. He examined the superconducting (SC) state in the two-dimensional Hubbard model. In the streamlining Monte Carlo technique, the SC state turns out to be to be sure steady when the Coulomb collaboration U is huge to be of the request for the data transmission.

Takashi displayed the minimum energy as a component of the superconducting boundary Δ in Figure 14.9 (left). The basic BCS wave work anticipated the chance of superconductivity in the Hubbard model, and the improved wave work additionally shows a security of the SC state. He likewise investigated the SC and antiferromagnetic (AF) order boundaries as a component of U in Figure 14.10. The AF order boundary has a pinnacle when U/t is at 10, which is of the request for the data transfer capacity, and the SC one additionally has a top at U c that is more prominent than the data transfer capacity.

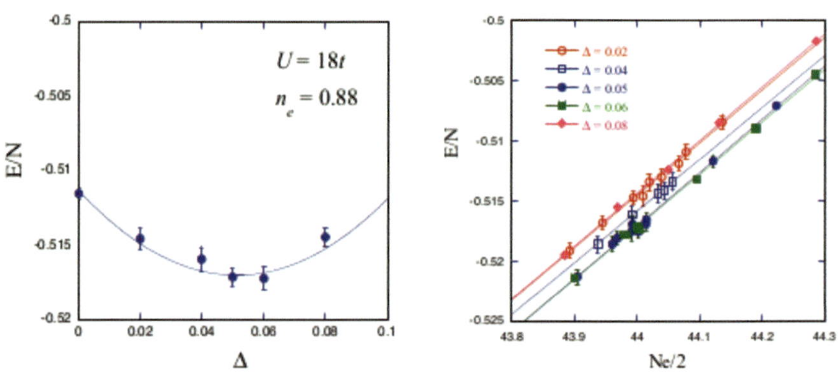

Figure 14.9 The ground-state energy for wave function ψ λ with superconducting gap Δ for the Hubbard model [71].

14.4.5 Phase transition

The relationship between electron–electron, the electron cross-section, orbital, and spin associations can likewise have comparable energy scales and all become an integral factor in correlated oxides. In light of the transaction between different levels of freedom in connected materials, various sorts of stage changes can happen in various materials, and the subtleties of the transition properties and instrument are, generally, subject to the materials. The temperature-changes stage progress in VO_2 is outstanding amongst other known MITs, and it has drawn huge interest for innovative technique, somewhat on the grounds that it happens near room temperature with a drastic change in the material's conductivity and optical properties. At room temperature, VO_2 is semiconducting with an optical bandgap of ~0.6 eV and is profoundly straightforward at microwave frequencies. Over the basic progress temperature of ~68 °C, it goes through a stage transition into a metallic stage. In single-precious stone VO_2, the resistivity drop occurs over an extremely thin temperature range. Then again, the resistance changes all

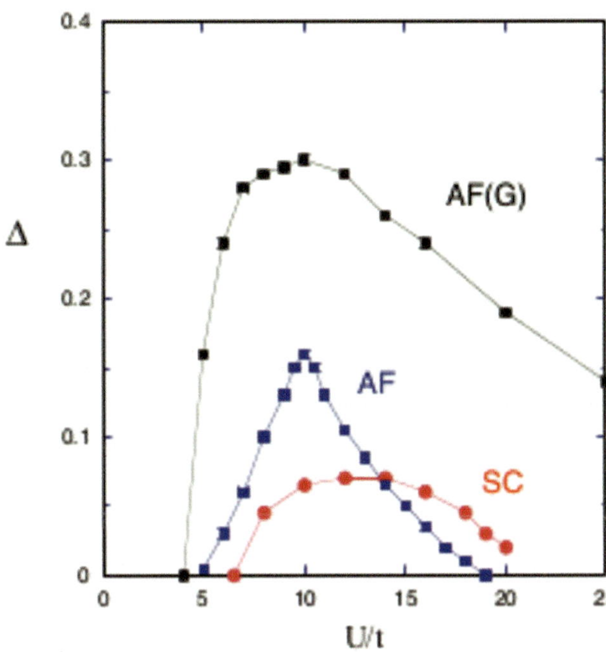

Figure 14.10 Variation of antiferromagnetic and strongly correlated materials as a function of U/t [69].

the more continuously with temperature in VO_2 flimsy movies with various spaces, making a temperature range where both the protecting stage and metallic stage exist together in the material. The adjustment of the conductivity across the phase transition can be because of an adjustment of either the free carrier density or the transporter portability.

14.5 Conclusion

The field of strongly correlated electronic systems is expansive and complex. Despite our theoretical insights, exploring this domain remains challenging. This review should be regarded as a "progress report," highlighting fundamental aspects of this rapidly evolving field. It focuses on key elements, including the intricate electron-electron interactions, as described by researchers.

References

[1] E. Dagotto, Complexity in Strongly Correlated Electronic Systems, Science 2005, 309, 257.
[2] E. Dagotto, Y. Tokura, Strongly Correlated Electronic Materials: Present and Future, MRS Bull. 2008, 33, 103
[3] Cohen AJ, Mori-S anchez P, Yang W (2012) Challenges for density functional theory. Chem Rev 112:289–320.
[4] Burke K (2012) Perspective on density functional theory. *J Chem Phys* 136:150901.
[5] Becke AD (2014) Perspective: Fifty years of density-functional theory in chemical physics. *J Chem Phys* 140:18A301.
[6] W. Kohn and L. J. Sham, Self-Consistent Equations Including Exchange and Correlation Effects Phys. Rev., 1965, 140, A1133
[7] Shen, Z.X.; Dessau, D. Electronic structure and photoemission studies of late transition-metal oxides—Mott insulators and high-temperature superconductors. Phys. Rep. **1995**, 253, 1–162.
[8] Lee, P.A.; Nagaosa, N.; Wen, X.G. Doping a Mott insulator: Physics of high-temperature superconductivity. Rev. Mod. Phys. **2006**, 78, 17–85.
[9] Bednorz, J.G.; Müller, K.A. Possible highT superconductivity in the Ba-La-Cu-O system. Z. Für Phys. B Condens. Matter **1986**, 64, 189–193
[10] Aryasetiawan, F.; Gunnarsson, O. The GW method. Rep. Prog. Phys. **1998**, 61, 237.
[11] Anisimov, V.I.; Lichtenstein, A.I. LDA+U Method: Screened Coulomb Interaction in the Mean-Field Approximation. In Strong Coulomb

Correlations in Electronic Structure Calculations. Beyond the Local Density Approximation; Anisimov, V.I., Ed.; Gordon and Breach: New York, NY, USA, 2001; Volume 1, pp. 97–165.

[12] Shen, Z.X.; Dessau, D. Electronic structure and photoemission studies of late transition-metal oxides—Mott insulators and high-temperature superconductors. Phys. Rep. **1995**, 253, 1–162.

[13] Imada M, Fujimori A and Tokura Y Metal-insulator transitions, Rev. Mod. Phys. 1998 70 1039

[14] T. L. Gilbert, Hohenberg-Kohn theorem for nonlocal external potentials Phys. Rev. B 1975, 12, 2111.

[15] N. N. Lathiotakis, N. Helbig, and E. K. U. Gross, Open shells in reduced-density-matrix-functional theory Phys. Rev. A 2005, 72, 030501.

[16] N. N. Lathiotakis, N. Helbig, and E. K. U. Gross, Performance of one-body reduced density-matrix functionals for the homogeneous electron gas Phys. Rev. B 2007, 75, 195120.

[17] S. Sharma, J. K. Dewhurst, N. N. Lathiotakis, and E. K. U. Gross, Reduced density matrix functional for many-electron systems Phys. Rev. B 2008, 78, 201103.

[18] N. Helbig, N. N. Lathiotakis, and E. K. U. Gross, Discontinuity of the chemical potential in reduced-density-matrix-functional theory for open-shell systems Phys. Rev. A 2009, 79, 022504.

[19] N. Helbig, N. N. Lathiotakis, M. Albrecht, and E. K. U. Gross, Europhys. Lett. 2007, **77**, 67003.

[20] S. Sharma, S. Shallcross, J. Dewhurst, and E. K. U. Gross, Spectrum of extended systems from Reduced Density Matrix Functional Theory arxiv:0912.1118.

[21] A. C. Hewson, The Kondo Problem to Heavy Fermions (Cambridge University Press, Cambridge, 1997).

[22] U. Fano, Effects of Configuration Interaction on Intensities and Phase Shifts Phys. Rev. 1961, 124, 1866.

[23] V. Madhavan et al., Tunneling into a Single Magnetic Atom: Spectroscopic Evidence of the Kondo Resonance Science 280, 567 (1998); L. Vitali et al., Phys. Rev. Lett. 101, 216802 (2008); N. Neet'l et al.,Phys. Rev. Lett. 101, 266803 (2008).

[24] N. Net'el et al., Conductance and Kondo Effect in a Controlled Single-Atom Contact Phys. Rev. Lett. 2007, 98, 016801.

[25] J. Kondo, Resistance Minimum in Dilute Magnetic Alloys Prog. Theor. Phys. 1964, 32, 37.

[26] A. Schiller and S. Hershfield, Theory of scanning tunneling spectroscopy of a magnetic adatom on a metallic surface Phys. Rev. B 2000, 61, 9036.

[27] J. J. Parks et al., Mechanical Control of Spin States in Spin-1 Molecules and the Underscreened Kondo Effect Science, 2010, 328, 1370.

[28] M. Karolak, D. Jacob, and A. I. Lichtenstein, Orbital Kondo Effect in Cobalt-Benzene Sandwich Molecules, PRL, 2011, 107, 146604.

[29] G. Kotliar, D. Vollhardtm, Strongly Correlated Materials: Insights From Dynamical Mean Field Theory, Physics Today 2004, 57, 53

[30] R Logemann, A N Rudenko, M I Katsnelson and A Kirilyuk, Exchange interactions in transition metal oxides: The role of oxygen spin polarization, J Phys.: Condens. Matter, 2017 29, 335801

[31] Zs.Rak and D.W. Brenner Exchange interactions and long-rangemagnetic order in the (Mg,Co,Cu,Ni,Zn)Oentropy-stabilized oxide: A theoreticalinvestigation, J. Appl. Phys. 2020, 127, 185108.

[32] J. B. Goodenough, An interpretation of the magnetic properties of the perovskite type mixed crystal $La_{1-x}Sr_xCoO_{3-X}$ J. Phys. Chem. Solids, 1958, 6, 287

[33] J. Kanamori, Superexchange interaction and symmetry properties of electron orbitals Journal of Physics and Chemistry of Solids, 1959, 2-3, 87-98

[34] Z. Szotek, W. M. Temmerman, and H. Winter, Application of the self-interaction correction to transition-metal oxides Phys. Rev. B 1993, 47, 4029.

[35] C. Rödl, F. Fuchs, J. Furthmüller, and F. Bechstedt, Quasiparticle band structures of the antiferromagnetic transition-metal oxides MnO, FeO, CoO, and NiO Phys. Rev. B 2009, 79, 235114.

[36] A. I. Liechtenstein, V. I. Anisimov, and J. Zaanen, Density-functional theory and strong interactions: Orbital ordering in Mott-Hubbard insulators Phys. Rev. B. 1995, 52, R5467.

[37] F. Bultmark, F. Cricchio, O. Grå näs, and L. Nordström, Multipole decomposition of LDA+U energy and its application to actinide compounds Phys. Rev. B 2009, 80, 035121.

[38] V. I. Anisimov, J. Zaanen, and O. K. Andersen, Band theory and Mott insulators: Hubbard U instead of Stoner I Phys. Rev. B 1991, 44, 943.

[39] A. Floris, S. de Gironcoli, E. K. U. Gross, and M. Cococcioni, Vibrational properties of MnO and NiO from DFT +U-based density functional perturbation theory Phys. Rev. B. 2011, 84, 161102.

[40] F. Essenberger, S. Sharma, J. K. Dewhurst,C. Bersier, F. Cricchio, L. Nordström E. K. U. Gross Magnon spectrum of transition-metal oxides: Calculations including long-range magnetic interactions using the LSDA+U method Phys. Rev. B. 2011, 84 174425

[41] Fisher P, Lebech B, Meier G, Rainford B D and Vogt O 1978 *J. Phys. C: Solid State Phys.* **11** 345 Meier G, Fisher P, Haly W, Lebech B, Rainford B D and Vogt O 1978 *J. Phys. C: Solid State Phys.* **11** 1173

[42] Norman M R and Koelling D D Fermi surface of field-induced ferromagnetic CeSb *Phys. Rev.* B. 1986, 33 6730

[43] Kasuya T, Sakai O, Tanaka J, Kitazawa H and Suzuki T Electronic structures in cerium monopnictides 1987 *J. Magn. Magn. Mater.* 63–64 9

[44] Tokura Y and Nagaosa N, Orbital Physics in Transition-Metal Oxides Science, 2000 288 462

[45] Dagotto E, Hotta T and Moreo A Colossal magnetoresistant materials: the key role of phase separation2001 Phys. Rep. 344 1

[46] Hotta T and Dagotto E Colossal Magnetoresistive Manganites ed T K Chatterji (Amsterdam: Kluwer) 2004 pp 207–62

[47] Tokura Y (ed) *Colossal Magnetoresistance Oxides* (New York: Gordon & Breach) 2000, pp 1–52

[48] Zener C Interaction between the d-Shells in the Transition Metals. II. Ferromagnetic Compounds of Manganese with Perovskite Structure Phys. Rev.1951, 82 403

[49] Cieplak M, Localized model for systems with double-exchange coupling Phys. Rev.B.1978 18 3470

[50] Wollan E O and Koehler W C, Neutron Diffraction Study of the Magnetic Properties of the Series of Perovskite-Type Compounds [(1−x)La, xCa]MnO3 Phys. Rev.1955 100 545

[51] Konishi Y, Fang Z, Izumi M, Manako T, Kasai M, Kuwahara H, Kawasaki M, Terakura K and Tokura Y Orbital-state-mediated phase-control of manganites J. Phys. Soc. Japan, 1999, 68 3790

[52] Fang Z, Solovyev I V and Terakura K Phase Diagram of Tetragonal Manganites Phys. Rev. Lett.2000 84 3169

[53] J. H. de Boer and E. J. W. Verwey, Semi-conductors with partially and with completely filled $_3$d-lattice bands Proc. Phys. Soc. 1937 49(4S), 59

[54] W. Metzner and D. Vollhardt, Correlated Lattice Fermions in d=∞ Dimensions Phys. Rev. Lett. 1989 62, 324.

[55] J. Kunes and V. I. Anisimov, Various scenarios of metal-insulator transition in strongly correlated materials, Ann. Phys. (Berlin) 2011, 523, No. 8 – 9, 682 – 688 / DOI 10.1002/andp.201100027

[56] C. G. Shull, W. A. Strauser, and E. O. Wollan, Neutron Diffraction by Paramagnetic and Antiferromagnetic Substances Phys. Rev. 1951, 83, 333.

[57] A. Fujimori, Photoemission satellites and electronic structure of Fe2O3 Phys. Rev. B 1986, 34, 7318.

[58] S. Mochizuki, Electrical conductivity of α-Fe_2O_3 Phys. Status Solidi A 1977, 41, 591-594

[59] J . Kuneš et al., Pressure-Driven Metal-Insulator Transition in Hematite from Dynamical Mean-Field Theory Phys. Rev. Lett. 2009,102, 146402.

[60] E. K. Li, K. H. Johnson, D. E. Eastman, and J. L. Freeouf, Localized and Bandlike Valence-Electron States in FeS2 and NiS2 Phys. Rev. Lett. 1974, 32, 470; W. Folkerts et al., J. Phys. C 1987 20, 4135; A. Fujimori et al., Phys. Rev. B 54, 16329 (1996).

[61] J. Kuneš, V. I. Anisimov, A. V. Lukoyanov, and D. Vollhardt, Local correlations and hole doping in NiO: A dynamical mean-field study Phys. Rev. B 2007, 75, 165115; J. Kunešetal, Phys. Rev. Lett. 2007, 99, 156404.

[62] J. Kunešetal, Metal-insulator transition in NiS2−xSex Phys.Rev.B 2010, 81, 035122.

[63] A. Fujimori, F. Minami, and S. Sugano, Multielectron satellites and spin polarization in photoemission from Ni compounds Phys. Rev. B 29, 5225 (1984); G. A. Sawatzky and J. W. Allen, Phys. Rev. Lett. 1984, 53, 2339

[64] Emery, V. J.,Theory of high-Tc superconductivity in oxides Phys. Rev. Lett. 1987, 58, 3759.

[65] Varma, C. M., S. Schmitt-Rink, and E. Abrahams, 1987, Charge transfer excitations and superconductivity in “ionic” metals Solid State Commun. 62, 681.

[66] S.D. Bader, Colloquium: Opportunities in nanomagnetism, Rev. Mod. Phys., VOLUME 2006, 1, 78

[67] V. N. Ant onov ,L. V. Bekenov, and A. N. Yaresko, Electronic Structure of Strongly Correlated Systems, dvances in Condensed Matter Physics, 107 2011, (2011)

[68] Dagotto Elbio, Correlated electrons in high-temperature superconductors, Reviews of Modern Physics, 1994, 66, 3

[69] Takashi Yanagisawa, Mechanism of High-Temperature Superconductivity in Correlated-Electron Systems, Condens. Matter 2019, 4, 57

15

Density Functional Theory Calculations of the Electronic Structure and Sensing Applications of Strongly Correlated Materials

Saju Joseph and Seetha Lakshmy

International and Inter University Centre for Nanoscience and Nanotechnology, Mahatma Gandhi University, India

Abstract

Understanding the physical, chemical, and electronic properties of strongly correlated electronic systems (SCES) and the consequent application of their properties in technological applications is the main focus of condensed matter physics. For the past several decades, there has been a vast experimental and theoretical effort to unravel the complicated behavior of strongly correlated systems. Calculations based on the density functional theory (DFT) have been widely utilized to analyze and successfully characterize the features of many condensed matter systems. Transition metal dichalcogenides (TMDs), an important class of condensed matter systems, whether single molecules, crystalline solids, or nanostructures, exhibit a variety of complex physical phenomena, and, thus, have attracted a great deal of interest from the scientific community. The strongly coupled electrons occupying in the partially filled d or f shells are responsible for their complicated behavior. In this book chapter, we have applied DFT to study the electronic structure and interfacial properties, in particular, sensing applications, of strongly correlated TMD-based nanostructures.

Keywords: Electronic Structure, Sensors, Transition metal dichalcogenides, Density functional theory

DOI: 10.1201/9781003337102-15

15.1 Introduction

The electronic structure, remarkable properties, and novel applications of correlated electron systems are a complex problem in condensed matter and materials. Materials with strong electronic correlations are novel substances and are defined as the movement of one electron strongly depending on the positions and movements of all other electrons due to the long-range electron–electron Coulomb interaction. Highly correlated electron systems such as transition metal dichalcogenides (TMDs) and other materials have been a focus of condensed matter physics for decades because of their remarkable electronic properties, which originate from strong electron correlation. Hence, we have seen remarkable advances in the physics, chemistry, and materials science of strongly correlated electronic systems (SCES) over the last few decades [1, 2, 3, 4, 5].

Strongly correlated electronic materials play a key role in physics, e.g., metal–insulator transitions (related to measurable changes in electrical conductivity), colossal magnetoresistance (tuned magnetic field), high temperature superconductivity, and many more [4, 5], as well as important areas of chemistry [1, 2, 3, 6] like catalysis, and sensing applications. In addition, recent rapid developments in experimental and theoretical techniques have led to an enhanced understanding of the molecular and bulk properties of SCES materials. Recently, one of the most active research fields in theoretical science and condensed matter physics is developing efficient first-principles-based density functional theory (DFT) approaches that can treat SCES efficiently and accurately.

The understanding of the electronic, magnetic, and optical properties of SCES is key to the main branches of the natural sciences, e.g., physics, chemistry, material science, and biology. Theoretical electronic structure computations based on DFT have grown in popularity in highly correlated electronic systems during the past few decades. Nowadays, DFT is one of the most widely used approaches for calculating the electronic structure of correlated electron systems. It is usually called the first-principle method or *ab initio* method, since it allows individuals to compute and determine numerous atomic and molecular properties of an SCES by simply providing some basic atomic and molecular structure information. DFT also offers an alternative to the traditional experimental method and pure theoretical method based on quantum theory for investigating and comparing the properties of the SCES. As a result, DFT has evolved into a powerful tool used by both experimentalists and theorists to better understand the properties of SCES, as well as to make specific theoretical predictions of experimentally observable phenomena for real materials and to design new materials.

DFT is a method in which all the properties of a system can be expressed as unique functionals of the ground state density (Thomas–Fermi model [7]) rather than the many-body wavefunction. The many-electron problem in DFT is transformed to a fictitious system consisting of non-interacting particles. The original interacting system is thus approximated by an effective single-particle problem. This approach reduces the problem dimension from $3n$-dimensional to three-dimensional spatial problem, where n is the number of particles. The following sections provide an overview of the many-body problem, and DFT methods.

15.2 Theoretical Background

15.2.1 Electronic Structure Problem

Strongly correlated materials are made from atoms that contain electrons and nuclei. The interactions between these give rise to various materials and their novel properties. The physical, chemical, and electronic properties of matter can be determined by solving the time-independent Schrödinger equation [8, 9, 10]

$$\hat{H}\Psi_i = E_i\Psi_i, \tag{15.1}$$

where H is the molecular Hamiltonian operator, Ψ_i is the total wavefunction of the i-th electronic state, and E_i is the corresponding energy eigenvalue of the system.

The Hamiltonian, $\hat{H}$, can be split up

$$\hat{H} = \hat{T}_e + \hat{T}_N + \hat{V}_{eN} + \hat{V}_{ee} + \hat{V}_{NN}, \tag{15.2}$$

where $\hat{T}_{\mathrm{N}}$ and $\hat{T}_{\mathrm{e}}$ are the kinetic energy operators for nuclei and electrons, respectively, with the general form:

$$\hat{T}_N = -\frac{1}{2}\sum_{I=1}^{N}\frac{1}{M_I}\nabla_I^2 \tag{15.3}$$

$$\hat{T}_e = -\frac{1}{2}\sum_{i=1}^{n}\nabla_i^2 \tag{15.4}$$

and $\hat{V}_{\mathrm{NN}}$ and $\hat{V}_{\mathrm{ee}}$, are the Coulombic potential energy operators for the inter-nuclear and inter-electron repulsions, and the term $\hat{V}_{\mathrm{eN}}$ representing the attraction between nuclei and electrons:

$$\hat{V}_{NN} = \sum_{I<J}^{N} \frac{Z_I Z_J}{|R_I - R_J|} \tag{15.5}$$

$$\hat{V}_{ee} = \sum_{i<j}^{n} \frac{1}{|r_i - r_j|} \tag{15.6}$$

$$\hat{V}_{Ne} = \sum_{I<1}^{N} \sum_{i<1}^{n} \frac{Z_I}{|R_I - r_i|}. \tag{15.7}$$

The above Schrödinger equation is deceptively simple but enormously complex to solve. The only molecular system that can be solved exactly by the Schrödinger equation is the hydrogen atom, which contains one proton and one electron, multielectronic systems; even for the helium atom, it is very difficult to solve the problem exactly. In systems like condensed matter physics and solid-state problems, it becomes even more difficult, since the large number of particles, electrons, and nuclei and, consequently, different levels of approximations have to be done.

15.2.2 The Born–Oppenheimer Approximation

The very first important approximation is obtained by decoupling the dynamics of the electrons and the nuclei, which is known as the Born–Oppenheimer approximation. [11, 12] The theorem states that the motion of the electrons and the nuclei can be considered decoupled. Because the mass of the nuclei is much heavier than that of the electrons, for example, a proton, mass is approximately 2000 times larger than the electron. As a result, the nuclei are said to be frozen in their equilibrium states. This means that nuclei can be thought of as stationary, and the interaction between nuclei and electrons can be looked upon as an external potential, $\hat{V}_{\text{ext}}$, to the electrons. Therefore, when characterizing the electrons and their contribution to the overall energy of the system, the interaction between nuclei is not a concern in most physical and chemical problems. The nuclear–nuclear interaction energy can be added as a constant later. The Hamiltonian (electronic Hamiltonian) of the electrons can be written as:

$$\hat{H}_e = -\frac{1}{2}\sum_{i=1}^{n} \nabla_i^2 + \sum_{i<j}^{n} \frac{1}{|r_i - r_j|} + \hat{V}_{\text{ext}}. \tag{15.8}$$

The resulting electronic Hamiltonian operator can be expressed in terms of electrons in the presence of frozen positions of nuclei. The electron-electron interaction, on the other hand, is a difficult problem to solve. Many other approaches have been introduced; these include multiconfiguration self-consistent field theory (MCSCF), multireference configuration interaction theory (MRCI), perturbation theory (MP2, MP3, and MP4 methods), and coupled cluster (CC) methods [13]. In addition, DFT provides an efficient and accurate method to study electron–electron interactions, which we have used here. The DFT method reduces the electronic Hamiltonian problem from a 3n-dimensional to a three-dimensional spatial problem. In the following sections, we will provide a brief overview of DFT-based techniques.

15.2.3 Density Functional Theory

15.2.3.1 Hohenberg–Kohn theorems

In the past decades, we have seen DFT achieve excellent results in describing the properties of a large variety of systems, in particular, solid-state physics, quantum chemistry, and material science. According to the Hohenberg–Kohn theorem [14], the ground–state electronic properties of a system can be described as a unique functional of the ground–state electron density. Hohengerg and Kohn developed two theorems that developed modern-day DFT.

Theorem I: There exists a one-to-one correspondence between external potential $\hat{V}_{\text{ext}}(r)$ and electron density $n(\mathrm{r})$.

Theorem II: The ground–state electron density can be found by using a variational principle.

The first theorem asserts that the ground–state electron charge density determines the ground state of an interacting system. The second Hohenberg–Kohn theorem states that for any external potential, the ground total energy assumes the following form:

$$E[n] = F_{HK} + \int V_{ext}(\mathrm{r})\mathrm{n}(\mathrm{r})\mathrm{dr}, \tag{15.9}$$

where $F_{HK}[n]$ is the universal functional, and $V_{\text{ext}}(\mathrm{r})\mathrm{n}(\mathrm{r})\mathrm{dr}$ is the external potential. The exact ground–state density is the density that minimizes the functional. However, the precise form of this functional is still not known.

15.2.3.2 Kohn–Sham ansatz

The Kohn–Sham approach [14] replaces the fully interacting N-particle interacting problem with N one-particle non-interacting one. In the Kohn–Sham ansatz, [14] the ground state electron density of the interacting system is considered to be the same as that of the non-interacting system exerted to two potentials, the so-called external potential V_{ext}, and an exchange correlation potential [15],V_{xc}.

$$E_0[n(\mathrm{r})] = T[n(\mathrm{r})] + \int V_{\text{ext}}(\mathrm{r})n(\mathrm{r})dr \quad + \frac{1}{2}\iint \frac{n(\mathrm{r})n(\mathrm{r}')}{|r-r'|}d\mathrm{r}'d\mathrm{r} + E_{\text{XC}}[n(\mathrm{r})] + E_{\text{NN}} \tag{15.10}$$

where the first term, $T[n(\mathrm{r})]$ is the the non-interacting kinetic energy of electrons, the second term, $V_{\text{ext}}(\mathrm{r})n(\mathrm{r})$, is the electrostatic potential due to the nuclei, the third term is the Hartree energy, the fourth term, $E_{\text{xc}}[n]$, is the exchange-correlation energy that takes into account all non-classical interactions between electrons and the last term and E_{NN}, is the nuclei–nuclei interaction energy. The Hartree term is the interaction energy due to the classical electron charge density. The Coulombic electrostatic repulsion in the Hartree energy term is, however, overestimated, and is taken care of in the exchange–correlation term.

Using the fact that the total number of electrons is conserved,

$$n(r) = \sum_{i<1}^{N} |\Psi_i(r)|^2, \tag{15.11}$$

and applying the variational principle leads to the one-particle Kohn-Sham equations,

$$\left(-\frac{\hbar^2}{2m_e}\nabla^2 + V_{\text{KS}}\right)\Psi_i(r) = \epsilon_i\Psi_i(r), \tag{15.12}$$

where $\Psi_i(r)$ is the Kohn–Sham orbitals and V_{KS} is,

$$V_{\text{KS}} = V_H + V_{\text{ext}} + V_{\text{XC}}. \tag{15.13}$$

The Kohn–Sham equation can be solved in a self–consistent manner to attain the correct ground–state electron density. The exchange-correlation potential, V_{XC}, is the only part we do not know how to describe and obtain precisely. In the next section, we will look at two important exchange–correlation functional approximations widely used in solid-state physics, quantum chemistry, and material science.

15.2.3.3 Exchange–correlation functionals

15.2.3.4 Local density approximation (LDA)

In DFT, the exchange–correlational (E_{xc}) functional plays a significant role. The two most widely used approximations for the exchange–correlation energy are the local density approximation (LDA) [14] and the generalized gradient approximation (GGA) [16].

In LDA, E_{xc}, energy is approximated with,

$$E_{\text{XC}}^{\text{LDA}}[n] = \int n(r)\epsilon_{\text{XC}} n(r)dr. \tag{15.14}$$

The LDA is the simplest approximation of E_{xc}, which is a very crude approximation. In this approximation, the electron density varies slowly (uniform electron density) in space. Although this is a very crude approximation, LDA is successful particularly where the electron density is uniform. Despite the fact that this is a fairly basic approximation, LDA works well in situations when the electron density is uniform.

15.2.3.5 Generalized gradient approximation (GGA)

In an attempt to improve the LDA and the description of exchange–correlation energy, the GGA has been developed. The GGA functional provides an enhancement in predicting the total exchange energies compared to the one obtained by the LDA., including density and its gradient. In LDA, the ϵ_{XC} term depends only on the value of the density at each point, but in GGA, it also depends on the gradient of the electron density [17, 15].

$$E_{\text{XC}}^{\text{GGA}}[n] = \int n(r)\epsilon_{\text{XC}}[n(r) \mid \nabla n(r) \mid]dr. \tag{15.15}$$

There are numerous attempts to find a good GGA functional to treat SCES. Some popular GGA exchange–correlation functionals include Perdew, Burke, and Ernzerhof (PBE) [15], Perdew and Wang 91 (PW91) [18], and BLYP [19]. There are various theoretical methods based on the level of approximations made to solve the many-electron Schrodinger equation. The Hartree-Fock (HF) method was among the first approaches to deal with the many-body problem in molecules and materials. This is due to the simplicity of the HF method. Accurate quantum mechanical extensions to address correlational effects [13], such as configuration interaction (CI) theory, perturbation theory, and (CC) methods, have been used to study molecules. For condensed matter and solid state systems, DFT has been considered as the most popular and widely used one.

15.3 An Overview of Strongly Correlated Electron Systems

Electron correlations play a critical role in determining the emergent properties of SCES. Common to correlated electron materials is that the valence electrons originate from partially filled $3d$ or $4f$ shells, including lanthanides and actinides ($3d$ or $4d$ electrons of the transition metals), lanthanides ($4f$ electrons), and actinides ($5f$ electrons). The above group of materials are collectively called âĂIJstrongly correlated systems,âĂİ where valance electrons are relatively localized around the atomic sites. When correlations are considerably strong, electrons become entangled and novel physical and chemical properties emerge.

First-principles calculations based on DFT and its extensions for strongly correlated systems agree reasonably well with experimental data, especially with better approximations for the general exchange–correlation energy functionals. In general, DFT methods such as LDA and GGA are widely applicable in highly correlated systems, even though there are still difficulties in using LDA/GGA-DFT to properly describe many body effects in strongly correlated systems. Also, the theoretical and computational complexity and costs were relatively low compared to expensive many-electron wavefunction calculations, such as HF theory and post-HF methods [20]. In wavefunction-based quantum mechanical calculations, there are many theoretical approaches developed that can treat strong correlation in a systematic and efficient way, including (MCSCF), (MRCI), Perturbation theory (MP2, MP3, and MP4 methods), and (CC) methods [13]. But these quantum chemical approaches need extremely high computational cost, and it is very difficult to apply them to condensed matter systems.

Recent efforts to combine model representations of strongly coupled electronic systems with DFT have opened the door to computing and predicting the electronic structure of materials using first-principles calculations [21, 20]. These include, the physical structure, electronic transport properties, tunable band gap properties, magnetic properties, phase transitions, sensing, catalysis, and many more. In the next section, we will discuss the sensing applications of TMDs based on correlated electron systems.

15.4 Modeling of Highly Correlated Materials for Sensing Applications

Sensors are essential devices used for the speedy detection of the given analyte molecules. Biosensors are sensors that mainly identify biological elements or biomolecules and have gained broad recognition due to their applicability in numerous fields such as medicine [22], industry, agriculture, and many more [23]. In most circumstances, the development of biosensors was to meet social concerns; for example, the sensors for identifying harmful biomolecules like nitrobenzene and catechol. [23, 24] For instance, glucose monitoring sensors or glucometers are devices used for the fast detection of glucose level in the blood, particularly in diabetic patients [25].

Highly correlated materials (HCMs) play a vital role in biosensing applications. The most investigated amongst them includes graphene [26], transition metal oxides [27], 2D (TMDs) [28] because of their high surface to volume ratio, high surface tunability, and other unusual fundamental properties. They have in-plane covalent bonding, and out-of-plane van der Waals (vdW) interactions that make them structurally stable. The critical property that aids them in sensing is that their surface activity can be tailored by a number of methods like doping, defect inducing, applying an electric field, and selective molecular adsorption [29].

The biosensing process involves the interaction of biomolecules with substrate materials. The interaction can cause the generation of a measurable signal, and the nature of these signals determines the sensor's design. The adsorption of biomolecules on the substrate surface happens largely due to the charge transport, which induces the signal. Molecular adsorption causes variations in the physical, electrical, and optical properties of the substrate materials.

It is challenging for experimentalists to choose a suitable sensing material from the vast availability. Moreover, it is very difficult for them to examine and explain the underlying process during biomolecule adsorption. In this regard, computational simulations can elaborate on the phenomenon of biomolecule interaction with the substrate materials. The binding energy, best adsorption sites, stable geometric configuration, charge transfer, orbital interactions, etc, can be well explained through simulations. These features unlock the doors for experimental researchers to design and devise highly selective and customized biosensors.

15.4.1 2D-Transition Metal Dichalcogenide Based Sensors

Popular 2D HCMs like graphene and TMDs (e.g., MoS_2) are theoretically predicted and experimentally verified as highly sensitive biosensing materials. The investigation of biomolecule interaction via computational calculation primarily requires a model of the preferred 2D-HCM satisfying the periodic planar boundary conditions. The bond parameters of this model should be highly compatible with their original structure. Sufficient vacuum should be created perpendicular to the planar structure to evade the periodic interactions. Then the model structure is allowed to optimize geometrically. The DFT optimized 4×4 supercell [30] of the 2H-MoS_2 monolayer is given in Figure 15.1(a). The DOS of the corresponding MoS_2 structure is shown in Figure 15.1(b). The incorporation of hybrid calculation (HSE06) resulted in a band gap that is in close agreement with the experimental value. Higher K point values are generally adopted to obtain the DOS structure. The next step is to introduce the preferred biomolecule at several adsorption sites of the HCMs and is again relaxed. The configuration with the merest binding energy is identified as the most promising structure. Figure 15.2 endows four configurations of biomolecule L-leucine interacting with the graphene surface [31].

The interactions between analyte and substrates in sensing processes are weak, i.e., the biomolecules get physisorbed on the surface of the 2D-HCMs with weak adsorption energy and at large bond distances. Since the LDA and GGA exchange-correlation functionals are inadequate to calculate the adsorption energy [32], the exchange–functional is added up with the dispersion corrections like DFT-D2 and DFT-D3 [33] employed.

The electronic characteristics and sensing abilities of these 2d-HCMs can be improved in several ways, like defect engineering, the introduction

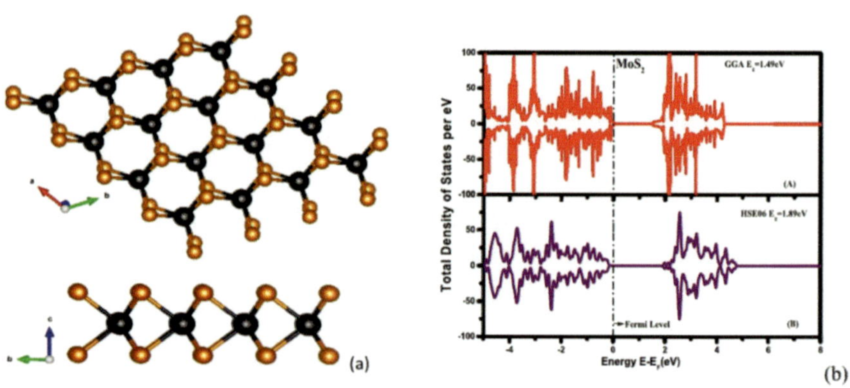

Figure 15.1 (a) Optimized configuration of pristine MoS_2. (b) Total DOS plot of MoS_2 calculated with GGA+U (upper panel) and hybrid functional (lower panel). The band gap variation is also shown in the plot. [30]

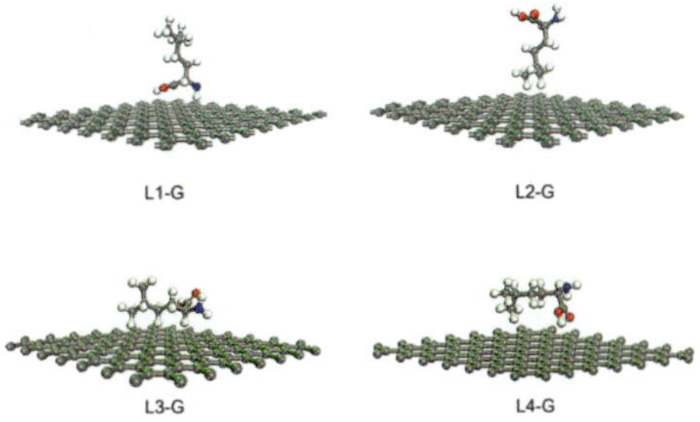

Figure 15.2 Diagrammatic representation of different configurations of interaction of L-leucine with graphene monolayer: L1-G, vertical interaction of the functional group with the surface; L2-G, methyl interacting vertically to the surface; L3-G, parallel interaction of the fundamental chain containing the amino group of L-leucine; and L4-G, parallel interaction of the fundamental chain containing carboxyl of L-leucine with the graphene surface. [31]

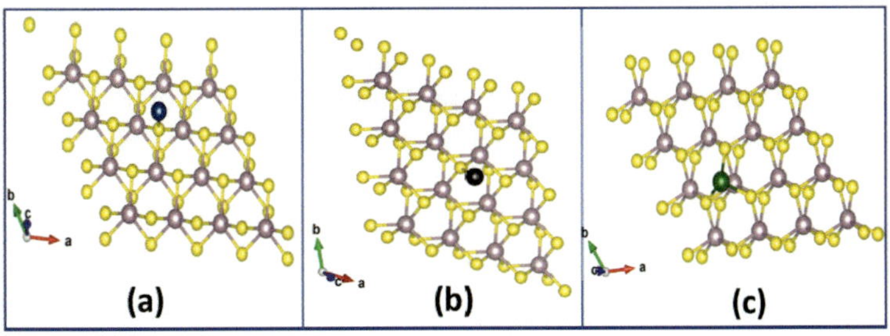

Figure 15.3 Optimized structures of metal doped MoS_2: (a), MoS_2+Pd; (b), MoS_2+Ag; (c), MoS_2+ Ti, where the Ti, Pd, and Ag atoms are indicated by the spheres with green, blue, and black, respectively. [34]

of strains, doping or decorating with transition metals, and the formation of hybrid structures [35, 36, 37, 38, 34]. These alterations can create the surface sites of HCMs more reactive with large surface energies compared to their pristine form. Figure 15.3, denotes the transition metal decorated structures of 2D monolayered MoS_2.

15.4.2 Interaction of Biomolecules with the Defected/Doped 2D-HCMs

The DFT relaxed structure of the biomolecule under consideration is placed over the defected or doped 2D-HCMs. The final structure is further relaxed with a keen selection of suitable exchange–correlation functionals and dispersion corrections. The binding energies of the biomolecule on the 2D-HCMs can be evaluated mathematically by the following expression:

$$E_{\mathrm{B}} = E_{(\mathrm{2D-HCM+Biomolecule})} - (E_{\mathrm{2D-HCM}} + E_{\mathrm{Biomolecule}}), \quad (15.16)$$

where $E_{(\mathrm{2D-HCM+Biomolecule})}$ denotes the energy of the biomolecule adsorbed in the 2D-HCM system, $E_{\mathrm{2D-HCM}}$ is the energy of defected or doped HCM, and $E_{\mathrm{Biomolecule}}$ is the energy of the desired biomolecule. The negative binding energy suggests that the adsorption process is exothermic, i.e., the process is energetically favourable.

The willingness of the biomolecule to get adsorbed on the 2D-HCMs can be well realized by exploring the electronic structure. The glucose or catechol biomolecule sensors exhibit a flow of charge from the biomolecule to the substrate material. In the transition metal doped MoS_2, the charge transfer occurs chiefly from the oxygen $2p$ orbitals to the partially filled d orbitals of the transition metal [30]. The Mullikan [39] or Bader charge analysis [40] techniques are commonly employed to study the charge transfer mechanism. Additionally, the partial density of states (PDOS) and charge density difference plots also reveal the charge transfer process. The PDOS plots for the valence of the interacting atoms give better insights into the binding mechanisms.

The binding energy and charge transfer analysis are not the sole requisites for these materials to be efficient biosensors. The chosen highly correlated substrate material should be highly stable, durable, sensitive and should exhibit a low recovery time when exposed to various environmental conditions like temperature, and pressure [29]. Since the DFT calculations are carried out at 0 K, the stability of the sensor materials to higher temperatures is highly essential. The molecular dynamic simulation snapshots of catechol adsorbed on Ti-doped MoS_2 at 300 K after 5 ps is shown in Figure 15.4. Another important property of the substrate material to be an efficient sensor

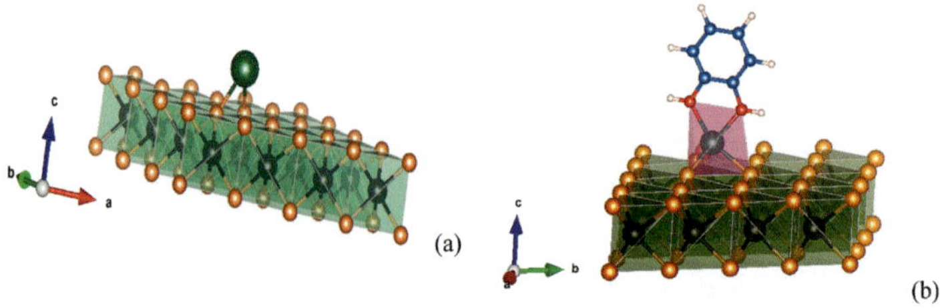

Figure 15.4 Molecular dynamics snapshots of (a)MoS_2+Ti and (b) MoS_2+Ti + catechol after 5 ps at 300 K.

is reversibility; i.e., the biomolecules need to be desorbed from the material surface to ensure reusability. All these features of the HCM sensors (graphene and MoS_2) could be well explored via theoretical calculations.

The theoretical results are often affected by certain parameters like the choice of exchange–correlational functional, energy cut-off, and dispersion corrections. Subtle change in the above parameters can vary the results. Hence, it is at most essential to specify these parameters precisely, and the selection should be justified for the system under study.

15.5 Conclusion

In this book chapter, it is discussed that different aspects of the 2D TMD-based HCMs as efficient sensors, in particular biosensors. Currently, the most common sensors are graphene and 2D TMDs, because of their high surface-to-volume ratio and high surface tunability. Also, the other fundamental properties, including particle size, in-plane covalent bonding, and out-of-plane vdW interactions make them structurally stable. These effects can influence the efficient charge transfers and adsorptions of the sensing properties. By understanding the various aspects that affect the sensing performance of the 2D TMD-based HCMs, it is possible to find ways to improve upon the performance of existing sensors. Those resembling graphene and 2D TMD-based HCMs could be developed to achieve sensitive detection of harmful molecules like nitrobenzene and catechol. First-principles-based DFT calculations can be used to further understand the sensing mechanism and potentially increase the possibilities for optimizing the novel properties of sensors.

References

[1] Dagotto, E.; Tokura, Y. *MRS Bulletin* **2008**, *33*, 1037–1045.

[2] Paier, J.; Penschke, C.; Sauer, J. *Chem. Rev.* **2013**, *113*, 3949–3985.

[3] Meng, Y. S.; Arroyo-de Dompablo, M. E. *Energy Environ. Sci.* **2009**, *2*, 589–609.

[4] Malrieu, J. P.; Caballol, R.; Calzado, C. J.; de Graaf, C.; Guihéry, N. *Chem. Rev.* **2014**, *114*, 429–492.

[5] Lee, P. A.; Nagaosa, N.; Wen, X.-G. *Rev. Mod. Phys.* **2006**, *78*, 17–85.

[6] Dagotto, E. *Science* **2005**, *309*(5732), 257–262.

[7] Thomas, L. H. *Mathematical Proceedings of the Cambridge Philosophical Society* **1927**, *23*(5), 542–548.

[8] Schrödinger, E. *Ann. Physik* **1926**, *79*, 361.
[9] Schrödinger, E. *Ann. Physik* **1926**, *79*, 489.
[10] Schrödinger, E. *Ann. Physik* **1926**, *79*, 734.
[11] Born, M.; Oppenheimer, J. R. *Ann. Physik* **1927**, *84*, 457.
[12] Hurley, A. C. *Introduction to the Electron Theory of Small Molecules;* Academic Press: London, 1976, 1.
[13] I. Shavitt, R. J. Bartlett, Many-Body Methods in Chemistry and Physics: MBPT and Coupled Cluster Theory; Cambridge University: Cambridge,2009.
[14] Kohn, W.; Sham, L. J. *Phys. Rev.* **1965**, *140*, A1133.
[15] Perdew, J. P.; Burke, K.; Ernzerhof, M. *Phys. Rev. Lett.* **1996**, *77*, 3865–3868.
[16] Becke, A. D. *Phys. Rev. A* **1988**, *38*, 3098–3100.
[17] Langreth, D. C.; Perdew, J. P. *Phys. Rev. B* **1980**, *21*, 5469–5493.
[18] Perdew, J. P.; Chevary, J. A.; Vosko, S. H.; Jackson, K. A.; Pederson, M. R.; Singh, D. J.; Fiolhais, C. *Phys. Rev. B* **1992**, *46*, 6671–6687.
[19] Lee, C.; Yang, W.; Parr, R. G. *Phys. Rev. B* **1988**, *37*, 785–789.
[20] Jiang, H. *Int. J. Quantum Chem.* **2015**, *115*(11), 722–730.
[21] Fulde, P. *Electron Correlations in Molecules and Solids;* Springer: Berlin, 2002.
[22] Haleem, A.; Javaid, M.; Singh, R. P.; Suman, R.; Rab, S. *Sensors International* **2021**, *2*, 100100.
[23] Chen, R. J.; Choi, H. C.; Bangsaruntip, S.; Yenilmez, E.; Tang, X.; Wang, Q.; Chang, Y.-L.; Dai, H. *J. Am. Chem. Soc.* **2004**, *126*(5), 1563–1568.
[24] Chen, R. J.; Bangsaruntip, S.; Drouvalakis, K. A.; Wong Shi Kam, N.; Shim, M.; Li, Y.; Kim, W.; Utz, P. J.; Dai, H. *Proc. Natl. Acad. Sci.* **2003**, *100*(9), 4984–4989.
[25] Oliver, N. S.; Toumazou, C.; Cass, A. E. G.; Johnston, D. G. *Diabetic Medicine* **2009**, *26*(3), 197–210.
[26] Pumera, M. *Mater. Today* **2011**, *14*(7), 308–315.
[27] Agnihotri, A. S.; Varghese, A.; M, N. *Applied Surface Science Advances* **2021**, *4*, 100072.
[28] Pumera, M.; Loo, A. H. *TrAC - Trends in Analytical Chemistry* **2014**, *61*, 49–53.
[29] Cho, B.; Kim, A. R.; Park, Y.; Yoon, J.; Lee, Y.-J.; Lee, S.; Yoo, T. J.; Kang, C. G.; Lee, B. H.; Ko, H. C.; Kim, D.-H.; Hahm, M. G. *ACS Appl. Mater. & Interfaces* **2015**, *7*(4), 2952–2959.

[30] Lakshmy, S.; Sanyal, G.; Vaidyanathan, A.; Joseph, S.; Kalarikkal, N.; Chakraborty, B. *Appl. Surf. Sci.* **2021**, *562*, 150216.

[31] Qin, W.; Li, X.; Bian, W.; Fan, X.; Qi, J. *j.biomaterials* **2010**, *31(5)*, 1007.

[32] Chakraborty, B. *Electronic structure and theoretical aspects on sensing application of 2D materials. Fundamentals and Sensing Applications of 2D Materials, Woodhead Publishing,* **2019**, p 145.

[33] Nguyen, C. V.; Hieu, N. N.; Nguyen, D. T. *Nanoscale Res. Lett.* **2015**, *10*(1), 433.

[34] Vaidyanathan, A.; Lakshmy, S.; Sanyal, G.; Joseph, S.; Kalarikkal, N.; Chakraborty, B. *Appl. Surf. Sci.* **2021**, *550*, 149395.

[35] Tedstone, A. A.; Lewis, D. J.; OÃćâĆňâĎćBrien, P. *Chem. Mater.* **2016**, *28*(7), 1965–1974.

[36] Toh, R. J.; Poh, H. L.; Sofer, Z.; Pumera, M. *Chem. Asian J.* **2013**, *8*(6), 1295–1300.

[37] Banhart, F.; Kotakoski, J.; Krasheninnikov, A. V. *ACS Nano* **2011**, *5*(1), 26–41.

[38] Aksimsek, S.; Jussila, H.; Sun, Z. *Opt. Commun.* **2018**, *428*, 233–239.

[39] Carbó-Dorca, R.; Bultinck, P. *J. Math. Chem.* **2004**, *36*, 231–239.

[40] Arnaldsen, A.; Tang, W.; Henkelman, G. *Bader Charge Analysis* , http://theory.cm.utexas.edu/bader/ (accessed 15 September 2012).

16

Strongly Correlated 2D Materials and Oxides

Muhammed Swalihu[2,5], **Parvathy Nancy**[1,4], **Sabu Thomas**[1,2,3], **and Nandakumar Kalarikkal**[1,4]

[1]School of Energy Materials, Mahatma Gandhi University, India
[2]International and Inter University Centre for Nanoscience and Nanotechnology, Mahatma Gandhi University, India
[3]School of Nano Science and Nano Technology, Mahatma Gandhi University, India
[4]School of Pure and Applied Physics, Mahatma Gandhi University, India
[5]Department of Physics, Catholicate College, India

Abstract

Highly correlated materials are excellent materials whose electronic properties are governed by strong correlations between electrons present in it. 2D layered materials, whose valence band structure is composed of one flat band and two Dirac bands, can host a broad range of tunable topological and strongly correlated electronic phases. The present chapter outlines strong correlations in magic-angle twisted bilayer graphene, complex metal oxide systems, transition metal dichalcogenides, etc. In such type of correlated electron materials, the interactions between the electronic charges, orbitals, spin, and the lattice produce a rich variety of electronic phases and self-organization. The cooperation among these correlated electronic phases can steer the emanation of surprising electronic phenomena and also of interesting functionalities through their responses to external fields.

Keywords: 2D materials, strongly correlated electrons, graphene, transition metal dichalcogenides

DOI: 10.1201/9781003337102-16

16.1 Introduction

Modern solid-state physics can easily explain the physical characteristics of numerous materials, such as basic metals, semiconductors, and insulators. However, it is harder to understand the properties of materials with open d and f electron shells because the electrons there occupy small orbitals. In transition metals, such as yttrium, vanadium, scandium, iron, chromium, and their oxides, for example, due to their spatial confinement in such orbitals, electrons encounter severe Coulomb repulsion. It is impossible to explain such strongly interacting (correlated) electrons as being embedded in a static mean field produced by the other electrons [1]. Simply put, an electron has too much of an impact on the other particles for them to be handled separately [2].

The effect of correlations has an intensive impact on material properties. The interplay of the d and f electrons' internal degrees of freedom – spin, orbital moment, and charge – can exhibit a whole zoo of mysterious ordering events at low temperatures. That interplay makes strongly correlated electron systems incredibly sensitive to even the tiny fluctuations in external parameters, such as pressure, temperature, or doping. One such exciting phenomenon was the metal–insulator phase transition detected in many transition metal oxides (TMOs) with variation in temperature, pressure, and doping concentration [3, 4].

Dramatic effects can include extremely high transition temperatures in superconductors with copper-oxygen planes, significant volume changes across phase transitions in lanthanides and actinides, and huge changes in resistivity across the metal–insulator transition in TMOs. In heavy fermions, larger effective mass of charge carriers when compared with free electron mass results in many interesting phenomena such as phase transition between magnetic orders, appearance and disappearance of local magnetic moments, superconductivity, etc [5]. Another such compound is magnetite (Fe_3O_4), which exhibits the electronic structure with a highly correlated ground state [6]. Strongly correlated materials can exhibit a variety of properties, including a significant thermoelectric response in certain cases and a high sensitivity to changes in an applied magnetic field in others, a phenomenon known as colossal magnetoresistance. The potential for creating applications from correlated-electron materials is exciting because of these characteristics. But their experimental and analytical analysis is made more demanding by the diversity of the phenomena and the remarkable sensitivity to microscopic details. In this chapter, we discuss the basics regarding highly correlated systems and, in addition to that, we discuss the high correlation in graphene, transition metal oxides, and transition dichalcogenides.

16.2 Graphene

16.2.1 Structure, properties, and disorder

Graphene is a single-layer carbon allotrope that exhibits extraordinary properties. They came up with SP^2 hybridization and a molecular bond length of 0.142 nm. As the thinnest compound, which is known to mankind, graphene shows a very high strength. The thermal conductivity at room temperature varies from 4.84 ± 0.44 to 5.30 ± 0.48 $Wm^{-1}k^{-1}$. And it is also popular as a good electric conductor. Also its potential stability for use in spin transport makes it a popular material for wide-ranging applications [7].

16.2.2 Structure of graphene

Carbon is marked as one of the most remarkable elements known to mankind. Due to its structure, properties, crystal alignment, and chemical bonding, carbon provides a wide range of applications. $1S^2$ $2S^2$ $2P_x{}^1$ $2P_y{}^1$ $2P_z{}^0$ is the carbon ground state configuration, which is shown in Figure 16.1 (b). The unique properties of carbon atoms arise due to their ability to form different hybridizations using four valence electrons. The carbon atom can form three types of hybrid configurations, such as SP, SP^2, and SP^3. Graphene forms

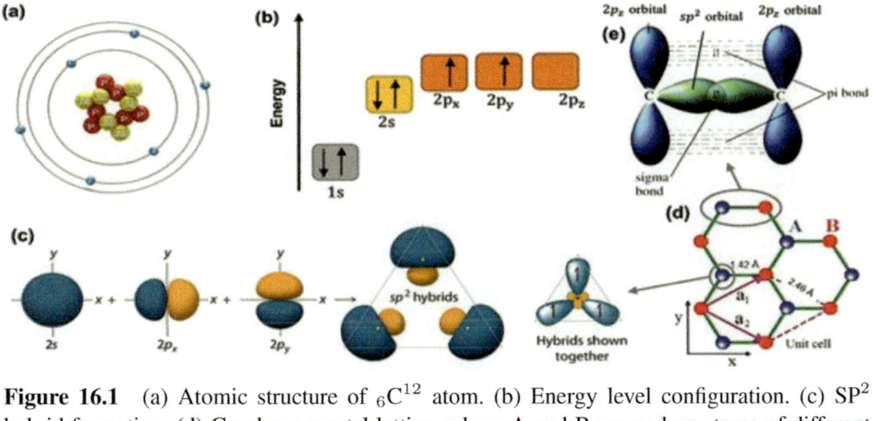

Figure 16.1 (a) Atomic structure of $_6C^{12}$ atom. (b) Energy level configuration. (c) SP^2 hybrid formation. (d) Graphene crystal lattice, where A and B are carbon atoms of different sub-lattices. (e) σ-bond and π-bond formed by SP^2 hybridization (reused with permission from [8]).

SP^2 hybridization as shown in Figure 16.1 (c) with a honeycomb-type planar structure and a bond length of 0.142 nm. The crystal graphene unit cell along with unit cell vectors a_1 and a_2 is shown in Figure 16.1 (d). The planar ring stability is due to the electron resonance and delocalization.

In graphene, carbon atom forms three sigma (σ) bonds with neighboring carbons (Figure 16.1 (e)) forming a stable hexagonal structure. The short interatomic covalent sigma bond of bond length is -1.42 Å, which results in remarkable mechanical properties. The P_z electron, which makes the π-bond is located vertical to the planar structure, contributes to graphene's electrical property.

16.2.3 Properties of graphene

Graphene shows outstanding properties in terms of mechanical strength, electrical conductivity, optical transparency, and thermal conductivity. The planar density of graphene is 0.77 mg/m^2, which makes it a super light material. The graphene unit structure is a hexagonal carbon ring with a 0.052 nm^2 area. The hexagonal carbon ring consists of two carbon atoms and each carbon atom is shared with three unit rings. Its ultrathin and super thin features are also contributed by its one-layer atomic thickness. The graphene shows a transparency of about 97.7%; it absorbs 2.3% of the visible spectrum [9]. The difference of single-layer and bilayer graphene transparency as well as between a substrate and single-layer graphene are both 2.3% [10]. Hence, the number of graphene layers can be effectively indicated using transparency. The non-interacting Dirac fermion theory simulation confirms it [11].

The room temperature conductivity of graphene is very high. It shows a conductivity of 10^6 S/m and sheet resistance of 31 omega/sq results in high mobility of 2×10^5cm^2/Vs [12]. The high conductivity is the result of the π-bond, which is free to move at room temperature. Abscence of band gap between conduction band with the valence band makes it a semi-metal [13]. The electron transport in graphene is generally governed by the Dirac equation and shows the anomalous quantum Hall effect [14, 15].

The high tensile strength of graphene and its elastic modulus is 125 GPa and 1.1 TPa, respectively,[16] which makes it the hardest and toughest material. Compared to steel with the same thickness, graphene is approximately 100 times stronger [17]. The outstanding thermal conductivity has a major impact in various applications. At room temperature, thermal conductivity is about 5×10^3 W/mK [18]. Compared to copper, it is approximately 10 times higher.

16.2.4 Lattice defects in graphene

Lattice defects alter the performance of graphene-based devices. However, this departure from optimal behavior can be used in a number of ways. The defects help to tailor new properties and thereby achieve new functionalities. Such structural imperfections modulate graphene's magnetic, electrical, optical, thermal, and mechanical properties. In general, lattice defects in graphene are grouped as intrinsic and extrinsic. Intrinsic defects are the perturbation of crystalline order without the presence of foreign particles. The extrinsic defect is the perturbation of crystalline order due to non-carbon atoms.

Intrinsic defects are subdivided into five categories, which are stone-wales defects [21], single vacancy defects [22, 23], multiple vacancy defects, carbon adatoms, and line defects. The stone-wales defect (Figure 16.2) results from the rotation of a single carbon layer. The hexagon ring is transformed into heptagons and pentagons. This point defect has the formation energy of 5 eV. Through the use of electron radiation or by rapidly cooling at high temperatures, this defect can be created into the lattice. The single vacancy defect is represented in Figure 16.3. It is the absence of a single carbon atom in the lattice site. The graphene undergoes a distortion, which results in the minimization of energy. The distortion is Jahn–Teller, which helps in the saturation of dangling bonds toward the missing atom. The formation energy, E_F, of a single vacancy defect is high. The formation energy is approximately $E_F = 7.5$ eV [16, 17].

The introduced defect in graphene is classified as substitutional impurities and foreign adatoms. In the strong oxidation method or chemical vapor deposition method, oxygen-containing functional groups or metal atoms are introduced into the surface of graphene. This defect is termed as foreign

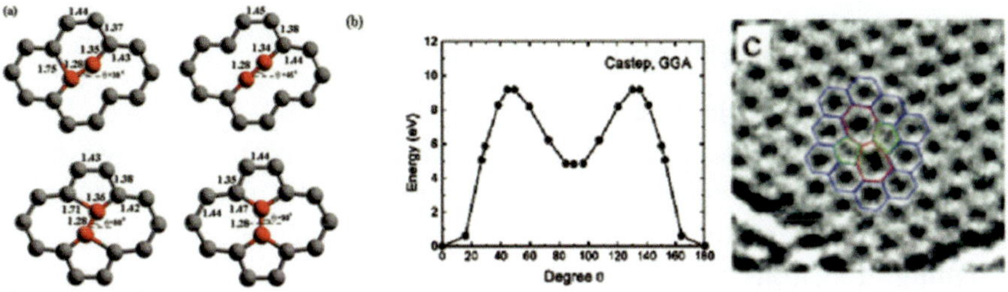

Figure 16.2 Stone-wales defect. (a) Theoretical atomic structure from DFT calculations. (b) Formation energy as a function of angle of rotation Θ. (c) TEM image of the stone-wales defect ((a) and (b) reprinted with permission from [19] and (c) reprinted with permission from [20]).

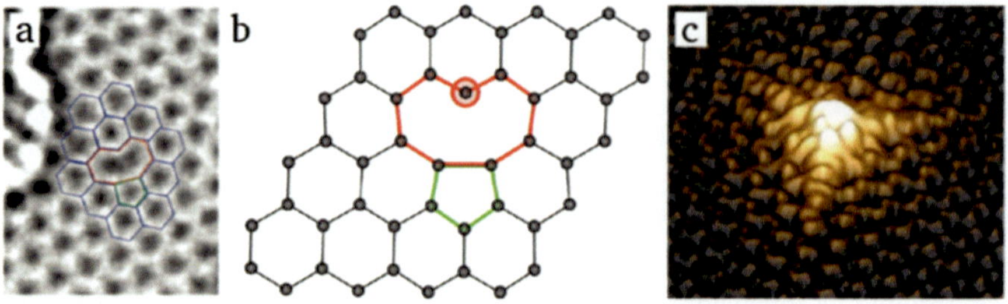

Figure 16.3 Single vacancy defect. (a) Experimental TEM image (reprinted with permission from [21]). (b) Atomic structure from DFT calculation (reprinted with permission from [20]). (c) Experimental STM image (reprinted with permission from [24]).

adatoms. The foreign non-carbon atom affects the properties of graphene. This is reliant on the bonding between graphene and the introduced atom. The introduced foreign atom can form covalent or Van der Waals bond with carbon neighbors. If the interaction is weak, the Van der Waals bonding between carbon and non-carbon atom leads to physisorption. The strong interaction leads to chemisorption due to the covalent bond.

In substitutional impurities, one or two carbon atoms are replaced by foreign impurities. Atoms like boron and nitrogen have the ability to form three chemical bonds; hence, they can take carbon's place in graphene. Even though nitrogen and boron have one electron more or less than carbon, they have approximately the same atomic radius. In graphene, boron and nitrogen can exist independently or they can be introduced at the same time using method control. The boron-doped or nitrogen-doped graphene shows excellent properties in terms of conductivity and catalytic activity. Replacing carbon using boron and nitrogen not only causes a change in the Fermi level but moreover changes the electrical structure [25].

16.2.5 Strong correlation in graphene

16.2.5.1 Magic angle graphene

Single graphene sheets are fascinating in and of themselves because they exhibit exceptional material strength and incredibly effective electrical conductivity. But when graphene sheets are stacked, things become even more intriguing. Electrons start interacting with electrons in the neighboring graphene layer in addition to those in the sheet they are currently in. These interactions alter as the angles of the sheets change, leading to intriguing quantum phenomena like superconductivity, electronic nematicity, etc.

Although the superconducting behavior of twisted graphene was unexpected, the possibility that something exciting might occur was not. When two graphene sheets are layered at angles more than a few degrees, they often act independently. The misaligned lattices, however, can produce a "super lattice" at smaller angles in which electrons can travel across layers. The super lattice's underlying structure would radically alter the behavior of electrons at certain minor twists, or "magic angles," slowing them down and allowing them to interact in ways that altered the material's electronic capabilities [26–28].

Bilayer graphene undergoes a transition from a Fermi liquid with little correlation to a two-dimensional electron system with strong correlation at a magic twist angle. This transition results in properties that are incredibly sensitive to carrier density as well as to controllable environmental factors like the proximity of nearby gates and twist-angle variation. Electronic nematicity, superconductivity, linear-in-temperature low-temperature resistivity, interaction-induced insulating states, magnetism, and quantized anomalous Hall states are only a few of the phenomena that magic-angle twisted bilayer graphene exhibits [29].

16.2.5.2 Condensation of electron–hole pairs in a two-layer graphene system

Josephson effects, super fluidity, and other anomalous electromagnetic phenomena in coupled semiconductor quantum wells were first suggested as probable causes by the pairing of spatially separated electrons and holes brought on by their Coulomb attraction. In dense electron–hole systems, this pairing is comparable to electron–hole pairing in excitonic insulators and Cooper pairing of electrons in superconductors. Strong Coulomb electron–hole attraction was thought to keep the pairing's critical temperature at a high level, while spatial isolation of the paired electrons and holes was thought to preclude interlayer tunneling, which would cause their recombination and fix the condensate phase. Experimental proof of a superfluid transition in two-layer semiconductor devices in a strong magnetic field (pairing of composite fermions) and in the absence of a magnetic field was discovered several decades later [30].

16.3 Transition Metal Oxides

As we all know, elements in the periodic table are arranged into three classes. The main group includes the elements with the last electron filled in the s or p orbital. Other groups are transition elements and inner transition elements. In transition elements, we have three series based on the last electron in 3d, 4d, or 5d orbit. The inner transition elements are classified as lanthanides and actinides; they are series in which the last electron fills in the 4f and 5f states. Transition metal oxides are compounds with oxygen as anion and transition metal as a cation. As a result of incompletely filled d and f shells of cation, this metal oxide becomes the most versatile and technologically inspiring class of materials with structural diversity and exhibits a variety of electric, magnetic, and color properties. Due to their exponentially rich physical as well as chemical properties, they show a broad range of potential applications in electronics, optics, biosystems, catalysis, and energy storage [31–34].

Oxygen plays a crucial role in the striking properties on the surface of TMOs. Due to the high polarizability of O^{2-} ions, surface properties dominate over material behaviour [35, 36]. This high polarizable nature enables planar transition metal oxides to exhibit non-uniform and non-linear charge distributions, which generate exceptional interfacial as well as local surface properties and results in specific energy states on the surface. These energy states are very different when compared with their bulk counterpart and induce a significant desire in 2D TMOs for Coulomb interactions with adjacent ions [37]. If two surfaces of TMOs are brought in the neighborhood of one another, the interface potentials arise, which results in the shifting of Fermi levels of the surface planes and influence the electronic properties. Another interesting property arises if d-orbitals of 2D TMOs are aligned normal to their planes, the orbital and spin ordering may get frustrated and the TMO surface becomes electronically activated. This results in unconventional semiconducting properties. The magnetic property shown by transition metal oxide is the result of holes in the 2p band of oxygen. Also, the cation arrangement results in magnetic frustrations and leads to Kondo effects and anti-ferromagnetic patterns [38].

Wide varieties of transition metal oxides can be formed using different transition metals. These cation species as well as their flexibility for changing oxidation state controls most of the fundamental properties. Especially in two-dimensional transition metal oxides, the cations can adopt different binding configurations and charge states, which stabilize large varieties of structures [39, 40]. The flexibility in changing oxidation states gives various electronic effects that vary from insulating to metallic behavior in the same TMO at different stoichiometry. Also, the d-electronic states contribute to the distinct transition from metal to insulator as a function of pressure and temperature.

Different varieties of TMOs are used in a wide range of applications. Materials such as MoO_3,41–43 V_2O_5,44,45 WO_3, and Ga_2O_3 were found as natural layered crystals in the hydrate or anhydrate phase. These TMOs can be exfoliated down to nanosheets using the gas or liquid phase techniques with oxygen-terminated basal surfaces. Many oxides such as zinc oxides and titanium oxides lack naturally layered crystal phases. They are exfoliated using layer by layer techniques.

16.3.1 Vanadium dioxide

At 67 °C, a temperature that works well for numerous applications, VO_2 is known for its dramatic metal–insulator transition with conductivity changes of up to four orders of magnitude [46]. Above 67 °C, the lattice changes into a poor metal with a rutile structure, whereas below 67 °C, VO_2 is a semiconductor with a gap of roughly 0.7 eV and with a monoclinic crystal structure [47, 48]. This modification of the lattice structure points to a first-order phase transition in which the vanadium chain dimerizes, doubling the size of the unit cell. This supports a Peierls-like transition from the band theory viewpoint. But some evidence points to a Mott transition that is fueled by powerful electron–electron interactions. These factors include the fact that an intermediate monoclinic "M2" phase, which can be made stable by doping [49] or stress [50], is insulating despite having undimerized vanadium chains [51], the fact that insulator band gap cannot be obtained using band structure calculations, the unusual low conductivity of the metallic state, and in optical studies the dependence of characteristics on excitation power, indicating sensitivity to excited carrier density.

16.4 Transition Metal Dichalcogenides

Transition metal dichalcogenides or TMDCs are emerging materials that belong to the family of two-dimensional nanosheets. They provide broad and rich opportunities for exploring highly correlated systems. They are semiconductors in the form of MX_2, where M belongs to transition elements and X belongs to chalcogens. This wealthy class of materials exhibits a wide range of physical characteristics, ranging from metals (VSe_2, NbS_2,

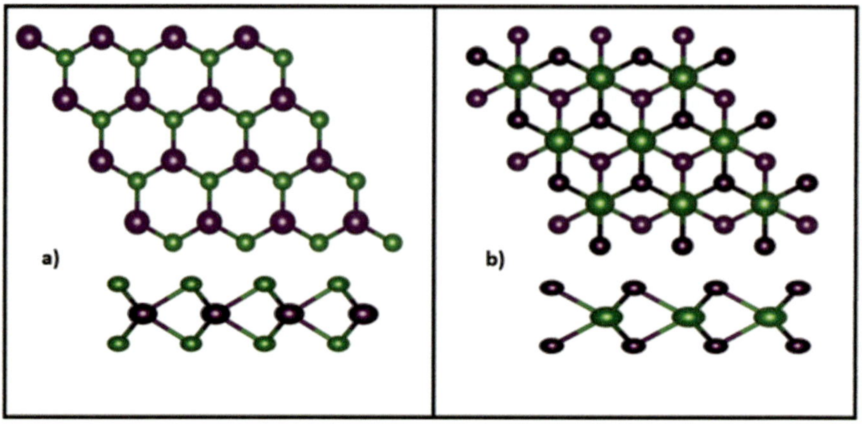

Figure 16.4 Top and side views of 2D MoS_2 structure in (a) 2H and (b) 1T phases.

etc.), semi-metals (TcS_2, WTe_2, etc.), insulators (HfS_2), and semiconductors (MoS_2, PtS_2, etc.). The extraordinary properties make them useful for a broad range of applications, such as nanoelectronics, nano-photonics, storage devices, sensing, etc.

Several structural phases of TMDCs exist. The common structural phases are either characterized by octahedral (1T) or trigonal prismatic (2H) coordination of metals (Figure 16.4). Based on their electrical band structure, the TMDs range from metals such as NbS_2 and TaS_2, to semi-metals such as VS_2 and WTe_2, to semiconductors such as $MoSe_2$ and MoS_2 [52]. Furthermore, the researchers found that chemical and physical properties have a strong layer dependency (Figure 16.1(g)). For example, the monolayer MoS_2 exhibits a direct bandgap of 1.9 eV, but the bulk possesses an indirect bandgap of 1.3 eV (Figure 16.1(c)–(e)).

The trigonal prismatic (2H) and octahedral (1T) phases can be viewed as different stacking orders. The 2H phases are characterized by an ABA stacking, that is, the chalcogen atoms occupy the same position A in different atomic planes and are placed on top of each other in a direction perpendicular to the layer. The 1T phases correspond to an ABC stacking order. Based on the combination of chalcogen and transition metal elements, 2H and 1T are thermodynamically the most stable phases, but other phases can also be obtained as metastable states.

Due to diversity in structural phases and chemical compositions, TMDCs exhibit a broad span of electronic properties from the perspective of metallic or insulating band structure character. The materials such as $MoSs_2$, MoS_2,

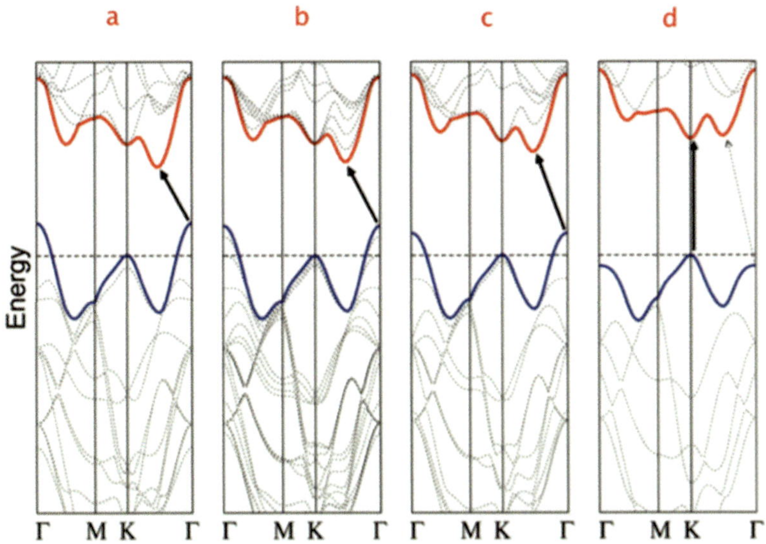

Figure 16.5 The relation between thickness and band structure of trigonal prismatic MoS_2 (reprinted with permission from [52]).

WSe_2, and WS2 are semiconductors at their thermodynamically stable trigonal prismatic phase. This property makes these TMDCs a potential player for multiple electronic applications. The first principle DFT calculation of trigonal prismatic MoS_2 band structure shows that as the thickness reduces the monolayer from bulk, the position of the conduction and valence band edges changes (Figure 16.5). Also, the indirect bandgap in bulk materials is replaced by the direct bandgap in monolayer. The theoretically calculated band gap value is 0.88 eV for bulk material and 1.71 eV for monolayer MoS2. The experimental value is 2.16 eV. Importantly, the conduction band minimum and valence band maximum is located at K/ and K. Here, K/ and K are two non-equivalent high symmetry points, which correspond to the hexagonal Brillouin zone corners [54].

The monolayer trigonal prismatic TMDCs lack inversion symmetry. Electronic bands spin apart in the lack of inversion symmetry, which is caused by spin–orbit interaction. Also, the spin degeneracy of the valence band and conduction band extrema at K/ and K are lifted. This is due to the fact that these points do not match the time-reversal invariant momenta. This effect in the valence band is particularly strong. In the valence band of monolayer MoS_2 (2H), the spin splitting value is 0.15 eV, and in WSe_2, it is 0.46 eV [50].

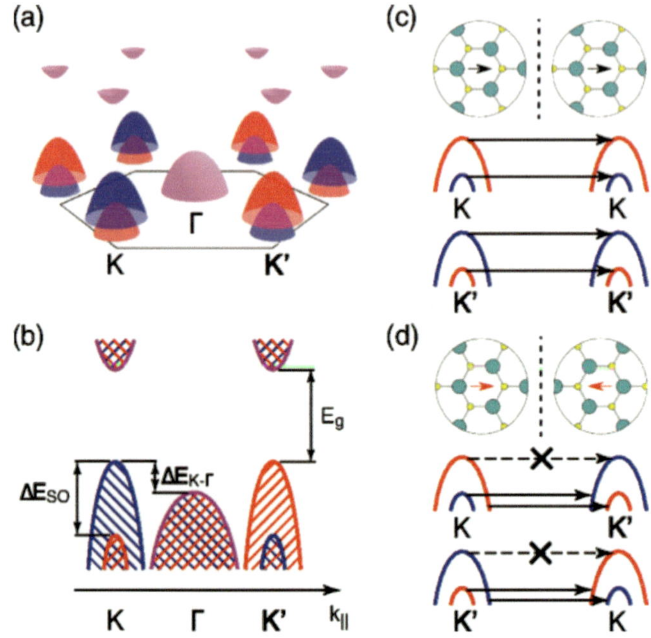

Figure 16.6 (a) Schematic representation of the electronic band structure of monolayer trigonal prismatic MoS_2 materials showing the spin-split valence bands at points K/ and K of the Brillouin zone and nearly spin-degenerate conduction bands at points K/ and K and the valence band at Γ. (b) Projection of the 2D band structure shown in Figure 16.6(a). (c and d) Allowed transmission channels for the hole charge carriers in the K/ and K valleys across line defects that preserve inversion domain boundaries and lattice orientation (reprinted with permission from [53]).

This effect is understood if we consider that the interaction between spin and orbit is relativistic. Also, this pattern is more prominent in heavier elements. Despite its very minimal magnitude, the spin splitting value in the conduction band cannot be ignored. Because of time-reversal symmetry, the spin splittings at K/ and K are opposite. Figure 16.6 illustrates the MoS_2's resulting band structure. This characteristic, known as the spin valley coupling, denotes the translation of valley polarization of charge carriers to spin polarization. This property in TMDCs has a greater impact on the design of spintronic devices using non-magnetic materials.

Most transition metal dichalcogenides exhibit single or perhaps even several charge density wave (CDW) transitions, which are inherent to their nearly two-dimensional (2D) structure. These periodic changes in the density of electrons result from the need to keep the overall electronic energy as low as possible when Coulomb repulsion is less than the energy gained from the gap widening as the new periodicity is introduced. Superconductivity (SC), which is a cooperative electronic state similar to the charge density wave (CDW), develops as a result of an electron–phonon interaction and a Fermi surface instability. Therefore, it may not come as a surprise that CDW and SC frequently engage in competition at low temperatures in a variety of TX_2 compounds. This competition has been thoroughly investigated as a central idea in condensed matter physics, and the use of materials design as a tuning mechanism has been a major help in this: electron doping (in TaS_2), pressure (in $NbSe_2$, TaS_3, $(TaSe_4)$ 2I, and $NbSe_3$), or both (in $Lu_5Ir_4Si_{10}$) suppress the CDW transition to low temperatures, where there is an increase in the critical temperature of an existing superconducting state.

References

[1] Pothier, H., Guéron, S., Birge, N. O., Esteve, D. & Devoret, M. H. Energy Distribution Function of Quasiparticles in Mesoscopic Wires. Phys. Rev. Lett. 79, 3490–3493 (1997).

[2] Kotliar, G. & Vollhardt, D. Strongly Correlated Materials: Insights From Dynamical Mean-Field Theory. Phys. Today 57, 53–59 (2004).

[3] Morosan, E., Natelson, D., Nevidomskyy, A. H. & Si, Q. Strongly Correlated Materials. Adv. Mater. 24, 4896–4923 (2012).

[4] Wei, J. & Natelson, D. Nanostructure studies of strongly correlated materials. Nanoscale 3, 3509 (2011).

[5] Wirth, S. & Steglich, F. Erratum: Exploring heavy fermions from macroscopic to microscopic length scales. Nat. Rev. Mater. 1, 16066 (2016).

[6] Ahn, C. H. et al. Electrostatic modification of novel materials. Rev. Mod. Phys. 78, 1185–1212 (2006).

[7] Allen, M. J., Tung, V. C. & Kaner, R. B. Honeycomb Carbon: A Review of Graphene. Chem. Rev. 110, 132–145 (2010).

[8] Yang, G., Li, L., Lee, W. B. & Ng, M. C. Structure of graphene and its disorders: a review. Sci. Technol. Adv. Mater. 19, 613–648 (2018).

[9] Sheehy, D. E. & Schmalian, J. Optical transparency of graphene as determined by the fine-structure constant. Phys. Rev. B 80, 193411 (2009).

[10] Nair, R. R. et al. Fine Structure Constant Defines Visual Transparency of Graphene. Science 320, 1308–1308 (2008).

[11] Ni, Z. H. et al. Graphene Thickness Determination Using Reflection and Contrast Spectroscopy. Nano Lett. 7, 2758–2763 (2007).

[12] Bolotin, K. I. et al. Ultrahigh electron mobility in suspended graphene. Solid State Commun. 146, 351–355 (2008).

[13] Son, Y.-W., Cohen, M. L. & Louie, S. G. Half-metallic graphene nanoribbons. Nature 444, 347–349 (2006).

[14] Novoselov, K. S. et al. Two-dimensional gas of massless Dirac fermions in graphene. Nature 438, 197–200 (2005).

[15] Zhang, Y., Tan, Y.-W., Stormer, H. L. & Kim, P. Experimental observation of the quantum Hall effect and Berry's phase in graphene. Nature 438, 201–204 (2005).

[16] Lee, C., Wei, X., Kysar, J. W. & Hone, J. Measurement of the Elastic Properties and Intrinsic Strength of Monolayer Graphene. Science 321, 385–388 (2008).

[17] Stankovich, S. et al. Graphene-based composite materials. Nature 442, 282–286 (2006).

[18] Balandin, A. A. et al. Superior Thermal Conductivity of Single-Layer Graphene. Nano Lett. 8, 902–907 (2008).

[19] Li, L., Reich, S. & Robertson, J. Defect energies of graphite: Density-functional calculations. Phys Rev B 72, 184109 (2005).

[20] Banhart, F., Kotakoski, J. & Krasheninnikov, A. V. Structural Defects in Graphene. ACS Nano 5, 26–41 (2011).

[21] Meyer, J. C. et al. Direct Imaging of Lattice Atoms and Topological Defects in Graphene Membranes. Nano Lett. 8, 3582–3586 (2008).

[22] Krasheninnikov, A. V., Lehtinen, P. O., Foster, A. S. & Nieminen, R. M. Bending the rules: Contrasting vacancy energetics and migration in graphite and carbon nanotubes. Chem. Phys. Lett. 418, 132–136 (2006).

[23] El-Barbary, A. A., Telling, R. H., Ewels, C. P., Heggie, M. I. & Briddon, P. R. Structure and energetics of the vacancy in graphite. Phys Rev B 68, 144107 (2003).

[24] Ugeda, M. M., Brihuega, I., Guinea, F. & Gómez-Rodríguez, J. M. Missing Atom as a Source of Carbon Magnetism. Phys Rev Lett 104, 096804 (2010).

[25] Boukhvalov, D. W. & Katsnelson, M. I. Chemical Functionalization of Graphene with Defects. Nano Lett. 8, 4373–4379 (2008).

[26] Cao, Y. et al. Unconventional superconductivity in magic-angle graphene superlattices. Nature 556, 43–50 (2018).

[27] Yankowitz, M. et al. Tuning superconductivity in twisted bilayer graphene. Science 363, 1059–1064 (2019).

[28] Lu, X. et al. Superconductors, orbital magnets and correlated states in magic-angle bilayer graphene. Nature 574, 653–657 (2019).

[29] Cao, Y. et al. Strange Metal in Magic-Angle Graphene with near Planckian Dissipation. Phys Rev Lett 124, 076801 (2020).

[30] Lozovik, Yu. E., Ogarkov, S. L. & Sokolik, A. A. Condensation of electron-hole pairs in a two-layer graphene system: Correlation effects. Phys Rev B 86, 045429 (2012).

[31] Naguib, M. & Gogotsi, Y. Synthesis of Two-Dimensional Materials by Selective Extraction. Acc. Chem. Res. 48, 128–135 (2015).

[32] Novoselov, K. S. et al. Two-dimensional atomic crystals. Proc. Natl. Acad. Sci. 102, 10451–10453 (2005).

[33] Osada, M. & Sasaki, T. Exfoliated oxide nanosheets: new solution to nanoelectronics. J. Mater. Chem. 19, 2503–2511 (2009).

[34] Wolfram, T. Two-Dimensional Character of the Conduction Bands of *d*-Band Perovskites. Phys Rev Lett 29, 1383–1387 (1972).

[35] Mannhart, J. & Schlom, D. G. Oxide Interfaces—An Opportunity for Electronics. Science 327, 1607–1611 (2010).

[36] Goniakowski, J., Finocchi, F. & Noguera, C. Polarity of oxide surfaces and nanostructures. Rep. Prog. Phys. 71, 016501 (2007).

[37] Chimene, D., Alge, D. L. & Gaharwar, A. K. Two-Dimensional Nanomaterials for Biomedical Applications: Emerging Trends and Future Prospects. Adv. Mater. 27, 7261–7284 (2015).

[38] Rienks, E. D. L. et al. Local zero-bias anomaly in tunneling spectra of a transition-metal oxide thin film. Phys Rev B 75, 205443 (2007).

[39] Surnev, S., Ramsey, M. G. & Netzer, F. P. Vanadium oxide surface studies. Prog. Surf. Sci. 73, 117–165 (2003).

[40] Wu, Q.-H., Fortunelli, A. & Granozzi, G. Preparation, characterisation and structure of Ti and Al ultrathin oxide films on metals. Int. Rev. Phys. Chem. 28, 517–576 (2009).

[41] Balendhran, S. et al. Two-Dimensional Molybdenum Trioxide and Dichalcogenides. Adv. Funct. Mater. 23, 3952–3970 (2013).

[42] Alsaif, M. M. Y. A. et al. Tunable Plasmon Resonances in Two-Dimensional Molybdenum Oxide Nanoflakes. Adv. Mater. 26, 3931–3937 (2014).

[43] Alsaif, M. M. Y. A. et al. High-Performance Field Effect Transistors Using Electronic Inks of 2D Molybdenum Oxide Nanoflakes. Adv. Funct. Mater. 26, 91–100 (2016).

[44] Terán-Escobar, G., Pampel, J., Caicedo, J. M. & Lira-Cantú, M. Low-temperature, solution-processed, layered V2O5 hydrate as the

hole-transport layer for stable organic solar cells. Energy Environ. Sci. 6, 3088–3098 (2013).
[45] Rui, X. et al. Ultrathin V2O5 nanosheet cathodes: realizing ultrafast reversible lithium storage. Nanoscale 5, 556–560 (2013).
[46] Morin, F. J. Oxides Which Show a Metal-to-Insulator Transition at the Neel Temperature. Phys Rev Lett 3, 34–36 (1959).
[47] Verleur, H. W., Barker, A. S. & Berglund, C. N. Optical Properties of VO_2 between 0.25 and 5 eV. Phys Rev 172, 788–798 (1968).
[48] Wei, J., Wang, Z., Chen, W. & Cobden, D. H. New aspects of the metal–insulator transition in single-domain vanadium dioxide nanobeams. Nat. Nanotechnol. 4, 420–424 (2009).
[49] Marezio, M., McWhan, D. B., Remeika, J. P. & Dernier, P. D. Structural Aspects of the Metal-Insulator Transitions in Cr-Doped VO_2. Phys Rev B 5, 2541–2551 (1972).
[50] Pouget, J. P., Launois, H., D'Haenens, J. P., Merenda, P. & Rice, T. M. Electron Localization Induced by Uniaxial Stress in Pure VO_2. Phys Rev Lett 35, 873–875 (1975).
[51] Rice, T. M., Launois, H. & Pouget, J. P. Comment on 'VO_2: Peierls or Mott-Hubbard? A View from Band Theory'. Phys Rev Lett 73, 3042–3042 (1994).
[52] Splendiani, A. et al. Emerging Photoluminescence in Monolayer MoS2. Nano Lett. 10, 1271–1275 (2010).
[53] Pulkin, A. & Yazyev, O. V. Spin- and valley-polarized transport across line defects in monolayer MoS_2. Phys Rev B 93, 041419 (2016).
[54] Lakshmy, S. et al. Catechol detection in pure and transition metal decorated 2D MoS2: Acumens from density functional theory approaches. Appl. Surf. Sci. 562, 150216 (2021).

Index

About the Editors

Dr. Parvathy Nancy, PhD, is the Research Officer at the International Centre for Ultrafast Studies (ICUS), Mahatma Gandhi University. She was the former Assistant Professor in the School of Energy Materials (SEM), Mahatma Gandhi University, and was a former post-doctoral fellow at the International and Inter University Centre for Nanoscience and Nanotechnology (IIUCNN). She obtained her PhD from the School of Pure and Applied Physics (SPAP), Mahatma Gandhi University, Kottayam -686560, Kerala, India.

Dr. Saju Joseph, PhD, is an Assistant Professor in the School of Nano Science and Nano Technology (SNSNT). He was a former Postdoctoral fellow at the International and Inter-University Centre for Nanoscience and Nanotechnology (IIUCNN), Mahatma Gandhi University, Kottayam-686560, Kerala, India. He obtained his PhD from the Department of Chemistry, University of Coimbra, Portugal.

Prof. Wilfred Prellier, PhD, is Director of the CRISMAT Laboratory (CNRS UMR 6508), ENSICAEN, 6 Bd MarAl'chal JUIN, France.

Prof. Sabu Thomas, D.Sc., PhD, FRSC, is the former Vice-Chancellor, Mahatma Gandhi University, Kottayam, Kerala, India, and the Founder Director, International and Inter-University Centre for Nanoscience and Nanotechnology. He was also a senior professor, School of Chemical Sciences (SCS), and Director of the School of Energy Materials (SEM) & School of Nano Science and Nano Technology (SNSNT), Mahatma Gandhi University, Kottayam-686560, Kerala, India.

Prof. Nandakumar Kalarikkal, PhD, is the Director, International Centre for Ultrafast Studies (ICUS), Mahatma Gandhi University, Kerala, India. He was the former Director & Chair, Senior Professor, School of Pure and Applied Physics (SPAP), and International and Inter University Centre for Nanoscience and Nanotechnology (IIUCNN), Mahatma Gandhi University, Kottayam -686560, Kerala, India.

For Product Safety Concerns and Information please contact our
EU representative GPSR@taylorandfrancis.com Taylor & Francis
Verlag GmbH, Kaufingerstraße 24, 80331 München, Germany